The Effect of Hydrology on Soil Erosion

The Effect of Hydrology on Soil Erosion

Editors

Jesús Rodrigo-Comino
José María Senciales-González
José Damián Ruiz-Sinoga

MDPI • Basel • Beijing • Wuhan • Barcelona • Belgrade • Manchester • Tokyo • Cluj • Tianjin

Editors
Jesús Rodrigo-Comino
University of Valencia
Spain

José María Senciales-González
Universidad de Malaga
Spain

José Damián Ruiz-Sinoga
Universidad de Malaga
Spain

Editorial Office
MDPI
St. Alban-Anlage 66
4052 Basel, Switzerland

This is a reprint of articles from the Special Issue published online in the open access journal *Water* (ISSN 2073-4441) (available at: https://www.mdpi.com/journal/water/special_issues/Hydrology_Soil).

For citation purposes, cite each article independently as indicated on the article page online and as indicated below:

LastName, A.A.; LastName, B.B.; LastName, C.C. Article Title. *Journal Name* **Year**, *Article Number*, Page Range.

ISBN 978-3-03936-644-6 (Hbk)
ISBN 978-3-03936-645-3 (PDF)

Cover image courtesy of Jesús Rodrigo-Comino.
This photo depicts the vineyards of Pepe Gámez, Almáchar (Málaga, Spain).

Contents

About the Editors

Jesús Rodrigo-Comino received his Ph.D. at the University of Málaga (Spain). He received his M.S. in land planning and GIS in 2013 in Granada and Málaga Universities. Since 2015, he has written two books related to soil geography, presented several oral conferences and posters in international meetings, and published his investigations about soil erosion, soil geography, and land degradation processes. He is editor-in-chief of *Air, Soil and Water Research* (SAGE), and also works as an associate editor for the *Hydrological Science Journal* (Taylor and Francis), *Journal of Mountain Science* (Springer) and *Hydrology* (MDPI). He is also a reviewer for more than 110 international indexed journals. At the time of publication, he is working on an Interreg project about light pollution (Smart Light-HUB) at the Trier University (Germany) and COST-Action Firelinks (CA18135) as a grant holder at the University of Valencia (Spain).

José María Senciales-González (Ph D. 1995) was Associate Lecturer from 1996 to 2003. He is now a Senior Lecturer in Physical Geography (since 2003). He has authored different scientific works including more than 20 indexed articles, book chapters, and complete monographs. He has also contributed to several scientific projects concerning regional environments, environmental impacts, soils, flood hazards, and urban risks, among others. He is a reviewer for several scientific journals and associate editor for *Water* (MDPI). His main expertise is focused on river morphology, weather types, and climate change.

José Damián Ruiz-Sinoga is a Full Professor of Physical Geography since 2011. His main research topics are the analysis of geomorphological processes and soil–water–vegetation relationships, at different scales, from the pedon to the small basin scales in the context of Global Change. This uninterrupted activity within the university environment has allowed the publication of numerous scientific articles (146, of which 72 are ISI-JCR, and 47 Q1), books or book chapters (53), participation in congresses, scientific meetings (154) and conferences, both nationally and internationally (63), as well as carrying out different projects in the autonomous, national, and international spheres (36).

Editorial

The Effect of Hydrology on Soil Erosion

Jesús Rodrigo-Comino [1,2,*], José María Senciales-González [3] and José Damián Ruiz-Sinoga [3]

1   Physical Geography, Trier University, 54286 Trier, Germany
2   Soil Erosion and Degradation Research Group, Department of Geography, University of Valencia, 46010 Valencia, Spain
3   Department of Geography, Málaga University, Campus of Teatinos s/n, 29071 Málaga, Spain; senciales@uma.es (J.M.S.-G.); sinoga@uma.es (J.D.R.-S.)
*   Correspondence: rodrigo-comino@uma.es

Received: 11 March 2020; Accepted: 15 March 2020; Published: 16 March 2020

Abstract: In this Special Issue, we have tried to include manuscripts about soil erosion and degradation processes and the accelerated rates due to hydrological processes and climate change. We considered that the main goal was successfully reached. The new research focused on measurements, modelling, and experiments under field or laboratory conditions developed at different scales (pedon, hillslope, and catchment) were submitted and published. This Special Issue received investigations from different parts of the world such as Ethiopia, Morocco, China, Iran, Italy, Portugal, Greece and Spain, among others. We are happy to see that all papers presented findings characterized as unconventional, provocative, innovative and methodologically new. We hope that the readers of the journal *Water* can enjoy and learn about hydrology and soil erosion using the published material, and share the results with the scientific community, policymakers and stakeholders new research to continue this amazing adventure, featuring plenty of issues and challenges.

Keywords: hydrological processes; soil erosion; different scales; models; experiments

---

1. Introduction

Soil erosion is one of the most important environmental issues in natural and anthropized territories [1,2]. Understanding the key parameters and factors of soil erosion will enable the conservation of soil system goods, services and resources, and will avoid the damage outside of fields caused by transported and accumulated sediments and water [3,4]. The dynamics of erosive processes are changing along with new trends of pluviometric patterns due to well-identified climate change, which is becoming an extra factor in soil degradation [5,6]. However, in several parts of the world, this problem is currently not well-understood because of the lack of information, or simply because it is ignored. This is the case in several parts of Africa or Asia or for land uses such as urban areas, vineyards or abandoned plots [7–9].

In this Special Issue, we attempted to discuss and address the state-of-the-art of soil erosion and degradation processes because of their accelerated rates due to hydrological processes and climate change but also due to non-controlled human impacts. To fill these gaps, studies focused on measurements, modelling, and experiments under field and laboratory conditions developed at different scales (pedon, hillslope, and catchment) were particularly welcome for submission. Understanding the effects of hydrology is complex, and a combination of several natural and anthropic mechanisms must be understood such as climate change, tillage effects, rill and interrill occurrence, gully formation, soil compaction and water losses, runoff generation thresholds, soil salinization, organic matter depletion, and so on.

2. Summary of This Special Issue

Two papers were published related to laboratory research. To understand soil erosion processes, each pedological, climate, geomorphological and hydrological parameter must be assessed. In Fernández-Raga et al. [10], the authors highlighted the need to choose the most correct measurement device during the experimental design; for example, in the case of splash erosion. They evaluated the hydrological response under different intensities of simulated rainfall of different devices under the same conditions. On the other hand, Meshkat et al. [11] assessed the geometry of hillslopes (plan and profile), which can affect soil erosion under rainfall–runoff processes. They studied the impact of surface roughness coefficients and complex hillslopes on runoff variables viz. their time of generation, time of concentration, and peak discharge value, with a total of 81 experiments under laboratory conditions.

Soil erosion also affects anthropized areas. In this Special Issue, three interesting manuscripts were published providing new insights about peri-urban areas, agricultural fields and soil conservation measures in forestry areas. Ferreira et al. [12] confirmed that understanding sediment dynamics in peri-urban catchments constitutes a research challenge because of the spatiotemporal complexity and variability of the land-uses involved. This research group analyzed differences in the concentration of total sediments and suspended sediments in the small peri-urban Mediterranean Ribeira dos Covões catchment (40% urban area) in central Portugal. The results provided key data about sediment dynamics to help stakeholders to establish strategies to reduce risks and reduce the impacts on urban aquatic ecosystems.

Capello et al. [13] illustrated that soil erosion in vineyards continues to be affected by soil erosion. This study case evaluated the impacts of rainfall temporal patterns and intensity variability when machine traffic is implemented with particular intensity from late spring to harvest. The results showed that soil management generated soil compaction, which likely affects soil hydraulic properties, runoff, and soil erosion. The authors stated that there is a real need to limit tractor traffic and to reduce negative effects due to soil compaction, especially in tilled inter-rows.

Another clear example of human disturbance is the research published by Mhiret et al. [14] about land degradation and soil conservation measures carried out to reduce soil loss in the Ethiopian Highlands. This territory suffers from severe land degradation, including erosion, and in response, the Ethiopian government has implemented soil and water conservation practices considering the acreage of eucalyptus, which has been expanded. They examined the impacts of these land-use changes on stream discharge and sediment load in a sub-humid watershed, collecting a total of 867 storm events for nine years. The results indicated that the techniques are either inappropriate for this sub-humid watershed or require improved design and maintenance.

Kirchhoff et al. [15] presented exotic research carried out in a poorly-studied region of Northern Africa, where the endemic argan tree (*Argania Spinosa*) populations are highly degraded due to their use as a biomass resource in dry years and illegal firewood extraction. In total, 36 rainfall simulation experiments, as well as 60 infiltration measurements, were conducted to investigate the potential difference between tree-covered areas and bare inter-tree spaces. They concluded that argan trees have a great influence on the soil underneath; meanwhile, the soil in intertree areas is less protected and more degraded. They stated that it can be supposed that there will be further soil degradation when intertree areas extend further due to the lack of rejuvenation of argan trees.

Finally, the rest of the published papers were related to modelling techniques of Earth surface landforms (e.g., gullies) and processes (e.g., landslides) or the quantification of soil erosion rates. In particular, two papers published here showing different study cases from China are very interesting. Yan et al. [16] explained that since large-scale agricultural irrigation began in the 1980s on the South Jingyang Plateau, 92 landslides have occurred. As a result, numerous casualties and substantial property loss have been caused. In their manuscript, the authors evaluated the soil erosion and mechanical mechanism of these irrigated shallow loess landslides, defining the spatial distributions, types and developmental characteristics of loess landslides and surveying and monitoring seasonal agricultural irrigation features and groundwater changes.

Hu et al. [17] explained that the algorithm for calculating the soil erodibility factor (K) in the RUSLE (Revised Universal Soil Loss Equation) is somewhat limited, particularly in the context of China, which features highly diverse soil types. They proposed a modified algorithm addressing the piecewise function of gravel content and relative soil erosion for the first time to modify the soil erodibility factor. They tested the vital key role that gravel content plays in the soil erodibility factor in the Chaohu Lake Basin (CLB) in East China.

Gullies are ones of the most important manifestations of land degradation in a territory. Using Google Earth, Karydas and Panagos [18] carried out a preliminary study on the presence of ephemeral gullies in Greece by sampling representative cultivated fields in 100 sites which were randomly distributed throughout the country for the period 2002–2019. This paper represents the first attempt to use visual interpretation with Google Earth image time-series on a country scale, producing a gully erosion inventory which could be very useful in implementing soil conservation practices such as contour farming and terraces in agricultural areas.

Javidan et al., [19] applied GIS (Geographic Information System) techniques and the MARS (Multivariate Adaptive Regression Splines) algorithm to evaluate gully erosion susceptibility mapping in a specific section of the Gorganroud Watershed in Golestan Province (Northern Iran), covering 2142.64 km^2. They also used Google Earth images and field surveys in combination with national reports to generate a map consisting of 307 gully-headcut points. The results showed that the performance of MARS for modelling gully erosion susceptibility can be quite consistent; meanwhile, changes in the testing and validation specimens were investigated. On the other hand, Arabameri et al. [20] successfully tested a new hybrid model combining the index-of-entropy (IoE) model with the weight-of-evidence (WoE) model. They used remote sensing and GIS techniques to map gully-erosion susceptibility in the watershed of the Bastam district of Semnan Province in northern Iran. They considered a total of 303 gullies and eight topographical, hydrological, geological, and environmental conditioning factors. The hybrid model predicted that 38% of the territory was registered as either highly or very highly susceptible to gullying.

Funding: This research received no external funding.

Acknowledgments: Thanks to all of the contributions to the Special Issue, the time invested by each author, as well as to the anonymous reviewers and editorial managers who have contributed to the development of the articles in this Special Issue. All the guest editors are very happy with the review process and management of the Special Issue and offer their thanks.

Conflicts of Interest: The authors declare no conflict of interest.

References

1. Panagos, P.; Meusburger, K.; Ballabio, C.; Borrelli, P.; Alewell, C. Soil erodibility in Europe: A high-resolution dataset based on LUCAS. *Sci. Total Environ.* **2014**, *479*, 189–200. [CrossRef] [PubMed]
2. García-Ruiz, J.M.; Beguería, S.; Nadal-Romero, E.; González-Hidalgo, J.C.; Lana-Renault, N.; Sanjuán, Y. A meta-analysis of soil erosion rates across the world. *Geomorphology* **2015**, *239*, 160–173. [CrossRef]
3. Poesen, J.; Ingelmo-Sanchez, F.; Mucher, H. The hydrological response of soil surfaces to rainfall as affected by cover and position of rock fragments in the top layer. *Earth Surf. Process. Landforms* **1990**, *15*, 653–671. [CrossRef]
4. Cerdà, A. Relationships between climate and soil hydrological and erosional characteristics along climatic gradients in Mediterranean limestone areas. *Geomorphology* **1998**, *25*, 123–134. [CrossRef]
5. Nadal-Romero, E.; González-Hidalgo, J.C.; Cortesi, N.; Desir, G.; Gómez, J.A.; Lasanta, T.; Lucía, A.; Marín, C.; Martínez-Murillo, J.F.; Pacheco, E.; et al. Relationship of runoff, erosion and sediment yield to weather types in the Iberian Peninsula. *Geomorphology* **2015**, *228*, 372–381. [CrossRef]
6. Ruiz-Sinoga, J.D.; Gabarrón Galeote, M.A.; Martinez Murillo, J.F.; Garcia Marín, R. Vegetation strategies for soil water consumption along a pluviometric gradient in southern Spain. *Catena* **2011**, *84*, 12–20. [CrossRef]

7. Rodrigo-Comino, J. Five decades of soil erosion research in "terroir". The State-of-the-Art. *Earth-Sci. Rev.* **2018**, *179*, 436–447. [CrossRef]

8. Biratu, A.A.; Asmamaw, D.K. Farmers' perception of soil erosion and participation in soil and water conservation activities in the Gusha Temela watershed, Arsi, Ethiopia. *Int. J. River Basin Manag.* **2016**, *14*, 329–336. [CrossRef]

9. García-Ruiz, J.M.; Lana-Renault, N. Hydrological and erosive consequences of farmland abandonment in Europe, with special reference to the Mediterranean region—A review. *Agric. Ecosyst. Environ.* **2011**, *140*, 317–338. [CrossRef]

10. Fernández-Raga, M.; Campo, J.; Rodrigo-Comino, J.; Keesstra, S.D. Comparative Analysis of Splash Erosion Devices for Rainfall Simulation Experiments: A Laboratory Study. *Water* **2019**, *11*, 1228. [CrossRef]

11. Meshkat, M.; Amanian, N.; Talebi, A.; Kiani-Harchegani, M.; Rodrigo-Comino, J. Effects of Roughness Coefficients and Complex Hillslope Morphology on Runoff Variables under Laboratory Conditions. *Water* **2019**, *11*, 2550. [CrossRef]

12. Ferreira, C.S.S.; Walsh, R.P.D.; Kalantari, Z.; Ferreira, A.J.D. Impact of Land-Use Changes on Spatiotemporal Suspended Sediment Dynamics within a Peri-Urban Catchment. *Water* **2020**, *12*, 665. [CrossRef]

13. Capello, G.; Biddoccu, M.; Ferraris, S.; Cavallo, E. Effects of Tractor Passes on Hydrological and Soil Erosion Processes in Tilled and Grassed Vineyards. *Water* **2019**, *11*, 2118. [CrossRef]

14. Mhiret, D.A.; Dagnew, D.C.; Alemie, T.C.; Guzman, C.D.; Tilahun, S.A.; Zaitchik, B.F.; Steenhuis, T.S. Impact of Soil Conservation and Eucalyptus on Hydrology and Soil Loss in the Ethiopian Highlands. *Water* **2019**, *11*, 2299. [CrossRef]

15. Kirchhoff, M.; Engelmann, L.; Zimmermann, L.L.; Seeger, M.; Marzolff, I.; Aït Hssaine, A.; Ries, J.B. Geomorphodynamics in Argan Woodlands, South Morocco. *Water* **2019**, *11*, 2193. [CrossRef]

16. Yan, R.-X.; Peng, J.-B.; Huang, Q.-B.; Chen, L.-J.; Kang, C.-Y.; Shen, Y.-J. Triggering Influence of Seasonal Agricultural Irrigation on Shallow Loess Landslides on the South Jingyang Plateau, China. *Water* **2019**, *11*, 1474. [CrossRef]

17. Hu, S.; Li, L.; Chen, L.; Cheng, L.; Yuan, L.; Huang, X.; Zhang, T. Estimation of Soil Erosion in the Chaohu Lake Basin through Modified Soil Erodibility Combined with Gravel Content in the RUSLE Model. *Water* **2019**, *11*, 1806. [CrossRef]

18. Karydas, C.; Panagos, P. Towards an Assessment of the Ephemeral Gully Erosion Potential in Greece Using Google Earth. *Water* **2020**, *12*, 603. [CrossRef]

19. Javidan, N.; Kavian, A.; Pourghasemi, H.R.; Conoscenti, C.; Jafarian, Z. Gully Erosion Susceptibility Mapping Using Multivariate Adaptive Regression Splines—Replications and Sample Size Scenarios. *Water* **2019**, *11*, 2319. [CrossRef]

20. Arabameri, A.; Cerda, A.; Tiefenbacher, J.P. Spatial Pattern Analysis and Prediction of Gully Erosion Using Novel Hybrid Model of Entropy-Weight of Evidence. *Water* **2019**, *11*, 1129. [CrossRef]

Article

Comparative Analysis of Splash Erosion Devices for Rainfall Simulation Experiments: A Laboratory Study

María Fernández-Raga [1,*], Julián Campo [2], Jesús Rodrigo-Comino [3] and Saskia D. Keesstra [4,5]

1   Department of Applied Physics, University of Leon, 24071 León, Spain
2   Department of Environmental Quality and Soil, Desertification Research Centre-CIDE (CSIC, UV. GV),
    46113 Valencia, Spain; julian.campo@uv.es
3   Instituto de Geomorfología y Suelos, Department of Geography, University of Málaga, 29071 Málaga, Spain;
    rodrigo-comino@uma.es
4   Wageningen Environmental Research, Team Soil, Water and Land Use, P.O. Box 47, 6700AA Wageningen,
    The Netherlands; saskia.keesstra@wur.nl
5   Civil, Surveying and Environmental Engineering, The University of Newcastle,
    Callaghan, NSW 2308, Australia
*   Correspondence: maria.raga@unileon.es; Tel.: +34-987291000-5342

Received: 24 May 2019; Accepted: 10 June 2019; Published: 12 June 2019

Abstract: For the study of soil erosion it is important to set up the experiments well. In the experimental design one of the key factors is the choice of the measurement device. This is especially important when one part of the erosion process needs to be isolated, such as for splash erosion. Therefore, the main aim of this research is to list the general characteristics of the commonly used splash erosion devices and to discuss the performance, to be able to relate them, and make suggestions regarding their use. The devices we selected for this comparative comparison were: the splash cup, funnel, Morgan tray, Tübingen cup, tower, and the gutter. The devices were tested under the same conditions (rainfall characteristics, slope, and soil type) to assess their hydrological response under different intensities of simulated rainfall. All devices were installed on a sloping plot (10°) with sandy soil, and were exposed to 10 min. of simulated rain with intensities ranging from 60 to 172 mm/h to measure the splashed sediment, and to describe problems and differences among them. The results showed that the Tübingen cup was the best performing device to measure kinetic energy of the rain, but, because of its design, it is not possible to measure the detached splashed sediment under natural (field) conditions. On the other hand, the funnel device showed a significant relation with rain intensity because it loses little sediment to washing. In addition, the device is easy to use and cheap. Therefore, this device is highly recommended to estimated splash erosion. to the good performance measuring the actual splash erosion, because it loses little sediment by washing. The device is also cheap and easy to install and manage.

Keywords: splash erosion; environmental assessment; soil erosion; rainfall simulation

---

1. Introduction

Splash erosion is the process of detachment of soil particles due to the impact of falling raindrops, and the deposit of those particles on adjacent sites [1–3]. Splash erosion is the first mechanism of soil erosion, and it is the one that produces most of the detachment work [4], and it affects other processes, such as infiltration [5], soil water repellency [6], overland flow [7], roughness [8], and the final soil erosion rates [9], as it changes the soil surface characteristics. The splash erosion process initiates movements that are concentrated at the surface and it affects the composition of the material on the soil, such as minerals and organic particles [10]. As a result, the transported material could be even rich in organic carbon [11], with splash erosion being one of the key factors that are responsible for the redistribution of the minerals on the soil surface [12]; however, the microtopography should be also considered as the key factor for soil erosion on small scales [13,14].

In arid and semi-arid areas, splash erosion also plays a major role in forming the landscape due to the control that exerts on shaping the landforms [15,16]. This is because the bare soils allow for the raindrop impact to develop a crust, and then to enhance surface runoff generation, which is a key process in the Mediterranean and temperate areas, such as Spain [17], Italy [18,19], Israel [20], or Iran [21]. The kinetic energy of the raindrops is responsible for breaking up soil aggregates and finer materials, such as fine sand, silt, clay, and organic particles that are detached by raindrops are carried away by runoff, leaving larger sand grains, pebbles, and gravel behind [22]. On a flat surface, the fine particles may clog up pores and facilitate ponding [23]. However, if the area is on a steep slope, the water that did not infiltrate may flow downslope as sheet erosion [24,25], carrying the soil particles that have been loosened by the raindrop impacts away [26]. Hills and ridges that are shaped by splash erosion use tend to have gently rounded tops that are very different from the sharper profiles that are created by other forms of water erosion [27].

Research in erosion processes is time-consuming and expensive [28]. To be effective, the erosion measurements should be precise, controlled, and replicable, and erosion measurement devices should be properly designed, constructed, well calibrated, and they also should be operated by a trained researcher or technician to assure effectiveness, [29]. The equipment is constantly redesigned, making the measurements not comparable and furthermore, creating a lack of available standardized devices because researchers try to improve old devices [30]. This lack of standardization is especially important in the case of splash erosion, as it is, by nature, a complex mechanism that is key for understanding soil erosion processes [31]. Moreover, it is very difficult to isolate the measurements of splash erosion from those of sheet and rill surface erosion since they all interact [32]. In general, soil erosion rates that are measured under simulated and natural rainfall by means of plots, erosion pins, silt fences, or flumes [33,34] can measure the soil erosion processes, but they cannot detect the individual mechanisms: splash, sheet surface wash, and rill flow. The information that is generated from experiments that separate individual soil erosion mechanisms, such as the one developed here, will be very effective in producing soil erosion models to foresee the soil loss changes in time and space.

There are many different types of devices that claim to measure splash erosion, and the number increases due to the contribution of new ideas. The objective of these changes in the designs are improving the sensitivity of the device with a little amount of splash erosion or changes in erosivity and precision shown in the standard deviation of the measurements under the same conditions [3]. However, all of the designs show important measurement limitations, which are yet to be well quantified. As a result, the research on splash erosion involves a high variability that is related to the different locations of the studies (i.e., soil and weather characteristics) and the different devices used for measurement (different design for reception and accumulation of the sample), which makes any comparison very difficult.

Within the devices that are used to characterize splash erosion, the first classification that can be made is whether the device is unbounded or bounded, depending on which surface the splashed sediment originates [3]. Bounded devices have a specific surface that is used to assess the soil erodibility, while the unbounded devices measure the amount of sediment that is received in the form of splashed sediment from the (unbounded) surrounding area. The unbounded devices that were selected for this study were: the funnel [35], the cup [36,37], and the gutter [38]. Another unbounded device, the tower of funnels, was also included in this research, although it has never been used before due to the different structure and data that can be obtained. All of the unbounded devices only measure a cumulative amount of splashed sediment (g) over the total measuring period, originated from an unknown area, which means that no traditional rate of splash (g m^{-2}) can be determined. For the bounded devices, the most common devices selected were the Tübingen cup (Tcup) [9,39] and the Morgan tray [40,41], which allow for calculating the quantitative splash rates, because the amount of sediment can be related to a known surface.

Our hypothesis is that the measurements of the splash erosion that were obtained with any of these devices should be related by an equation to enable a good comparison of the data obtained with different splash erosion device. The main goal of this research was to compare the measurements that were obtained with six different splash erosion devices (using the same soil, slope angle, and rain characteristics), and to try to propose a relation among the obtained data that is able to help to make the results of earlier and new research comparable. Additionally, this study aims to describe the specific problems that a researcher needs to take into account when dealing with splash erosion, while paying special attention to the sensitivity, the cost, and the user-friendliness of the device.

2. Materials and Methods

2.1. Description of Selected Splash Devices

The selection of splash devices for this study intends to provide a wide representation of the main types of splash systems that are commonly used (Figure 1). Starting with the unbounded area devices, the splash cup (Figure 1a) is currently one of the most common devices to measure splash erosion [3]. It consists of an aluminium or PVC cylinder with a paper filter at 2 cm from the upper edge, placed between a metal mesh, which is fixed, and a coarser mesh on top of it. This cylinder has a standing height of 7 cm and a diameter of 10 cm, and the lower part is cut open to allow for runoff to flow freely underneath the cup. It is inserted in the ground until only 3 cm protrudes. When empty, this device usually acts as a receptor, but it could also be used to estimate erosion when it is initially filled with a known quantity of soil and the remaining amount of soil after the rain impact is measured [42]. It has a measuring area of 0.0079 m^2.

The funnel device (Figure 1b) was specifically designed for minimizing the loss of particles that were already accumulated within the device by the washing out caused by raindrops that fall inside the device after the deposit of some particles [35]. It consists of a couple of piled-up funnels, with a filter paper in between that should be weighed before and after the rainfall event. The diameter of funnels is 12 cm, and they are inserted in the ground until only 3 cm protrudes. It has a measurement surface of 0.0095 m^2.

The collection pipe or gutter [38] is a plastic gutter (Figure 1c) of 5.95 m long and 6.5 cm width with the upper part open, at a height of 3 cm, from the soil surface. This gutter is installed across the studied area and it allows for collecting the splashed soil particles in the lower part by filtering the water. It has a measurement area of 0.37 m^2.

The tower of funnels (Figure 1d) was designed for this experiment as a device that enables the determination of the vertical transport of splash erosion. This system or the Leguédois tray only obtain this information [43]. 13 funnels that are attached to a vertical bar with an angle of 45 degrees, and the lower funnel at 7.5 cm from the surface form it. There is a filter paper inside each funnel. Each of the 13 funnels has an area of 0.011 m^2.

In the group of bounded area, the Morgan tray and the Tübingen cup are the devices. The Morgan tray [44] collects all of the soil splashed from a well-defined eroding area, allowing for the calculation of the splash rate. This device (Figure 1e) consists of a 10 cm diameter cylinder that was placed directly on the soil, which is surrounded by plastic and closed circular plate of 30 cm in diameter. Raindrops impacting the bare soil surface in the inner circle detached soil particles, and later they jump over the rim and fall on the outer circle. The outer rim of the tray has a wall of 20 cm height to avoid the contamination of splashed material from the outside soil. The device has a source area of soil to be eroded of 0.0079 m^2.

The Tübingen cup or Tcup [45] consists of a plastic flask that is filled with a known quantity of sand of 212.50 μm with a carrier system attached (Figure 1f). It measures the difference in weight of the sand before and after the rain, and therefore measures the splash erosion that is generated by the applied rainfall for a given particle size of quartz. A silk cover separates the flask and the carrier systems, which prevents the loss of sand and guarantees free drainage of water between the cup and the carrier. It does not estimate splash erosion rates of natural soil because it only measures the loss of sand. Nevertheless, the measurement of this sand can be related somehow to the raindrop impact on the soil or to the kinetic energy affecting the area. It has a source measuring area of 0.002 m^2.

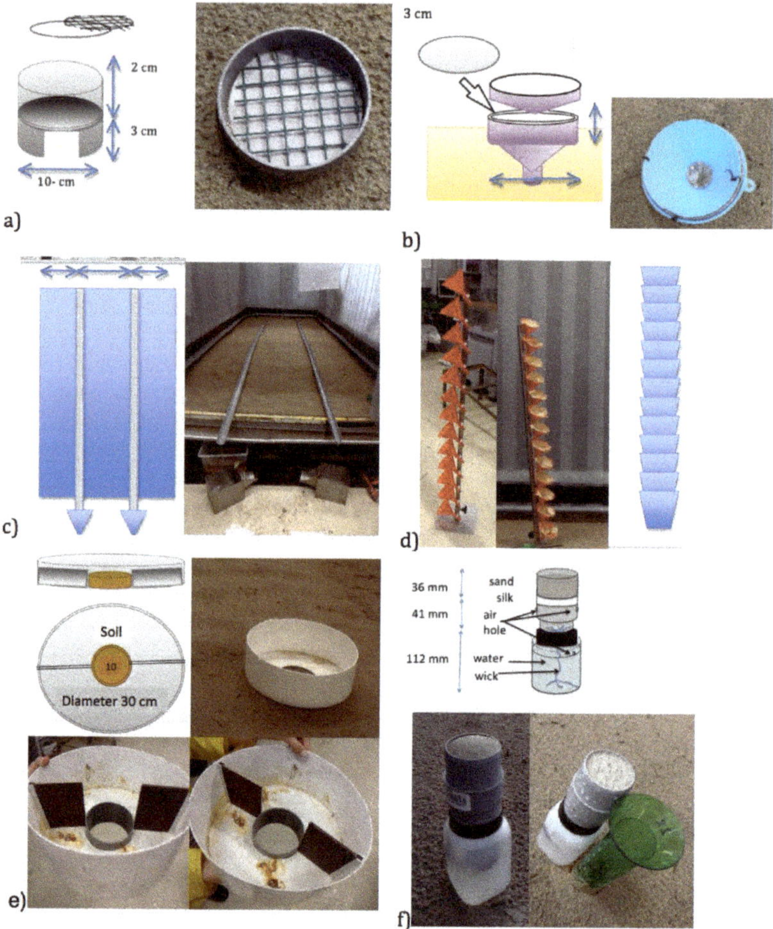

Figure 1. Splash devices used in this study: (**a**) cup, (**b**) funnel, (**c**) gutter, (**d**) tower of funnels, (**e**) Morgan tray, and (**f**) Tübingen cup (Tcup).

2.2. The Wageningen Rainfall Simulator

The rainfall simulator of Wageningen University and Research consists of a 6 m long and 2.5 m wide plot, with a 2.8 m high metal lateral frame (Figure 2). The rainfall is produced from two mobile central nozzles that are situated at 3.5 m height. The slope that can be adapted from 0 to 15.5° was fixed at 10° for this study. The bottom is freely drained and it is subdivided into six segments, which were covered by a permeable geotextile to allow infiltration. The plot was filled with air-dried sand 369 µm in diameter, which allows for a very constant infiltration (mini-disk infiltrometer tests done on 10 different positions showed an average value of K = 0.024 cm s^{-1} with a standard deviation of 0.0014) and can act as a homogenous soil source that can be splashed. The sides were covered with a plastic curtain, and a tube along the sides of the plot was installed to collect water that was dripping from the curtains.

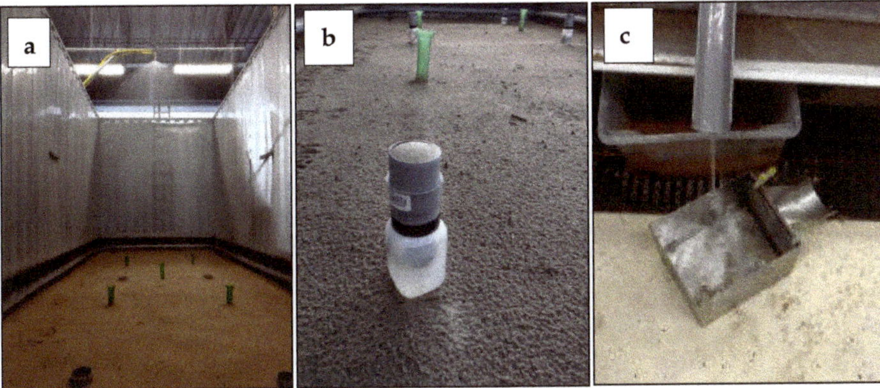

Figure 2. Rainfall simulator of Wageningen University and Research; (**a**) Rainfall simulator plot during one experiment; (**b**) splash cups; and, (**c**) sediment collector.

Lassu et al. [46] extensively described the characteristics of this rainfall simulator, which is regulated by changing the pressure of the pump, the flow of the water, and the type of nozzles. In their study, Lassu et al. [46] selected two Lechler nozzles (nr. 460.788 and nr. 461.008) to produce low and high-intensity rainfall, respectively, and a pressure of 2 bars as the best conditions to work with the Wageningen simulator. Their results showed rainfall characteristics that were steady in time and reproducible. Moreover, they also found that the intensity of the rainfall slightly varies along the area, and the position of the nozzles, which is crucial for the spatial distribution of the rainfall, concentrating higher intensities under the nozzles, and lower intensities on both sides (Figure 3). The size and speed of the raindrops that were produced by the Wageningen rainfall simulator were measured using a disdrometer Thies Laser Precipitation Monitor to determine that intensities from 38 to 160 mm/h produce steadily kinetic energy values from 25.7 to 29 J m^{-2} mm^{-1}. In the current study, we reproduced exactly the same conditions that in this above-mentioned research (pressure, position of the nozzles and slope), also measuring very similar intensities (60–160 mm/h with only one measurement done at 172 mm h^{-1}), which allows for us to assume that the kinetic energy values of a hypothetical study area under extreme rainfall conditions. This will allow for obtaining the maximum potential of each studied device.

Additionally, it is important to remember that fixing a value of pressure, water flow, and specific nozzle, in this simulator, there will be one value of kinetic energy that is related to each value of intensity, because the drop size distribution (DSD) will be constant under such conditions. Drop size distribution represents the number of drops that are measured as a function of diameter in one sample [47]. This is different in the natural rain, where it is possible to obtain different kinetic energies from the same intensity due to the different DSD of rains that were more convective or stratiform [37,48].

The Wageningen rainfall simulator is very sensitive to small changes in the flow, so the intensity values in every specific experimental spot of the area were checked before and during every simulation. Additionally, a modification of the Christiansen Uniformity Coefficient (CUC) was used to assign a value of intensity to every point of the rainfall simulator area. This is a commonly used statistic for the evaluation of sprinkling systems, and a value of 80 is usually considered to be acceptable in sprinkling systems [49]. It can be calculated with Equation (1).

$$CUC = 100 \times \left(1 - \frac{\sum_{i=1}^{n} \left| X_i - \overline{X} \right|}{n\overline{X}} \right) \qquad (1)$$

where *CUC* is the Christiansen's Uniformity Coefficient (%), X_i is the depth (mm) or the mass of water precipitated in each collector (g), $\overline{X}$ is the mean depth (mm) precipitated in the collectors, and *n* is the number of collecting container.

This method consists on recovering the rainfall of 10 min. with rain gauges being installed every 40 cm (96 containers along the area), obtaining the intensity in every point (Figure 3), and only in the experimental positions (places where the splash devices were settled), obtaining a mean value of uniformity of 78.9. This value was used as a reference to set the experimental spots in the experiments. However, in any case, the values of rainfall intensity in every experimental spot were measured prior using each splash device. The goal was to confirm the stability of the parameters of the simulated rainfall in these specific spots between different experiments with the rainfall simulator that is characterized by the same flux, pressure, position of the nozzles, and inclination. We measure a value directly related to the kinetic energy because the artificial drops produced by the rainfall simulator will show the same DSD and speed. The reason is that there will be one unique value of kinetic energy for each value of intensity if the rest of the parameters of the rainfall simulator stay constant (flux, pressure, the position of nozzles, and slope).

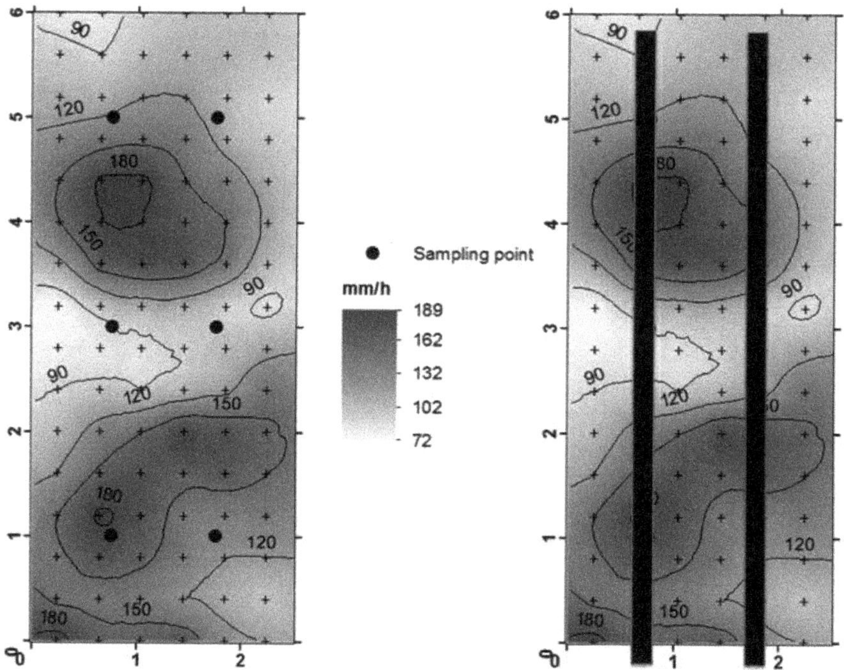

Figure 3. Schema of the plot used for rainfall simulation with the mean intensity distribution along the area, and the position of the splash devices individual on the left and gutter on the right. The symbol "+ "means the position of the gauges to measure the intensity of the rain.

2.3. Experimental Design

The measurement of the intensity of the rain in every experimental spot before each splash-measurement was considered as one of the most important steps as the aim of this study is to compare splash devices performance. Assuring the same rainfall intensity value, no matter what the intensity value is, compared all of the used devices. Additionally, we rely on the hypothesis that a certain relationship exists between rainfall intensity and the amount of splash that is recovered by each splash erosion device; therefore, this relationship will be shown in the results that are obtained with different devices.

Accordingly, the first important step was to install a set of six splash devices of the same type in well-defined positions of the rainfall simulator area (Figure 3) and to assign a rainfall intensity value to each position. We set containers in the six positions of the devices, and measured the intensities among repetitions, previously, to assure the stability of the rainfall intensity values prior to conducting each rainfall simulation. In the case of the gutter device, because of its different design, the determination of the rain intensity was done by measuring all the squared area with the gauges while taking into account that this area was situated at the same positions of the other devices (Figure 3). The difference in the area has been taken into account and the gutter results have been independently analysed. After 10 min of simulated rain, the splashed soil was collected from each device to evaluate the differences among the measurements. The distance between devices was always larger than 1 m to avoid interferences between them following the recommendations of Geißler et al. [39].

2.4. Comparison of Measurements

The experiments were repeated under controlled conditions, conducting at least 30 measurements with each device in order to compare the device measurements (Table 1), only varying the intensity of simulated rainfall values that were measured prior conducting each experiment in every spot. The measurements are organized in classes of 5 mm/h each, the average and deviation of splash erosion recovered in each interval of the intensity of rain can be also seen in Figure 4 (left graph).

Table 1. Rainfall experiments conducted with bounded and unbounded splash devices, including the number and intervals of coincidence of rainfall intensities.

Types	n	Funnel		Cup		Tcup		Morgan		Total	
		P n	I mm/h	P n	I mm/h	P n	I mm/h	P n	I mm/h	P n	I mm/h
Funnel	48	48	60–176	23	69–172	15	69–173	20	69–142	15	69–116
Cup	48	23	69–173	48	69–173	15	69–173	19	69–142	15	69–116
Tcup	40	15	69–173	15	69–173	50	66–204	20	66–144	15	69–116
Morgan	42	20	69–142	19	69–142	20	66–144	42	69–144	15	69–116

n: number of repetitions; P: points; I: intensity. Note: Gutter has 15 measurements recording 121 mm/h and 15 measurements with 135.86 mm/h. Tower of funnels has six measurements with intensities from 108 to 162.8 mm/h (13 levels of height per measurement).

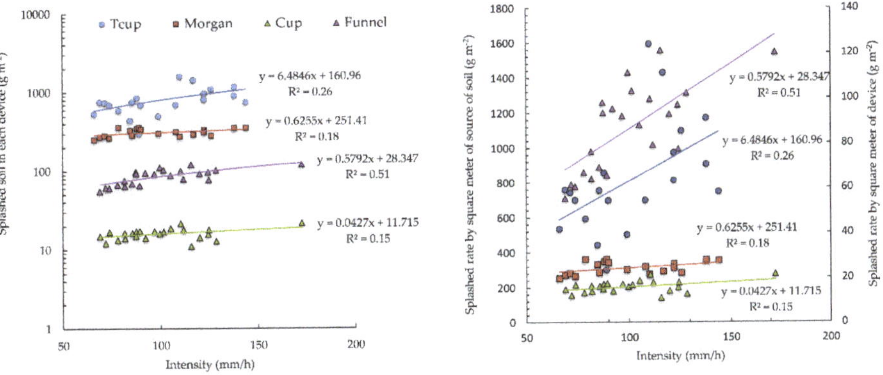

Figure 4. Comparison of the splashed soil recovered at different rainfall intensities using the funnel, cup, Morgan, and Tcup devices.

The data that were recovered from the devices were analysed and compared based on their sensitivity and reproducibility, following the methodology that was proposed by Thomsen et al. [50].

The sensitivity is related to the response of the device to changes in the rainfall intensity. The reproducibility compares the results that were obtained with each device after several simulations using the same intensity. It is important to keep in mind that only the devices operating in a similar way can be directly compared. Accordingly, unbounded devices are compared first and bounded ones afterwards.

It is still possible to obtain the same units ($g\ m^{-2}$) with both types of devices, being conscious of the difference in the performance of bounded and unbounded devices. Obviously, the same units do not represent the same physical process, but they allow for comparing the relation that they have with the changes in intensity.

The design of the bounded devices is generally divided into two parts: one inside, with a certain amount of soil susceptible of being eroded, and one outside, with a deposit to collect the splashed soil. This means that the soil that is collected in the outside area will come from the inner deposit, and absolutely all of the soil moved from the inside deposit to the outside deposit will be measured. Accordingly, the soil lost from the inner deposit (g), divided by the area of this deposit (less than $0.01\ m^2$, depending on the device), represents the splash rate ($g\ m^{-2}$).

In the unbounded devices, as it was explained in the introduction, the calculation of the splash rate is not possible. The soil source area is not precisely established (it is not possible to know the origin of the soil collected), and the devices will not collect a lot of mobilized splash sediment from this undefined area. However, to give the same units of bounded ones, the number of sediments collected in each unbounded device can be divided by its collection area. In both kinds of devices, an increase in the rain intensity should be logically related to an increase in the soil mobilized per area. However, it should be taken into account that the sample that was collected by an unbounded device will be much less than in the bounded one, which collects all of the mobilized soil.

In conclusion, the amount of splashed soil measured are very different, always being bigger in the bounded devices, but the relationship between splash and intensity of rain should be positive in both types, because both types of devices are measuring differently. These differences in absolute values should be not a problem for conducting statistical analysis, because both types of measurements are calculated while using the same units ($g\ m^{-2}$). However, this issue should be considered during the interpretation of the results, and the comparison of bounded and unbounded devices. Additionally, even while taking into account the different capacity that those devices have in recovering splashed soil, we have the possibility to compare the data by calculating the standard score or typifying data, by extracting the mean score from each score and dividing the result by the standard deviation.

Additionally, to study the consistency of the response of the studied devices under different rainfall intensities, the splash rate that was obtained with each splash device was also divided by the rainfall intensity used in each experiment in $g\ h\ m^{-2}\ mm^{-1}$ relating a splash rate per mm of rain. Table 2 shows the average obtained with all of the repetitions of the six types of devices used in the experiments.

The normal distribution of data was checked using the Shapiro–Wilk test. Parametric statistics and tests were used, as the null hypothesis was not rejected in most cases. Pearson's correlation coefficients were used to analyse possible linear relations between devices as well as Spearman's in the cases of non-parametrical distributions (Table 2). With these tests, we may confirm the linear relation between the compared data, although these analyses cannot measure the relationships in the data. To assess these relations, it is necessary a test that usually needs data that can be used as a reference of the real values, but, in this case, it is not known which data should be used as a reference device. Therefore, it was necessary to use the Bland–Altman plot test that is designed for a comparison between methods when it is not possible to know the reference base method. The Bland–Altman plot test defines the intervals of agreement within which 95% of the differences of a method fall as compared to another one. It does not confirm whether those limits are acceptable or not. Moreover, a regression analysis was conducted while using the rainfall intensity and each studied device as parameters. If the devices react with the intensity of the rain, some relation should be found, and if the devices correctly measured the splash erosion, some relation should be found with the results taken by several devices.

Finally, a qualitative analysis comparing the cost, the installation limitations, soil collection, and any extra information given by each splash device was also done by means of the Leopold matrix that is commonly used for the selection of methodologies and decision making [51,52].

3. Results

3.1. Splashed Soil Measured with the Different Devices

The values presented in Table 2 correspond to the average of the total amount of splashed soil, measured with each device, divided by the surface of the device and the rain intensity previously collected to every experiment in each spot. It is possible to observe the variability among some of these results. Funnel, cup, gutter, and tower are unbounded devices and they show similar values, though the tower device collected less soil per area. The tower device has a larger surface divided into different height levels; some of them (the upper ones) collect very little soil, reducing the average. Out of the other three devices with a similar design, the cup collected much less splashed soil than the funnel. Regarding the bounded devices, the Tübingen cup accumulated much more splashed soil than the Morgan tray (Figure 4—right graph).

Table 2. Splashed soil measured with different devices after 10 min of simulated rain with different intensities (60–150 mm h^{-1}).

Type	N	Average (g h mm^{-1} m^{-2})	S.D.	Maximum (g h mm^{-1} m^{-2})	Minimum (g h mm^{-1} m^{-2})	Normality Test (KS/SW) Statistic	P-value	P
Funnel	48	0.84	0.17	1.12	0.5	0.97	0.28	yes
Cup	48	0.17	0.04	0.25	0.09	0.97	0.26	yes
Gutter	30	0.38	0.07	0.52	0.25	0.96	0.28	yes
Tf	78	0.05	0.14	0.65	0	0.36	0	no
Tcup	50	9.80	3.51	19.73	3.33	0.11	0.09	yes
Morgan	42	3.18	0.66	4.69	2.25	0.93	0.01	no

Tf H (m)	N	Average (g h mm^{-1} m^{-2})	S.D.	Maximum (g h mm^{-1} m^{-2})	Minimum (g h mm^{-1} m^{-2})	Normality Test (KS/SW) Statistic	P-value	P
0.075	6	0.48	0.18	0.65	0.19	0.89	0.307	yes
0.185	6	0.07	0.03	0.12	0.03	0.89	0.316	yes
0.295	6	0.04	0.03	0.11	0.01	0.89	0.328	yes
0.405	6	0.02	0.02	0.06	0	0.92	0.476	yes
0.515	6	0.01	0.01	0.03	0	0.93	0.589	yes
0.625	6	0.01	0.01	0.03	0	0.92	0.504	yes
0.735	6	0.01	0.01	0.03	0	0.78	0.037	no
0.845	6	0.00	0.01	0.03	0	0.76	0.026	no
0.955	6	0.01	0.01	0.03	0	0.89	0.311	yes
1.065	6	0.01	0.01	0.02	0	0.81	0.068	yes
1.175	6	0.01	0.01	0.04	0.00	0.75	0.018	no
1.285	6	0.01	0.01	0.04	0	0.88	0.266	yes
1.395	6	0.01	0.01	0.04	0	0.90	0.392	yes

N: number of rainfall simulation experiments; S.D.: standard deviation; Tf: Tower of funnel; S: Kolmogorov Smirnov test (with correction of Lilliefors) performed for 50 or more samples; H: height; SW: Shapiro-Wilk test for less than 50 samples; P: Parametric.

The gutter is separately analysed because of the difference in the total area covered by this device. These results show consistently more splash erosion in the gutter on the left side of the rainfall simulator plot, where the rainfall intensity was slightly higher. Accordingly, there was 10% less of splashed soil than on the left side on the right side of the plot (Figure 5).

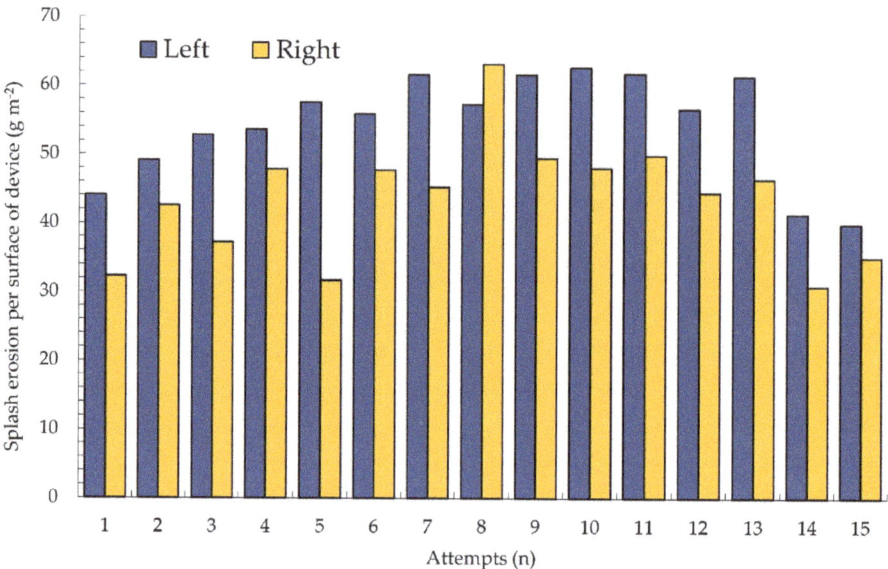

Figure 5. Splashed soil recovered in the gutter device (in relation to different mean intensities observed in the left and right side of the plot).

In Figure 6, the results that were obtained with the tower of funnels showed that, with rainfall intensities up to 150 mm h^{-1}, more than 90% of the total splash erosion reached up to 0.405 m height. At higher elevations, the arrival of splashed soil is scarce or even negligible. If the distribution of collected splashed soil at each height is analysed, it is possible to observe that 73% of the total sediment was collected at a height of 0.075 m, 10% at 0.185 m, around 5% at 0.295 m, and only 1% or less at higher heights.

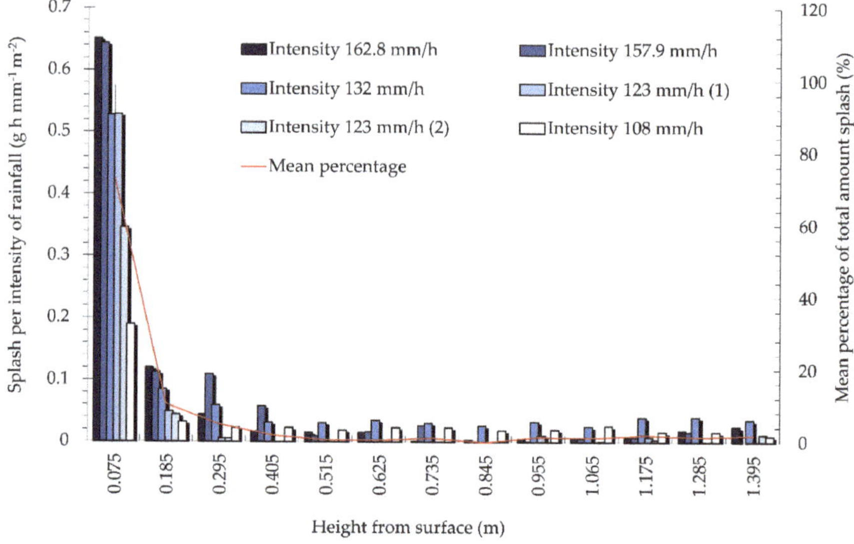

Figure 6. Splashed soil recovered with the tower of funnels using different rainfall intensities.

Table 2 also presents the standard deviation of the measurements of each device, data that may be used to calculate the standard deviation related to the average calculated for each device. Gutter presented the lowest value of (17.7%) and Tcup, the highest one (32.7%). The large percentage that was obtained for the splash tower (248.7%) responds to the intrinsic deviation of the measurements more than to the variation between experiments.

Correlation analyses were carried out in order to find the possible relations between device measurements and rain intensities (Table 3). The algorithm runs through all the possible combinations of independent variables (device measurements) to select the model with the best goodness of fit (highest correlation coefficient). Only the results from the Tcup, Morgan tray, and funnel showed significant correlations with the rain intensities, with the relations of the latter being significant at 99%. Similar relations can be observed when plotting the splashed soil measured with each device vs. the rainfall intensities (Figure 4). The highest regression coefficient was obtained for the funnel, followed by Tcup and Morgan tray (Table 3).

Table 3. Significant correlation coefficients between rainfall intensities and splash erosion measured with different devices.

Device	Index	Correlation Coefficient	P-Value
Funnel	Pearson	0.711 **	<0.001
Morgan	Spearman rho	0.478 *	0.033
Tcup	Pearson	0.509 *	0.022
Cup	Pearson	0.039	0.065

(**) Significant at a 99% confidence level and (*) significant at 95% confidence level.

3.2. Relations between Splash Erosion Devices

While searching for relations between devices (Table 4), there was a significant correlation between cup and gutter, which was likely because both of the devices collect splashed soil from an unbounded area, without any protection against the loss of sample by raindrops. Tcup and funnel also showed a significant correlation that may respond to the capacity of both to react influenced by the rain intensity (Table 3). The similar response to the rain intensity on both of the devices can be observed in Figure 4. On the other hand, the gutter was significantly correlated with the cup, Tcup, Morgan tray, and the tower. The relation that was found between the Morgan tray and the tower is very difficult to explain.

Table 4. Significant correlations at the 95% confidence level between the splash erosion measurements obtained with different devices.

Devices	Index	Correlation Coefficient	P-Value
Gutter/cup	Pearson	0.484 *	0.007
Gutter/Tcup	Pearson	0.444	0.014
Gutter/tower	Pearson	−0.443	0.014
Gutter/Morgan	Spearman rho	−0.383	0.037
Tcup/funnel	Pearson	0.547	0.012
Morgan/tower	Spearman rho	0.449 *	0.003

* Significant at a 99% confidence level.

When devices with a similar source of soil, like cup and funnel, were compared and the results the were obtained with them were different (Table 2 and Figure 7), it is necessary to think which one represented the reality better and whether the different results can be explained by the different sizes of the different design of the devices. Figure 7 showed the relations between the measurements of both devices and a linear relationship with a very low coefficient of determination (R^2) among them. This weak relation can also be observed in the Bland Altman plot, where the results showed an agreement interval that is extremely large (Figure 8—left), where a large bias between the two devices has been found, similarly to the result from the regression analysis (Figure 7).

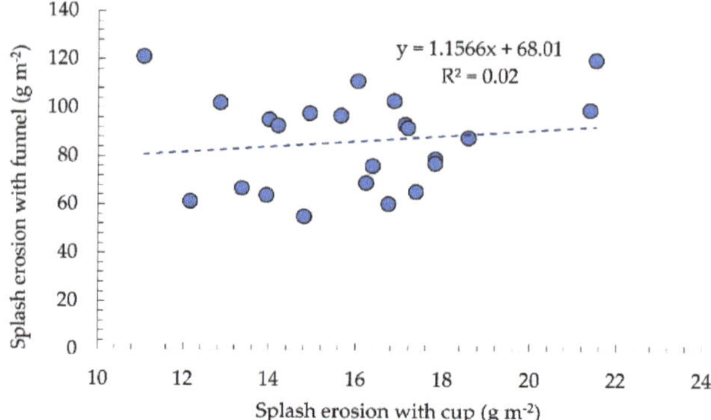

Figure 7. Relationships between splash erosion measured with cup and funnel devices (only the mean values per each intensity are presented for both devices).

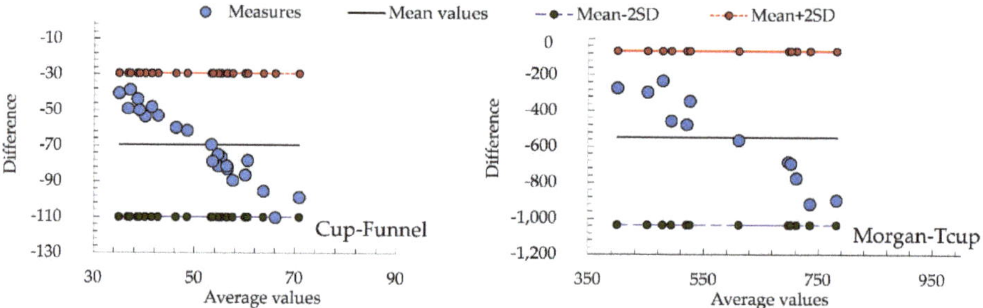

Figure 8. Bland Altman plot to represent a comparison between the average and the difference of cup and funnel measurements (**left part**) and Morgan and Tcup ones (**right part**).

Similarly to the results with unbounded devices, when the analysis is done on the splash erosion that was collected from bounded areas, using both the Morgan tray and the Tcup, the splash erosion increases with rainfall intensity, but in the Morgan tray very smoothly (Figure 9). The different value given by the two different measuring devices is due to their different peculiarities, and, if the tendency is the same, this cannot be considered as a relevant result. However, in addition, there are other differences, for example, the accuracy of the data that were obtained with the Tcup is not very high. This result, together with the large disagreement interval detected by the Bland Altman method (Figure 8—right), points out the great differences between methods.

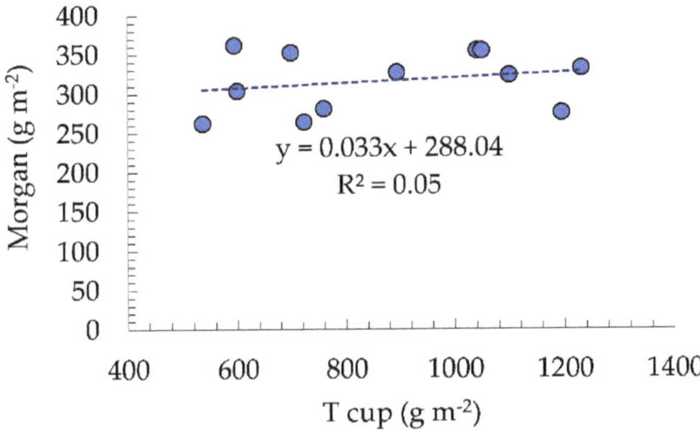

Figure 9. Correlation of splash erosion recovered with Tcup and Morgan system.

Linear regression analyses have been performed, with a confidence level of 95%, while looking for a model that is able to relate the splash results that were obtained with the different devices (Table 5). Significant relations with the rain intensity could only be established for the funnel, cup, Tcup, and Morgan tray, confirming the correlation results (Tables 3 and 4). However, the regression coefficients were, in the best case, slightly higher than 0.5, which means that there is a low agreement between the results that were obtained with the different devices and between the devices and the rain intensity.

Table 5. Linear regression equations relating to splash devices results and rainfall intensities. Factors proving such adjustment are also presented.

Equation	R^2	Adjusted R^2	P-Value	D-W
Funnel = 5.931 + 0.015Tcup + 0.702Int	0.591	0.543	0.001	2.298
Funnel = 5.189 + 0.842Int	0.540	0.514	<0.001	—
Funnel = 0.022 + 0.811Int + 0.025Morg	0.542	0.488	0.001	—
Funnel = 10.739 + 0.834Int + 0.028Morg − 0.869Cup	0.553	0.469	0.004	1.864
Funnel = 16.263 + 0.867Int − 0.854Cup	0.551	0.498	0.001	1.308
Funnel = −15.077 + 0.033Tcup + 0.229Morg	0.495	0.436	0.003	2.103
Tcup = 942.026 + 8.858Int − 3.21Morg	0.359	0.284		1.704
Tcup = 35.001 + 9.245Funnel	0.300	0.261	0.012	1.638
Tcup = 160.957 + 6.485Int	0.259	0.218		1.704
Morg = 243.293 + 0.739Int	0.257	0.216	0.023	1.096
Morg = 230.341 + 0.724Int + 0.919Cup	0.260	0.173	0.077	1.096

Int: mean intensity; Morg: Morgan; Tcup: Tübingen cup; D-W: Durbin Watson residues distribution.

3.3. Complementary Information about the Splash Devices

The difficulty of installation in the field is one of the factors that can determine the selection of a splash device (Table 6). All of the devices selected for this study can be used in the field, with the only limitation of the gutter, which is not advisable to use in fieldwork, because its large area can easily be polluted by particles that are transported by air when installed. The weight and size are also classified into categories, where the funnel, cup, and Tcup are the best devices, because their smaller sizes simplify the installation. There is also very little disturbance in the soil surface with the selected devices, except the hole required by the funnel in the soil, and the removal of some vegetation from the surface for the Morgan and the gutter devices (the disturbances will be minimal in bare soils). All of the chosen devices were very cheap to build (Table 6), except the tower of funnels that was a little more expensive because it demands a structure to maintain the funnels in their position.

Table 6. Complementary information about the splash devices scored from 1 (lowest value) to 3 (highest value).

(A)						
Comparison among Splash Devices						
Evaluated Parameter	**Funnel**	**Cup**	**Gutter**	**Tower of Funnels**	**Tcup**	**Morgan**
Possibility of been used in the field	3	3	2	3	3	3
Installation weight	3	3	1	1	2	2
Installation size	3	3	1	1	3	1
Disturbance of soil surface during installation	2	3	2	3	3	2
Price	3	3	2	1	2	2
Difficulty on the collection of samples	3	3	3	2	3	2
Possibility of losing sample when collected	3	2	2	3	3	2
It collects information that no other device does	1	1	1	3	1	2
Total	21	21	14	17	20	16

(B)		
Evaluated Parameter	**Points**	**Rubric**
Price	1	More than 20 euros per plot
	2	Between 10–20 euros per plot
	3	Less than 10 euros
Installation weigh	1	More than 1 kg
	2	From 0.2 to 1 kg
	3	Less than 0.2 kg
Installation size	1	Bigger than 200 cm^3
	2	From 200–100 cm^3
	3	Less than 100 cm^3
Disturbance of soil surface during installation	1	The installation produces some disturbance
	2	The installation produces very little disturbance
	3	The installation hardly produces disturbance
Difficulty on the collection of samples	1	It cannot be done by one person alone
	2	It can be done by one person but with more time
	3	It can be done by one person fast.
Possibility of losing sample when collecting	1	It is very likely to lose some sample during sampling
	2	It is possible to lose some sample during sampling
	3	It is impossible to lose sample in the sampling
Possibility of been used in the field	1	It is very difficult to install in the field (only rainfall simulations)
	2	It is difficult to install in the field
	3	It is easy to install in the field
It collects information that no other device does	1	It does not collect different information
	2	It collects different information
	3	It does collect special information

The easiness in retrieving the splashed soil and the probability of losing the sample during this process are also important factors. The funnel and the Tcup are the devices that better protect the sample as well as those of easiest use by one single researcher. Finally, the last characteristic that is useful to compare devices is the capacity for providing additional information. Thus, the tower of funnels can tell the height that a splashed particle can achieve, or the Morgan is able to differentiate between upslope and downslope splash erosion. In Table 6, all of these features are scored from 1 (lowest value) to 3 (highest value), which provides additional information regarding the performance of the different devices. According to this, the funnel and the cup are the best splash devices from a functionality point of view.

In any case, there is a lot of research done with the different splash devices, and a summary of the lack of standardization in the values and units reported in the published studies carried out until now can be observed in Table 7.

Table 7. Summary of splash rates reported in the literature using different splash devices.

Ref.	Device	Maximum Splash Detachment Rate			Minimum Splash Detachment Rate		
		R-U	RI	CR	R-U	RI	CR
Zhou et al. [53]	cup	0.78 g min^{-1}	38 mm h^{-1}	27 mm	0.1 g min^{-1}	38 mm h^{-1}	10 mm
Ma et al. [54]	cup	930.11 g m^{-2} h^{-1}	59.94 mm h^{-1}	n.a.	41.43 g m^{-2} h^{-1}	31.52 mm h^{-1}	n.a.
Fernández-Raga et al. [37]	cup	100 g m^{-2}	n.a.	131.7 mm	2.3 g m^{-2}	n.a.	1.5 mm
Fernández-Raga [48]	funnel	220 g m^{-2}	n.a.	131.7 mm	6 g m^{-2}	n.a.	1.5 mm
Jordán et al. [35]	funnel	21.74 g	360 mm h^{-1}	60 mm	1.29 g	360 mm h^{-1}	60 mm
Jomaa et al. [38]	flume	2000 g m^{-2}	80 l m^{-1}	n.a.	n.a.	n.a.	n.a.
Jomaa et al. [55]	flume	603 g	60 mm h^{-1}	n.a.	301 g	60 mm h^{-1}	n.a.
GeiBler et al. [39]	Tcup	77.42 g m^{-2}	n.a.	37 mm	9.93 g m^{-2}	n.a.	21 mm
Liu et al. [56]	Tcup	9000 g m^{-2}	7 mm/10 min^{-1}	55 mm	10 g m^{-2}	1 mm/10 min^{-1}	5 mm
Beguería et al. [40]	Morgan	2.77 g	92.9 mm h^{-1}	n.a.	0.22 g	92.9 mm h^{-1}	n.a.
Beguería et al. [40]	Morgan	5642.98 g m^{-2}	92.9 mm h^{-1}	n.a.	448.9 g m^{-2}	92.9 mm h^{-1}	n.a.
Angulo-Martinez et al. [41]	Morgan	6.06 g	n.a.	61.3 mm	0.97 g	n.a.	8.2 mm
Angulo-Martinez et al. [41]	Morgan	12,367.3 g m^{-2}	n.a.	61.3 mm	1979.6 g m^{-2}	n.a.	8.2 mm

Ref: reference; RI: rainfall intensity; n.a.: not available; CR: Cumulated rainfall; R-U: Rate and unit.

4. Discussion

4.1. The Rainfall Simulator and Laboratory Sensitivity

There are two potential causes of error in this comparative research—potential changes in the rainfall and errors with the sample collection. The rainfall simulator that was used in this study is a very stable instrument, with less than 5% variance between the average rainfall intensities among repetitions [46]. A great effort in getting similar intensities along the simulations with the different devices was done, but small differences between repetitions could happen (because of the simultaneous use of the rainfall simulator by other researchers). In any case, the intensity that was used on each device was always perfectly known.

Other sources of error can be the transport of the soil samples to the laboratory (to weigh them) where the loss of soil can occur. Additionally, the filter papers that are used for collecting the splashed soil can absorb some humidity from the air. Accordingly, after drying the filters in the oven, it is very important to keep them in a desiccator until they have an adequate temperature to operate with them. Filters were kept for 30 min. in the desiccator before weighing them, to ensure a stable weight of the samples in order to make the experiment reproducible.

4.2. Splash Devices Measurements: Lack of Standardization

It is surprising that it is not common for the comparison between device results to have been done so far. The reason may be based on the difficulty of designing a proper experiment that is able to compare data that are sensitive to soil samples, rainfall characteristics (simulated or real), duration of events, etc. in a way that supports the achievement of general conclusions.

Although, in the experiment, the main factors that are involved in the variability of measurements with splash devices were controlled (homogeneous soil (sand), reproducible rainfall and constant slope), the results obtained, presented a high variability (Table 2). This is in agreement with other studies [49,50], for example, devices like the cup used under the simulated rain or in field conditions with similar intensities (31.5 and 38 mm h^{-1}, respectively) showed very different results (Table 7).

These studies showed higher splash rates in funnels than in cups in spite of their similar design, similarly to our experiment.

4.2.1. Bounded vs. Unbounded Devices

One analysis was carried out comparing the devices working with an unbounded source of soil, and another one, with those performing with bounded areas. The bounded devices produced higher values of splash erosion than the unbounded ones. Devices like the Morgan and the Tcup may lose some sand by wash-off effect, because they are completely filled with sand and some grains of sand suspended in the water deposited on the top may fall from the device. This means that raindrops falling on the surface can pond and cause some particles to move with the surface flow instead of by splash transportation. On the contrary, devices like the cup and the funnel have a bound of 3 cm high to avoid the particles coming from processes different to splash erosion. In this case, only the splashed particles that are ejected higher than 3 cm will be able to enter into the device. Another effect that may appear is the rim effect or the lowering of the soil surface in relation to the solid rim of the cup, but this was not detected in this study.

4.2.2. Size of the Sampling Area and Design

In relation to unbounded devices, funnels, which have a similar design to cups, showed higher splash rates than cups, as was also reported in other studies [35,37]. The devices with a larger sampling area, like the gutter or collection through, seem to have high splash rates, but these values cannot be used for comparison, because the data about the intensity or total rainfall producing the splash are not described (only the flow is given). For the bounded devices, the data that were collected with Morgan or Tcup also showed large variability [39,40,56]. In this study, the Tübingen cup presented much more splashed soil than the Morgan tray. The reason for this difference could be related to the loss of sample from the Morgan device during its collection, bigger than the loss from the cup in Tcup. The loss of sample in both cases, together with the difference in sand-size (in Tcup sand is 212.50 μm, while in Morgan is 369 μm) increased the difference in the average of the loss sample between them. The analysis of these results confirms the difficulty in comparing data coming from different researches and/or devices. Data that were obtained from the tower of funnels showed that splashed soil above 0.5 m is negligible. However, we observed some particles to reach a height of 1.395 m.

The size of the device is an important factor to keep in mind when comparing the results in splash studies. On the one hand, and according to measurements of this research, devices with a larger size—like funnels (12 cm in diameter)—have shown higher splashed soil than smaller ones—cups (10 cm in diameter). These results are not in agreement with Poesen and Torri [42], who reported an inverse influence of device size in the reception of the splashed sample. Although, indeed, it is not a consequence of the size, but of the funnel design that avoids surface wash of the sample [37]. It was observed that many of the splashed particles that had already been splashed inside the cup were moved outside again after the impact of new raindrops. On the other hand, the Tcup with the smallest area (only 19.63 cm^2) is the device that collects the highest quantity of splashed soil (Table 2), probably because its measurement depending on the lost soil is more exact. For this reason, and because the Tcup is more sensitive to very low intensities [39], it has been selected for splash experiments under plants. However, the largest device used here, the gutter, showed splash values from 30 up to 60 g m^{-2}, which are between the values of splashed soil that were collected with cups and funnels. These quantities seem logic because the gutter can also be affected by surface wash processes, losing sample from re-splash of the soil deposited on the gutter. In any case, these values cannot be compared with those that were obtained with other devices because of gutter's design (6 m long), which make very difficult to determine the rain intensity over the whole surface of the gutter (particularly in cases as ours, in which the rainfall intensity was different along the surface of the studied plot, Figure 3).

Other factors affecting the splash erosion measurements, like soil moisture [5], acidic components in the rain [57] structure and size of soil aggregates [6], or the presence of rock fragments [58] are not considered in this study, because both the composition and the conditions of soil (sand) were homogeneous along the experiments. However, the duration of simulated rainfall could also have some influence on the outcomes when comparing the results that were obtained with those of other authors. Mermut et al. [59] simulated rainfall during 120 min. and observed a decrease in the splash erosion after 50 min. In this study, some trials were done simulating the rain for 20 min., but significant differences with regard to 10 min were not found (data not shown). This result contradicts that of Katebikord et al. [60], who found that even five minutes of change between simulations might influence the erosion processes and the gathering of sediments from splash erosion. After the first drop, there is a change in the conditions in the place impacted by the drop, because a micro-pool appears [5], and the changes in the humidity also have very little effect when there is saturation.

4.3. Influence of Rainfall Intensity on Measured Splash Erosion

The splash measurements that were obtained with the studied devices showed different relations with rain intensities, being only Tcup, Morgan tray, and funnel results in those presenting significant correlations (Figure 4). It seems that the response of the funnel to changes in the rain intensity is higher than cup's one, which is nearly unappreciable. Indeed, close observation of the cup showed that the amount of splashed soil is not changing after some time. This means that there is a threshold after which the cup always gives the same splash rate, regardless of the intensity, because it reaches a balance. The main reason for this balance is the removal of splashed particles deposited already on the filter paper when washed by new raindrops.

The funnel obtained the highest regression coefficient, followed by the Tcup and Morgan tray. These results indicate that both of the devices measuring splash erosion rates from a defined soil source represent the movement of soil particles that are produced by different rain intensities in a similar way, probably because they lost fewer sediments by surface wash. According to these results, the funnel is the best device that is able to measure the soil splashed at different rain intensities, because its lateral protection prevents surface wash of the sample reducing the loss of collected soil, as its low percentage of deviation from the mean confirms (Table 2).

4.4. Specific Problems and User Recommendations

Properties like weight and volume of the device are also very important when evaluating the possibility of using it in the field (Table 6). Some devices are small and easy to carry, while larger devices, like the Morgan or the gutter, need to be transported by car. Another important aspect to have in mind is the disturbance that is produced by the device over the experimental surface. In splash erosion research, measurements in highly sensitive areas, like burnt zones, agricultural terraces, or mines, are very common, and disturbances that are related to the measurement technique may produce inaccurate results. In general, the devices that were selected for this study do not generate negative impacts on the studied soil surface, although the funnel, gutter, and Morgan devices can have a higher impact than the others can.

The easiness in the sampling of the splashed soil collected, and the probability of losing the sample during this process, is another important factor in the comparison among the devices. It is indeed one of the most important features to take into account, because the success of the experiment depends on the feasibility of collecting the splashed soil, with the least personal and economic effort. In this aspect, the funnel and Tcup look at the best devices because one single researcher can do the sampling. The Morgan device is the most sensitive and difficult-to-handle apparatus.

For the future, the next mandatory step will be to carry out new attempting to assess device performance in the open or beneath crop or plant canopies in order to compare our results under real conditions when considering variable inclinations, different rainfall intensities (time-space), and roughness.

5. Conclusions and Future Recommendations

In this study, we compared six different devices that were designed to measure splash erosion. The analysis of the devices shows that the funnel and the cup are the best devices because of their functionality, which includes low-cost, easy installation, sampling, and the information that they provide. However, only the funnel shows a significant relation with the rain intensity. This can be related to the design of the funnel device that prevents the loss of sample by washing. For all of these reasons, the use of the funnel is highly recommended to estimate splash erosion. However, the use of the Tcup is advised to measure the kinetic energy of the rainfall, because it is a contrasting method, which can be used in isolated places as well as being very easy to transport and cost-effective. The Morgan tray and the tower of funnels present a more complex use and demand more work to collect the samples than the others do; however, the acquired data provide information about distance and potential erosivity power of the rain to move natural soil, which other methods do not provide.

The height that splash erosion is able to reach can be detected with the tower of funnels; however, this method does not properly measure the splashed soil and it should only be used as a complement of other devices, such as the funnel.

Another interesting conclusion is that the researcher should decide in advance whether the calculation of a splash rate is necessary for his or her research, selecting a bounded splash device if this is indeed needed. Unfortunately, in this study, it was not possible to establish a relation between the different splash devices, because of the high variability and low agreement of the results. For future research in splash erosion, the recommendation is to exclusively make comparisons with studies that use the same methodology, because the transformation of the results from one device to another is uncertain and the comparison is unclear.

Author Contributions: Conceptualization, M.F.-R., J.C., S.D.K.; methodology, M.F.-R., J.C., S.D.K.; validation, M.F.-R.; formal analysis, M.F.-R., J.C.; data curation, M.F.-R.; writing—original draft preparation, M.F.-R., J.C., S.D.K., J.R.-C.; writing—review and editing, M.F.-R., S.D.K., J.R.-C.; visualization, M.F.-R., J.R.-C.; supervision, S.D.K.

Funding: This research was funded by a José Castillejo Grant, a Program for Junior professors (CAS16/00385) of the Spanish Ministry of Education, Culture and Sports.

Acknowledgments: The authors want to thank Shmuel Assouline, for his help, advice and technical support, and Piet Peters for his help in the construction of the devices and Antonio Jordán for his advices.

Conflicts of Interest: The authors declare no conflict of interest.

References

1. Moreno-de las Heras, M.; Nicolau, J.M.; Merino-Martín, L.; Wilcox, B.P. Plot-scale effects on runoff and erosion along a slope degradation gradient. *Water Resour. Res.* **2010**, *46*, W04503. [CrossRef]
2. Nearing, M.A.; Yin, S.; Borrelli, P.; Polyakov, V.O. Rainfall erosivity: An historical review. *CATENA* **2017**, *157*, 357–362. [CrossRef]
3. Fernández-Raga, M.; Palencia, C.; Keesstra, S.; Jordán, A.; Fraile, R.; Angulo-Martínez, M.; Cerdà, A. Splash erosion: A review with unanswered questions. *Earth-Sci. Rev.* **2017**, *171*, 463–477.
4. Quansah, C. The Effect of Soil Type, Slope, Rain Intensity and Their Interactions on Splash Detachment and Transport. *J. Soil Sci.* **1981**, *32*, 215–224. [CrossRef]
5. Ryżak, M.; Bieganowski, A.; Polakowski, C. Effect of Soil Moisture Content on the Splash Phenomenon Reproducibility. *PLoS ONE* **2015**, *10*, e0119269. [CrossRef] [PubMed]
6. Fox, D.M.; Darboux, F.; Carrega, P. Effects of fire-induced water repellency on soil aggregate stability, splash erosion, and saturated hydraulic conductivity for different size fractions. *Hydrol. Process.* **2007**, *21*, 2377–2384. [CrossRef]
7. Ghahramani, A.; Ishikawa, Y.; Gomi, T.; Shiraki, K.; Miyata, S. Effect of ground cover on splash and sheetwash erosion over a steep forested hillslope: A plot-scale study. *CATENA* **2011**, *85*, 34–47. [CrossRef]
8. Römkens, M.J.M.; Helming, K.; Prasad, S.N. Soil erosion under different rainfall intensities, surface roughness, and soil water regimes. *Catena* **2002**, *46*, 103–123. [CrossRef]

9.	Liu, H.; Blagodatsky, S.; Giese, M.; Liu, F.; Xu, J.; Cadisch, G. Impact of herbicide application on soil erosion and induced carbon loss in a rubber plantation of Southwest China. *CATENA* **2016**, *145*, 180–192. [CrossRef]

10.	Marzen, M.; Iserloh, T.; Casper, M.C.; Ries, J.B. Quantification of particle detachment by rain splash and wind-driven rain splash. *CATENA* **2015**, *127*, 135–141. [CrossRef]

11.	Campo, J.; Gimeno-García, E.; Andreu Pérez, V.; González-Pelayo, Ó.; Rubio, J.L. Effects of experimental repeated fires in the soil aggregation and its temporal evolution. In Proceedings of the EGU General Assembly 2013, Vienna, Austria, 7–12 April 2013.

12.	Fukuyama, T.; Takenaka, C. Upward mobilization of 137Cs in surface soils of Chamaecyparis obtusa Sieb. et Zucc. (hinoki) plantation in Japan. *Sci. Total Environ.* **2004**, *318*, 187–195. [CrossRef]

13.	Remke, A.; Rodrigo-Comino, J.; Gyasi-Agyei, Y.; Cerdà, A.; Ries, J.B. Combining the Stock Unearthing Method and Structure-from-Motion Photogrammetry for a Gapless Estimation of Soil Mobilisation in Vineyards. *ISPRS Int. J. Geo-Inf.* **2018**, *7*, 461. [CrossRef]

14.	Di Stefano, C.; Ferro, V.; Palmeri, V.; Pampalone, V. Measuring rill erosion using structure from motion: A plot experiment. *CATENA* **2017**, *156*, 383–392. [CrossRef]

15.	Cerdà, A. Effects of rock fragment cover on soil infiltration, interrill runoff and erosion. *Eur. J. Soil Sci.* **2001**, *52*, 59–68. [CrossRef]

16.	Lesschen, J.P.; Cammeraat, L.H.; Nieman, T. Erosion and terrace failure due to agricultural land abandonment in a semi-arid environment. *Earth Surf. Process. Landf.* **2008**, *33*, 1574–1584. [CrossRef]

17.	Bergkamp, G.; Cerdà, A.; Imeson, A.C. Magnitude-frequency analysis of water redistribution along a climate gradient in Spain. *CATENA* **1999**, *37*, 129–146. [CrossRef]

18.	Baiamonte, G.; Minacapilli, M.; Novara, A.; Gristina, L. Time Scale Effects and Interactions of Rainfall Erosivity and Cover Management Factors on Vineyard Soil Loss Erosion in the Semi-Arid Area of Southern Sicily. *Water* **2019**, *11*, 978. [CrossRef]

19.	Biddoccu, M.; Opsi, F.; Cavallo, E. Relationship between runoff and soil losses with rainfall characteristics and long-term soil management practices in a hilly vineyard (Piedmont, NW Italy). *Soil Sci. Plant Nutr.* **2014**, *60*, 92–99. [CrossRef]

20.	Lavee, H.; Imeson, A.C.; Sarah, P. The impact of climate change on geomorphology and desertification along a mediterranean-arid transect. *Land Degrad. Dev.* **1998**, *9*, 407–422. [CrossRef]

21.	Sadeghi, S.H.; Kiani Harchegani, M.; Asadi, H. Variability of particle size distributions of upward/downward splashed materials in different rainfall intensities and slopes. *Geoderma* **2017**, *290*, 100–106. [CrossRef]

22.	Sanchez-Moreno, J.F.; Mannaerts, C.M.; Jetten, V.; Löffler-Mang, M. Rainfall kinetic energy–intensity and rainfall momentum–intensity relationships for Cape Verde. *J. Hydrol.* **2012**, *454–455*, 131–140. [CrossRef]

23.	Wang, L.; Shi, Z.H.; Wang, J.; Fang, N.F.; Wu, G.L.; Zhang, H.Y. Rainfall kinetic energy controlling erosion processes and sediment sorting on steep hillslopes: A case study of clay loam soil from the Loess Plateau, China. *J. Hydrol.* **2014**, *512*, 168–176. [CrossRef]

24.	Biddoccu, M.; Zecca, O.; Audisio, C.; Godone, F.; Barmaz, A.; Cavallo, E. Assessment of Long-Term Soil Erosion in a Mountain Vineyard, Aosta Valley (NW Italy). *Land Degrad. Dev.* **2018**, *29*, 617–629. [CrossRef]

25.	Vanwalleghem, T.; Amate, J.I.; de Molina, M.G.; Fernández, D.S.; Gómez, J.A. Quantifying the effect of historical soil management on soil erosion rates in Mediterranean olive orchards. *Agric. Ecosyst. Environ.* **2011**, *142*, 341–351. [CrossRef]

26.	Cerdà, A. Effect of climate on surface flow along a climatological gradient in Israel: A field rainfall simulation approach. *J. Arid Environ.* **1998**, *38*, 145–159. [CrossRef]

27.	Dunne, T.; Malmon, D.V.; Mudd, S.M. A rain splash transport equation assimilating field and laboratory measurements. *J. Geophys. Res. F Earth Surf.* **2010**, *115*. [CrossRef]

28.	Boix-Fayos, C.; Martínez-Mena, M.; Arnau-Rosalén, E.; Calvo-Cases, A.; Castillo, V.; Albaladejo, J. Measuring soil erosion by field plots: Understanding the sources of variation. *Earth-Sci. Rev.* **2006**, *78*, 267–285. [CrossRef]

29.	Stroosnijder, L. Measurement of erosion: Is it possible? *Catena* **2005**, *64*, 162–173. [CrossRef]

30.	Iserloh, T.; Fister, W.; Seeger, M.; Willger, H.; Ries, J.B. A small portable rainfall simulator for reproducible experiments on soil erosion. *Soil Tillage Res.* **2012**, *124*, 131–137. [CrossRef]

31.	Iserloh, T.; Ries, J.B.; Arnáez, J.; Boix-Fayos, C.; Butzen, V.; Cerdà, A.; Echeverría, M.T.; Fernández-Gálvez, J.; Fister, W.; Geißler, C.; et al. European small portable rainfall simulators: A comparison of rainfall characteristics. *Catena* **2013**, *110*, 100–112. [CrossRef]

32. Gholami, L.; Sadeghi, S.H.; Homaee, M. Straw mulching effect on splash erosion, runoff, and sediment yield from eroded plots. *Soil Sci. Soc. Am. J.* **2013**, *77*, 268. [CrossRef]

33. Rodrigo-Comino, J.; Senciales, J.M.; Sillero-Medina, J.A.; Gyasi-Agyei, Y.; Ruiz-Sinoga, J.D.; Ries, J.B. Analysis of Weather-Type-Induced Soil Erosion in Cultivated and Poorly Managed Abandoned Sloping Vineyards in the Axarquía Region (Málaga, Spain). *AirSoil Water Res.* **2019**, *12*, 1178622119839403. [CrossRef]

34. Cerdà, A.; Ackermann, O.; Terol, E.; Rodrigo-Comino, J. Impact of Farmland Abandonment on Water Resources and Soil Conservation in Citrus Plantations in Eastern Spain. *Water* **2019**, *11*, 824. [CrossRef]

35. Jordán, A.; Zavala, L.M.; Granged, A.J.P.; Gordillo-Rivero, Á.J.; García-Moreno, J.; Pereira, P.; Bárcenas-Moreno, G.; de Celis, R.; Jiménez-Compán, E.; Alanís, N. Wettability of ash conditions splash erosion and runoff rates in the post-fire. *Sci. Total Environ.* **2016**, *572*, 1261–1268. [CrossRef] [PubMed]

36. Ma, Y.; Li, X.; Guo, L.; Lin, H. Hydropedology: Interactions between pedologic and hydrologic processes across spatiotemporal scales. *Earth-Sci. Rev.* **2017**, *171*, 181–195. [CrossRef]

37. Fernández-Raga, M.; Fraile, R.; Keizer, J.J.; Varela Teijeiro, M.E.; Castro, A.; Palencia, C.; Calvo, A.I.; Koenders, J.; Da Costa Marques, R.L. The kinetic energy of rain measured with an optical disdrometer: An application to splash erosion. *Atmos. Res.* **2010**, *96*, 225–240. [CrossRef]

38. Jomaa, S.; Barry, D.A.; Brovelli, A.; Heng, B.C.P.; Sander, G.C.; Parlange, J.-Y.; Rose, C.W. Rain splash soil erosion estimation in the presence of rock fragments. *CATENA* **2012**, *92*, 38–48. [CrossRef]

39. Geißler, C.; Kühn, P.; Böhnke, M.; Bruelheide, H.; Shi, X.; Scholten, T. Splash erosion potential under tree canopies in subtropical SE China. *CATENA* **2012**, *91*, 85–93. [CrossRef]

40. Beguería, S.; Angulo-Martínez, M.; Gaspar, L.; Navas, A. Detachment of soil organic carbon by rainfall splash: Experimental assessment on three agricultural soils of Spain. *Geoderma* **2015**, *245–246*, 21–30.

41. Angulo-Martínez, M.; Beguería, S.; Navas, A.; Machín, J. Splash erosion under natural rainfall on three soil types in NE Spain. *Geomorphology* **2012**, *175–176*, 38–44.

42. Poesen, J.; Torri, D. The effect of cup size on splash detachment and transport measurements. Part I. Field measurements. *Geomorphic Process. Environ. Strong Seas. Contrasts* **1988**, *12*, 113–126.

43. Leguédois, S.; Planchon, O.; Legout, C.; Le Bissonnais, Y. Splash Projection Distance for Aggregated Soils. *Soil Sci. Soc. Am. J.* **2005**, *69*, 30–37. [CrossRef]

44. Morgan, R.P.C. Field studies of rainsplash erosion. *Earth Surf. Process.* **1978**, *3*, 295–299. [CrossRef]

45. Scholten, T.; Geißler, C.; Goc, J.; Kühn, P.; Wiegand, C. A new splash cup to measure the kinetic energy of rainfall. *J. Plant Nutr. Soil Sci.* **2011**, *174*, 596–601. [CrossRef]

46. Lassu, T.; Seeger, M.; Peters, P.; Keesstra, S.D. The Wageningen rainfall simulator: Set-up and calibration of an indoor nozzle-type rainfall simulator for soil erosion studies. *Land Degrad. Dev.* **2015**, *26*, 604–612. [CrossRef]

47. Jameson, A.R.; Kostinski, A.B. What is a Raindrop Size Distribution? *Bull. Amer. Meteor. Soc.* **2001**, *82*, 1169–1178. [CrossRef]

48. Fernández-Raga, M.; Castro, A.; Palencia, C.; Calvo, A.I.; Fraile, R. Rain events on 22 October 2006 in León (Spain): Drop size spectra. *Atmos. Res.* **2009**, *93*, 619–635. [CrossRef]

49. de Sousa Júnior, S.F.; Mendes, T.A.; de Siqueira, E.Q.; de Sousa Júnior, S.F.; Mendes, T.A.; de Siqueira, E.Q. Development and calibration of a rainfall simulator for hydrological studies. *RBRH* **2017**, *22*. [CrossRef]

50. Thomsen, L.M.; Baartman, J.E.M.; Barneveld, R.J.; Starkloff, T.; Stolte, J. Soil surface roughness: Comparing old and new measuring methods and application in a soil erosion model. *Soil* **2015**, *1*, 399–410. [CrossRef]

51. Valizadeh, S.; Hakimian, H. Evaluation of waste management options using rapid impact assessment matrix and Iranian Leopold matrix in Birjand, Iran. *Int. J. Environ. Sci. Technol.* **2018**, *16*, 3337–3354. [CrossRef]

52. Jebelli, J.; Madramootoo, C.; Gebru, G.T. A Simplified Matrix Approach to Perform Environmental Impact Assessment of Small-scale Irrigation Schemes: An Application to Mekabo Scheme in Tigray, Ethiopia. *Am. J. Environ. Eng.* **2017**, *7*, 21–34.

53. Zhou, H.; Peng, X.; Darboux, F. Effect of Rainfall Kinetic Energy on Crust Formation and Interrill Erosion of an Ultisol in Subtropical China. *Vadose Zone J.* **2013**, *12*. [CrossRef]

54. Ma, B.; Yu, X.; Ma, F.; Li, Z.; Wu, F. Effects of Crop Canopies on Rain Splash Detachment. *PLoS ONE* **2014**, *9*, e99717. [CrossRef] [PubMed]

55. Jomaa, S.; Barry, D.A.; Brovelli, A.; Sander, G.C.; Parlange, J.-Y.; Heng, B.C.P.; Tromp-van Meerveld, H.J. Effect of raindrop splash and transversal width on soil erosion: Laboratory flume experiments and analysis with the Hairsine–Rose model. *J. Hydrol.* **2010**, *395*, 117–132. [CrossRef]

56. Liu, W.; Luo, Q.; Li, J.; Wang, P.; Lu, H.; Liu, W.; Li, H. The effects of conversion of tropical rainforest to rubber plantation on splash erosion in Xishuangbanna, SW China. *Hydrol. Res.* **2015**, *46*, 168–174. [CrossRef]
57. Kavian, A.; Alipour, A.; Soleimani, K.; Gholami, L.; Smith, P.; Rodrigo-Comino, J. The increase of rainfall erosivity and initial soil erosion processes due to rainfall acidification. *Hydrol. Process.* **2019**, *33*, 261–270. [CrossRef]
58. Poesen, J.W.; Torri, D.; Bunte, K. Effects of rock fragments on soil erosion by water at different spatial scales: A review. *CATENA* **1994**, *23*, 141–166. [CrossRef]
59. Mermut, A.R.; Luk, S.H.; Römkens, M.J.M.; Poesen, J.W.A. Soil loss by splash and wash during rainfall from two loess soils. *Geoderma* **1997**, *75*, 203–214. [CrossRef]
60. Katebikord, A.; Khaledi Darvishan, A.; Alavi, S.J. Changeability of soil erosion variables in small field plots from different rainfall durations with constant intensity. *J. Afr. Earth Sci.* **2017**, *129*, 751–758. [CrossRef]

Article

Effects of Roughness Coefficients and Complex Hillslope Morphology on Runoff Variables under Laboratory Conditions

Masoud Meshkat [1], Nosratollah Amanian [1,*], Ali Talebi [2], Mahboobeh Kiani-Harchegani [2] and Jesús Rodrigo-Comino [3,4]

[1] Department of Civil Engineering, Faculty of Engineering, Yazd University, Yazd P.O. Box 89195-741, Iran; masoudmeshkat@ymail.com

[2] Department of Watershed Management, Faculty of Natural Resources, Yazd University, Yazd P.O. Box 89195-741, Iran; talebisf@yazd.ac.ir (A.T.); mahboobeh.kiyani20@gmail.com (M.K.-H.)

[3] Department of Physical Geography, University of Trier, 54296 Trier, Germany; rodrigo-comino@uma.es

[4] Soil Erosion and Degradation Research Group, Department of Geography, University of Valencia, Avda. Blasco Ibáñez, 28, 46010 Valencia, Spain

* Correspondence: namanian@yazd.ac.ir; Tel.: +98-9132-745-480

Received: 9 October 2019; Accepted: 29 November 2019; Published: 3 December 2019

Abstract: The geometry of hillslopes (plan and profile) affects soil erosion under rainfall-runoff processes. This issue comprises of several factors, which must be identified and assessed if efficient control measures are to be designed. The main aim of the current research was to investigate the impact of surface Roughness Coefficients (RCs) and Complex Hillslopes (CHs) on runoff variables viz. time of generation, time of concentration, and peak discharge value. A total of 81 experiments were conducted with a rainfall intensity of 7 L min^{-1} on three types of soils with different RCs (i.e., low = 0.015, medium = 0.016, and high = 0.018) and CHs (i.e., profile curvature and plan shape). An inclination of 20% was used for three replications. The results indicate a significant difference (p-value ≤ 0.001) in the above-mentioned runoff variables under different RCs and CHs. Our investigation of the combined effects of RCs and CHs on the runoff variables shows that the plan and profile impacts are consistent with a variation in RC. This can implicate that at low RC, the effect of the plan shape (i.e., convergent) on runoff variables increases but at high RC, the impact of the profile curvature overcomes the plan shapes and the profile curvature's changes become the criteria for changing the behavior of the runoff variables. The lowest mean values of runoff generation and time of concentration were obtained in the convex-convergent and the convex-divergent at 1.15 min and 2.68 min, respectively, for the soil with an RC of 0.015. The highest mean of peak discharge was obtained in the concave-divergent CH in the soil with an RC of 0.018. We conclude that these results can be useful in order to design planned soil erosion control measures where the soil roughness and slope morphology play a key role in activating runoff generation.

Keywords: rainfall-runoff processes; soil erosion; hillslope morphology; surface flow; roughness

---

1. Introduction

Soil erosion processes lead to soil depletion, reduced fertility, and consequently affect crop quality and crop yields [1–5]. At the pedon scale, soil erosion can be attributed to the separation of soil particles by raindrops and runoff during interrill erosion processes, so preventing this process is an important goal in the management of water and soil resources [6–8]. The amount of runoff is a function of navigation time, therefore the accurate estimation of the time of concentration and runoff threshold results in a more accurate hydrograph of flooding [9]. Designing methods and structures for soil water conservation requires accurate estimation of the amount and time of flood peak discharge and time of concentration. To achieve these goals, it is necessary to study the factors affecting the amount of runoff produced, the time to runoff generation, the time to the concentration of runoff and, consequently, the nature of the hydrograph [10,11].

It is well known that the runoff process consists of three parts: surface runoff, subsurface runoff, and baseflow. In arid and semi-arid areas, the soils show some inherent properties that produce an elevated irregularity of these hydrological mechanisms, generating mixed models of surface and subsurface flows [12]. A representative example of this kind of mixed and complex dynamic is the degraded soils of Iran. Since most soils in Iran have low permeability, precipitation exceeds infiltration, hence, surface runoff becomes dominant [9,13].

Research in recent years on the behavior and performance of surface runoff indicates the impact of various climatic and physiographic factors such as rainfall intensity, slope steepness, and surface Roughness Coefficient (RC) on surface runoff variables in different temporal and spatial scales (e.g., [14–23]). In this regard, Vermang et al. [24] studied the effect of RC on runoff and soil loss rate using a rainfall intensity of 50.2 mm h^{-1} on a silt loam soil. Soil particle sizes were divided into four size groups of 3–12, 12–20, 20–45 and 45–100 mm under a slope steepness of 5%. The results showed a significant decrease ($p \leq 0.05$) in the rate of runoff changes along with an increase in RC. However, in RC with particle sizes larger due to the impact of raindrops and breaking of aggregates and the formation of a seal layer on surfaces, the permeability decreased, and the runoff rate reached a steady state. Ding et al. [25] also measured runoff and sediment at different times using rainfall simulation under laboratory conditions on two types of soils with different RCs. The results showed that high RC delayed runoff generation and concentration of runoff, but no significant changes in the amount of runoff were observed. Another example can be found in Vaezi and Ebadi [26]. They recently investigated the effect of nine rainfall intensities from 10 to 90 mm h^{-1} under five slope steepness' from 0% to 40% as the most important factors in producing surface runoff under laboratory conditions. Their results show that the runoff threshold in interrill erosion occurred using rainfall intensities close to 20 mm h^{-1}. They also reported that the highest surface runoff occurred under a slope steepness of 20%, with no significant difference for runoff as the slope increased.

In recent years, Complex Hillslope (CH) (i.e., profile curvature and plan shape) has also been considered as an effective factor in surface and subsurface runoff dynamics. Once a mutual relation is proven between the shape of the hillslope and the hydrological processes, it can be both useful and efficient in managing natural and urban watersheds [27,28]. Hillslope shape is an effective measure in studying the complex impact of topography on the different runoff variables. Troch et al. [14,29] introduced nine different CHs through the combination of three plan shapes (convergent, divergent, and parallel) and three profile curvatures (convex, straight, and concave). Agnese et al. [30] showed that for constant plan shapes, the convex profile generates more surface runoff than the other profiles. For the constant profile curvatures, the converged plan obtained higher amounts of surface runoff than the parallel and divergent plans. In addition, Talebi et al. [31] also demonstrated that convex and divergent hillslopes could be generally more stable than other types of hillslopes and that concave and convergent hillslopes could be less stable. According to the results obtained by Geranian et al. [32] on the effect of CH plan and profile on surface runoff changes, the impact of plan on runoff generation and time of concentration of runoff could be much greater than profile curvature, and convergent hillslopes with surface runoff concentration had an earlier runoff activation time than divergent and parallel hillslopes. Moreover, the runoff thresholds on convex hillslopes could be lower than on concave hillslopes. In addition, they observed that the peak discharge in divergent and parallel hillslopes was much higher than for convergent hillslopes due to the greater width at the outlet of the hillslope. Thus, the effect of profile on the rate of discharge changes needs further investigation.

The straight hillslope reaches a peak discharge earlier than the convex hillslope, which reaches a peak discharge earlier than the concave hillslope. Sabzevari et al. [9] found that the geometry of the hillslopes could change the peak discharge of hydrograph several times since the divergent hillslopes had a higher peak discharge than the parallel and convergent hillslopes. In addition, their results indicated that the highest peak discharge could be found in concave-divergent hillslopes. In their investigation of the effect of hillslope geometry on surface and subsurface runoff, they also showed that the rate of change in the hydrograph of the divergent hillslopes was decreasing, and the concave curve was downward. In another study, Sabzevari et al. [33] further examined the effect of hillslope geometry on the temporal variables of runoff, such as lag time and time to equilibrium. The results showed a 33% increase in time to equilibrium in the divergent hillslopes as compared to the convergent hillslopes. Talebi et al. [18] examined the effect of CH plan and profiles on separation and deposition of sediment particles caused by sheet erosion. They stated that the rate of particle separation because of runoff on convex profiles is about 15 times higher than that in straight profiles. Regarding this, Fariborzi et al. [34] employed the subsurface time area model for the prediction of subsurface flow in CHs. To validate these results, a rainfall simulator on a sandy loam soil was tested, which was used under three rainfall intensities and three inclinations. Then, subsurface time area model results were compared with those of a laboratory of subsurface flow. The results showed an accurate estimation with a determination coefficient of 0.85 for the method of subsurface time area in CHs.

Although there is a great number of research on the individual effect of RC and CH on the performance and dynamics of surface runoff and subsurface flow, the combined effects of RC and CH morphology on runoff variables during rainfall-runoff processes in laboratory conditions is still unknown. The main aim of this study is to perform a comparative analysis of the individual and combined effects of RC and CH on runoff production and parameters related to runoff, such as start time, time of concentration, and runoff peak, by using hydrographs. To achieve this goal, a total of 81 rainfall simulations under laboratory conditions were conducted with a rainfall intensity of 7 L min^{-1} on three soils with different RCs (i.e., low = 0.015, medium = 0.016, and high = 0.018) and CHs (i.e., profile curvature and plan shape). An inclination of almost 20% was used in three replications per treatment.

2. Materials and Methods

2.1. Preparing Rainfall Simulator Conditions

In the current research, a laboratory model was used to investigate the rainfall-runoff processes in CHs under three different types of RCs. The longitudinal profiles and slope plans were designed based on the Evans model [35]. According to Troch et al. [14], nine different CHs were designed with the simultaneous change of plan and profile of the hillslopes (Table 1). To investigate the plan shapes (convergent, divergent, and parallel) and three profile curvatures (convex, straight, and concave), a three-dimensional geometric model of the hillslopes were considered [34,36]. To introduce a suitable function that can represent the geometry of the CHs, the model proposed by Troch et al. [14] was used.

In this study, a rainfall simulator was used with a uniform rainfall intensity of 7 l min^{-1} (210 mm/h). The simulator has a plot with a length of 2 m, a width of 1 m, and a depth of 0.8 m. The height of the surface of the plot to the precipitation nozzles is 20 cm. The simulator has a water tank with a capacity of 200 L beneath the unit where the water is pumped into the nozzles and rained down on the surface of the plot. There are two jacks below and on the left side of the plot for longitudinal slope adjustment. A slope steepness of almost 20% was used, as this value was considered a representative inclination of the most eroded areas in Iran [26] Surface runoff was also measured using the outlet pipes of P2 and P4 in outlet plot (Figure 1).

Table 1. Geometric characteristics of complex hillslopes (CHs).

No.	Longitudinal Profile	Plan Shape	CHs	H (m)	n (No Dimension)	L (m)	$\omega\ (m^{-1})$	A (m^2)
1		Convergent					+0.0997	1.8
2	Concave	Parallel		0.36	1.5	1.90	0.0000	2.4
3		Divergent					−0.0997	1.8
4		Convergent					+0.0997	1.5
5	Straight	Parallel		0.36	1	1.90	0.0000	2
6		Divergent					−0.0997	1.5
7		Convergent					+0.0997	1.8
8	Convex	Parallel		0.36	0.5	1.90	0.0000	2.4
9		Divergent					−0.0997	1.8

H: maximum heights relative to the baseline; L: total length of the hillslope; n: profile curvature parameter; ω: plan curvature parameter calculated [14].

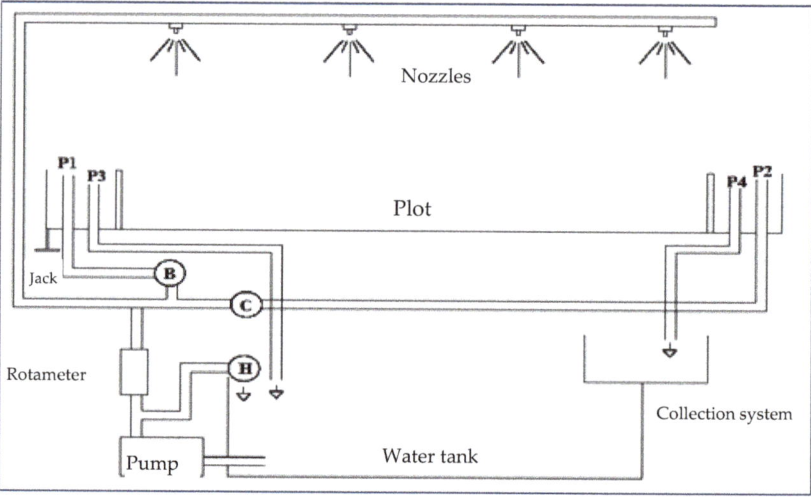

Figure 1. Schematic diagram of the water cycle within the rainfall simulator [32].

A rectangular weir was used to measure the time of concentration and amount of runoff generated in CHs. After preparing the rainfall simulator and adjusting the slope and rainfall intensity, a digital chronometer was turned on. Then, the start time of the runoff was recorded by observing runoff on the within plot and runoff travel from the plot most distant point to the outlet plot as the time of runoff generated and time of runoff concentration, respectively. In the following, runoff peak (L min^{-1} m^{-2}) was measured using a hydrograph, so that the maximum amount of runoff in each hydrograph was considered as the runoff peak. Then, the amount of runoff was measured in intervals of 1 min with a total duration of 15 min after runoff generation under different CH and RC conditions. Finally, using the rectangular weir information at different times, the peak discharge rate was also determined [26].

2.2. Determining the RC

Soil RC has a great impact on the surface storage, velocity, and direction of runoff [37]. In this research, in order to investigate the effect of RCs on the rainfall-runoff processes in CHs, soil particle diameter in the plot was considered as a factor influencing hydrological processes on RCs, and three different granulation types were used to investigate different RCs. Various relations have been proposed to determine the hydraulic RC through granulation. Therefore, in order to select the appropriate RC, the granulation curve of three soil types was determined by granulometry tests. The soil uniformity was then calculated using Equation (1) [38]:

$$C_u = \frac{d_{60}}{d_{10}}$$

(1)

where C_u is the soil uniformity and d_{60} and d_{10} represent the diameters at which 60% and 10% of each soil type are comprised of particles with a diameter less than those values, respectively. For C_u smaller than 5, the soil is uniform. For $5 < C_u < 15$, it is non-uniform, and for larger than 15, it is very non-uniform [38]. Finally, Soils 1, 2, and 3 (Figure 2) were uniform with a $C_u < 5$ (C_u: 1.5, 2.63, and 1.74, respectively). Thus, the Strickler equation (Equation (2)) was used to determine the RCs of Soils 1, 2, and 3. Their RCs, according to the granulation curve, were obtained to be 0.015, 0.016, and 0.018, respectively. In this equation, d_{50} (mm) is the median diameter of particles [39].

$$n = 0.0474 \, (d_{50})^{1/6}.$$

(2)

Figure 2. The three soil types with different roughness coefficients (RCs) used during the rainfall simulation experiments, (**1**): Cu: 1.5; (**2**): Cu: 2.63; (**3**) Cu: 1.74.

2.3. Statistical Analysis

After measuring the surface runoff, under different CHs and RCs, the data obtained from the experiments were categorized in Excel 2013 (Microsoft, USA). Before performing any statistical analysis, the normality of the data was tested through the Kolmogorov–Smirnov test. Regarding this, the non-normal data were changed to normal data using a transforming data method and this was then analyzed. In the following, the means of different groups were compared using Tukey's test. Two-way ANOVA tests were used to measure the individual and combined effects of CHs and RCs on runoff variations as was done by Kiani-Harchegani et al. [40].

3. Results and Discussion

3.1. Descriptive Statistics of Runoff Variables in RCs in CHs

Table 2 presents the descriptive statistics, including the mean, standard deviation (SD) and coefficient of variation (CV) of different runoff variables, including start time, time of concentration, and runoff peak in three soil types with different RCs under different CHs.

The results indicate that the slowest start time for Soils 1 and 2 was in the concave-parallel hillslopes (Hillslope No. 2). The slowest start time was observed in Soil 3 in the straight-parallel hillslopes (Hillslope No. 5), showing the longest start time of runoff amongst all CHs. The lowest start time of runoff for all soils was obtained in the convex-convergent hillslopes (Hillslope No. 7) and the earliest start time was Soil 1. The results, as shown in Table 2, also indicate that the highest time of concentration was in Soil 2 with convex-convergent hillslopes (Hillslope No. 7) and the lowest time of concentration in Soil 1 with straight-divergent hillslopes (Hillslope No. 9). Finally, the highest runoff peak was observed in Soil 3 in the concave-divergent hillslopes (Hillslope No. 3) and the lowest runoff peak in Soil 1 with the convex-convergent hillslopes (Hillslope No. 7).

3.2. Start Time of Runoff

According to the results of the two-way ANOVA presented in Table 3, there was a significant difference ($p \leq 0.001$) in the effect of different CHs and different RCs on the start time of runoff. The interaction between RCs and CHs also shows a significant difference ($p \leq 0.001$), indicating different effects of different RCs under CHs on the start time of runoff.

The results of the Tukey's test (Figure 3) also showed that runoff reaches the plot outlet in Soil 1 (n = 0.015) earlier under convex-convergent hillslopes (Hillslope No. 7) due to surface flow concentration, high water height at the plot outlet, and reduced permeability. The earliest start time of runoff in Soils 2 and 3 at the hillslope of No. 7 can be also accounted for the more effective simultaneous permeability effect and profile curvature. Therefore, because of an increased RC of the soil due to coarser soil particles increasing and a higher surface in convex profile, as well as the increased runoff height at the plot outlet, runoff occurs quickly.

By considering the soil profile and type constant, the effect of the plan shape on the runoff start time can be determined, such that the convergence reduces the start time of runoff and the divergence delays the start time of runoff. In addition, by considering the plan and type of soil constant, we can confirm the impact of profile on the start time of runoff, such that concave delays the start time of runoff and convex hillslopes have an earlier start time of runoff. These results are in line with the results obtained by Sabzevari et al. [9] and Geranian et al. [32].

Table 2. Summary of characteristics of runoff variables using different RCs and CHs (three replications).

N			1	2	3	4	5	6	7	8	9
Hillslopes No.				Concave			Straight			Convex	
Variables	**Soil Type**	**Profile Plan**	Converge	Parallel	Divergent	Converge	Parallel	Divergent	Converge	Parallel	Divergent
Runoff generation (min)	1	Mean	2.15	4.43	4.12	2.24	3.25	3.39	1.15	2.91	3.45
		SD	0.50	0.76	0.29	0.51	0.45	0.32	0.50	0.10	0.50
		CV	0.23	0.17	0.07	0.23	0.14	0.09	0.43	0.03	0.14
	2	Mean	2.53	6.17	3.70	4.08	5.43	3.67	2.20	4.12	4.47
		SD	0.58	0.58	0.00	0.29	0.58	0.58	0.10	0.29	0.58
		CV	0.23	0.09	0.00	0.07	0.11	0.16	0.05	0.07	0.13
	3	Mean	4.25	6.67	7.42	3.93	7.78	6.25	1.82	3.18	5.43
		SD	0.50	0.58	0.29	0.29	0.76	0.50	0.29	0.10	0.58
		CV	0.12	0.09	0.04	0.07	0.10	0.08	0.16	0.03	0.11
Time of runoff concentration (min)	1	Mean	4.00	4.86	3.15	3.33	3.53	2.77	5.28	3.37	2.68
		SD	0.20	0.11	0.09	0.15	0.15	0.12	0.03	0.06	0.16
		CV	0.05	0.02	0.03	0.05	0.04	0.04	0.01	0.02	0.06
	2	Mean	5.27	5.68	3.78	4.28	5.63	4.27	7.78	4.93	3.87
		SD	0.38	0.03	0.03	0.19	0.15	0.06	0.03	0.03	0.06
		CV	0.07	0.01	0.01	0.04	0.03	0.01	0.00	0.01	0.01
	3	Mean	5.68	6.68	5.47	6.65	6.75	4.47	7.45	3.52	4.73
		SD	0.16	0.16	0.15	0.00	0.05	0.15	0.13	0.10	0.21
		CV	0.03	0.02	0.03	0.00	0.01	0.03	0.02	0.03	0.04
Runoff peak (L min^{-1} m^{-2})	1	Mean	1.95	1.89	2.35	2.42	1.84	3.02	1.27	1.92	2.43
		SD	0.10	0.06	0.06	0.15	0.14	0.06	0.03	0.13	0.06
		CV	0.05	0.03	0.02	0.06	0.08	0.02	0.02	0.07	0.02
	2	Mean	2.02	1.89	2.52	3.02	2.10	2.98	1.48	1.72	2.02
		SD	0.15	0.06	0.06	0.49	0.01	0.15	0.15	0.06	0.15
		CV	0.08	0.03	0.02	0.16	0.01	0.05	0.10	0.03	0.08
	3	Mean	2.30	1.89	2.89	2.86	2.17	2.96	1.74	1.94	2.60
		SD	0.14	0.07	0.10	0.03	0.14	0.21	0.06	0.15	0.20
		CV	0.06	0.04	0.03	0.01	0.07	0.07	0.03	0.08	0.08

SD, standard deviation; CV, coefficient of variation.

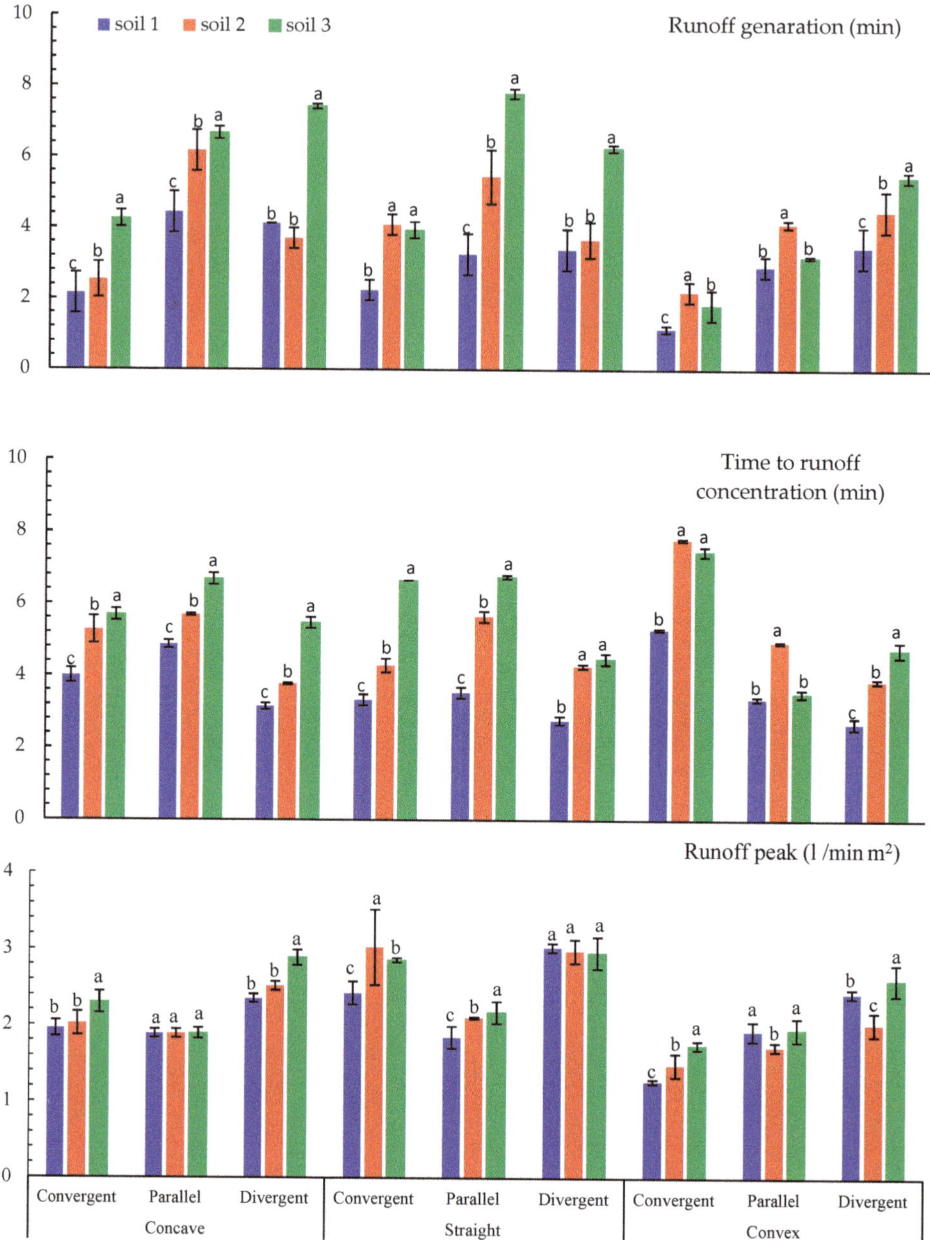

Figure 3. Effect of different RCs and CHs on runoff variables by Tukey's test. (a, b and c are statistically different at $p \leq 0.05$ on three soil types in each hillslope). Means followed by the same letters are not significantly different according to Tukey's HSD at $p \leq 0.05$.

Table 3. Effect of different RCs and CHs and their interaction on runoff variables using two-way ANOVA.

Variables	Factors	df	Mean Squared	F-Value	*p*-Value
	CH	8	55,945.03	2237.80	0.00
Start Time (min)	**RC**	2	114,643.11	4585.72	0.00
	CH × RC	16	8513.61	340.54	0.00
	CH	8	32,212.25	1288.49	0.00
Time of concentration (min)	**RC**	2	84,037.00	3361.48	0.00
	CH × RC	16	5248.25	209.93	0.00
	CH	8	82.36	82.40	0.00
Runoff peak (L min^{-1} m^{-2})	**RC**	2	25.44	25.40	0.00
	CH × RC	16	5.07	5.07	0.00

df: degrees of freedom; F-value: variation between gropes mean/variation within the gropes; *p*-value: *p*-value $\leq$ 0.01 and *p*-value > 0.05 indicates a significant variation within the group and a low variation within the group, which leads to the rejection and acceptance of the null hypothesis, respectively.

3.3. Time of Concentration

The results of the two-way ANOVA presented in Table 3 indicate a significant difference ($p \leq 0.001$) in the individual effect of different CHs and RCs on time of runoff concentration. The interaction between CHs and RCs also shows a significant difference ($p \leq 0.001$), indicating different effects of different RCs and CHs on the time of concentration.

The results of means of different groups using Tukey's test in Figure 3 also show that the divergent hillslopes decrease the time of concentration and concave hillslopes increase it, which results in an increased time of concentration on the concave-divergent hillslopes (Hillslope No. 3) with increasing RCs (n = 0.016 to n = 0.018). This shows that hillslope profiles have a greater effect than plan shape. Therefore, as soil permeability increases, more time is needed to accumulate runoff at the bottom of the plots, thus increasing the time of concentration [22]. In the convex-parallel hillslopes (Hillslope No. 8), with increasing soil RC, the time of concentration was first increased and then decreased. The reason may be that with the low slope of the convex profile at the bottom of the plot as well as water accumulation, the water is discharged earlier [41,42]. The results also show that the convergent plan, due to the concentration of surface and subsurface runoff, retains more water and takes longer to discharge. As well, the runoff path is curved to reach the exit plot outlet in the convergent plan, but in other plans, it follows a smooth path. This finding is consistent with those obtained by Geranian et al. [32].

In examining the CH profiles with increasing RCs, it was noticed that concave profiles with a lower slope angle at the end of the plot required a considerable runoff, thus an increasing trend was observed with RCs in different plan shapes. In straight profiles, due to the uniformity of the slope, time of concentration is affected by the RC and soil permeability [43].

The examination of the effect of the plan on time of concentration by considering soil profile and type constant indicates that time of concentration in divergent hillslopes is less than that in parallel and convergent hillslopes [21,44]. The reason may be due to the curved path of runoff in convergent plans. Considering the plan and soil type constant, concave hillslopes are also shown to have a higher time of runoff concentration than convex and straight hillslopes because, on the convex hillslopes, the water flow is faster than other profiles for the high slope and accumulation of runoff [45]. For example, the highest time of concentration for all soils was obtained for the convex-convergent hillslopes. The lowest time of concentration for Soil 1, 2 and 3 was observed in the convex-divergent, concave-divergent, and convex-parallel hillslopes, respectively. In most hillslopes, as the soil RC increases, the time of concentration is delayed, which is consistent with the results obtained by Geranian et al. [32]. However, from the results of simultaneous impact analysis of profiles, plans, and RCs, it is shown that in different RCs, the effect of the profile and plan on runoff variables varies. Moreover, in some hillslopes, by increasing permeability, the effect of profiles is observed more than the plan, and in other hillslopes, the opposite is the case. For example, in the concave-divergent hillslopes (Hillslope No. 3), we can observe the effect of profile (concavity) on the plan (divergence), so by increasing soil RC, the time of concentration and permeability is increased. A similar case can also be observed for Hillslope No. 8 (convex-parallel).

3.4. Runoff Peak

The results of the two-way ANOVA, presented in Table 3, indicate a significant difference ($p \leq 0.001$) for the individual influence of different CHs and RCs on the runoff peak rate. In addition, the interaction between soil RC and CHs shows a significant difference ($p \leq 0.001$), indicating the different effects of different soil RCs and CHs on runoff peak rate.

The results of Tukey's test (Figure 3) also show that in concave-convergent, concave-divergent, straight convergent, and straight-parallel hillslopes (Hillslopes No. 1, 3, 4, and 5, respectively), increasing soil RC relatively increases the runoff peak rate. In the concave-parallel and straight-divergent CHs (Hillslopes No. 2 and 6), the soil RC has no effect on the runoff peak rate. Meanwhile, in the convex-convergent CHs (Hillslope No. 7), increasing soil RCs significantly increases the runoff peak, and in CHs (Hillslopes No. 8 and 9), by increasing the soil RCs, the time of peak rate first decreases and then increases. This is due to the effect of the plan on the runoff peak rate in different RCs and therefore permeability [16,17]. The highest runoff peak rate was also observed in Soil 3 (n = 0.018) with concave-divergent hillslopes (Hillslope No. 3) and the lowest runoff peak in Soil 1 (n = 0.015) with convex-convergent hillslopes (Hillslope No. 7).

A different profile affects the speed of discharge rate with slight changes (Figure 3). But in the CHs plan, due to changes in the width of the down plot, great changes can be observed in the discharge rate. Divergent or parallel plans can pass all runoff but in the convergent plan, there is some runoff accumulation due to a lower width. The individual effect of the hillslope profiles can be attributed to the speed of changes in the discharge rate due to the same amount of input discharge for all hillslopes (assuming hillslope plan and soil type constant). The runoff peak in concave hillslopes has a lower increase than the other hillslopes due to runoff accumulation in the outlet and higher initial discharge [46]. Thus, the runoff peak of the convergent hillslopes is smaller in the outlet than the parallel and divergent plans, which also reflects the higher time to concentration (greater discharge time). This could be consistent with the results obtained by Tucker and Bras [47] and Bonetti et al. [48] if we consider the differences generated at different scales. According to the results of this study, in all soils with constant RC and different CHs, the highest runoff peak is for divergent hillslopes with straight or concave profiles and the lowest amount for convex-convergent hillslopes.

3.5. Hydrographs

Divergent hillslopes have a decreasing rate of change due to parallel and direct movements of runoff toward the outlet and the effect of the lower part of plot on the outlet runoff volume, while parallel hillslopes have a constant rate of head change due to the constant up and down widths of the plot [49]. However, in convergent hillslopes due to a greater area of upward of plot over time, the rate of increase of runoff peak increases [50,51]. The study of the CHs also shows that the start time of runoff and time to concentration increases with increasing soil RCs (Figure 4). But in convex profiles, due to the steep slope of the lower part of the plot, with increasing soil RCs and consequently increasing its permeability, the runoff is discharged earlier and the time of concentration is reduced. So, the hydrographs of Soil 1, 2, and 3 are almost overlapped. Similarly, in convergent hillslopes, the start time of runoff decreases because the convergent hillslope saturation is faster than parallel and divergent hillslopes. The results of this research and previous researches, such as Sabzevari et al. [9] and Geranian et al. [32], show that convergence and convexity of hillslopes accelerate the start time of runoff and divergence and concaveness delay it.

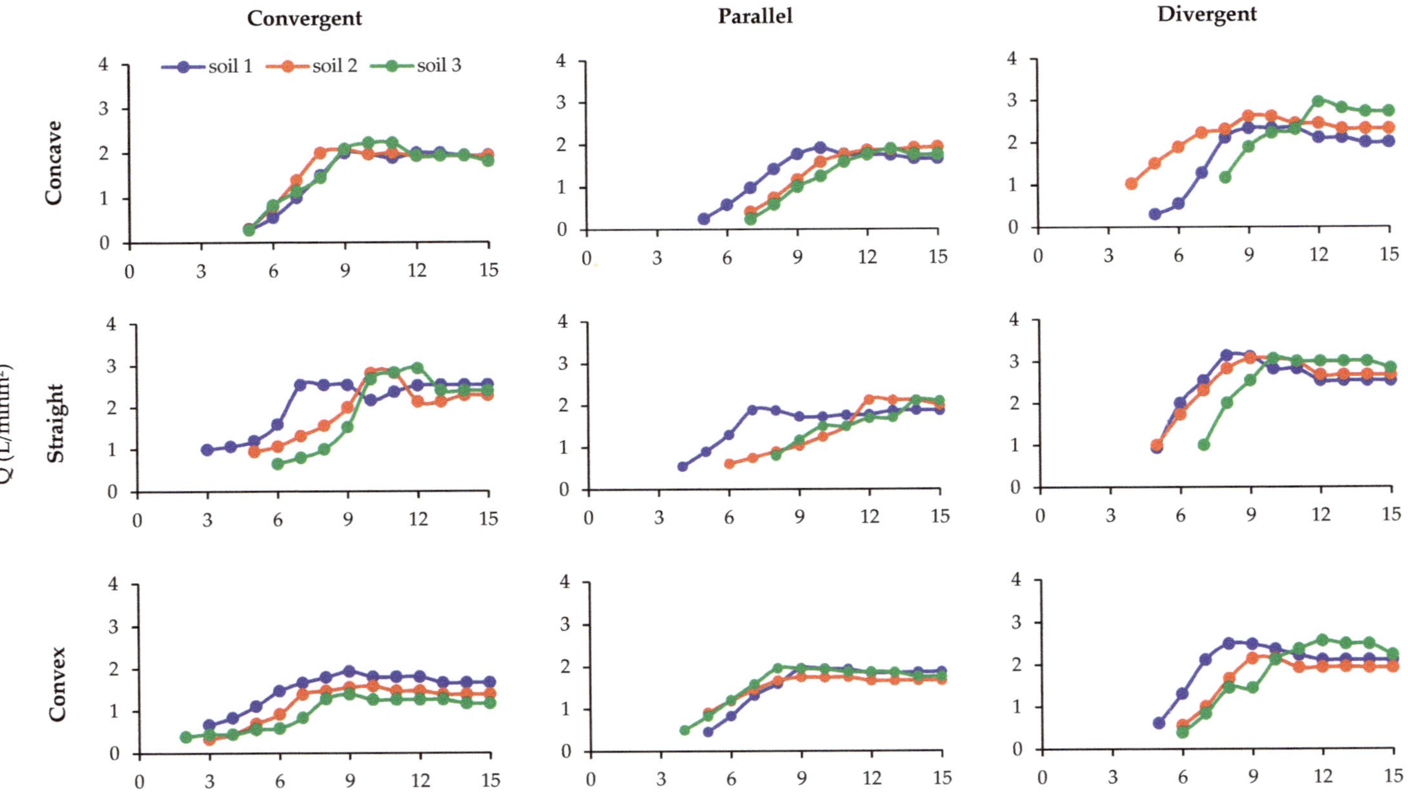

Figure 4. Hydrographs of rainfall-runoff processes at different RCs and CHs.

3.6. Recommendations, Challenges and Future Research Work

We recommend delaying the start time of runoff and its subsequent destructive consequences in one of the following ways in landscapes with high levels of convexity morphology. Firstly, increasing surface RC by changing soil granulation or by changing vegetation, as well as enhancing soil horizon formations, biological processes, and water retention capacity [52,53]. For this purpose, modified natural or synthetic coatings could increase the RC, thereby reducing the start time and time of concentration and runoff peak rate of the runoff. Secondly, it could be possible to introduce an embankment for diverging artificial paths or reducing the convexity of hillslopes with an appropriate stair levelling, which could increase the start time of runoff and relatively reduce the runoff peak rate. This is very common in cultivated fields where terraces are used to retain water, conserve the soil, and reduce hillslope inclination [54,55].

Another point to be discussed at this point as a challenge for the future should be the effect of other external factors to assess the connectivity processes at the pedon scale [56]. It is well known that laboratory experiments are not representative of natural conditions [10,57]; however, they can be very useful to detect the specific factors that condition the activated dynamics and processes. In the future, other related factors that affect the connectivity of the surface and subsurface processes should be considered, such as the number of rock fragments [58], changes in organic carbon contains and litter cover [59,60], or different parent materials [61].

4. Conclusions

The main purpose of this research was to compare the variability of a surface runoff between different soil RCs under different CHs, both individually and together during rainfall-runoff processes. The results showed that the effect of the CH plan due to the change in runoff storage volume at the bottom of the plot was greater than the CH profile. With this, the divergent hillslopes showed a slower start time than the convergent and parallel hillslopes, and with increasing soil RC, the start time of runoff increased. With increasing soil RC, start time on convex profile hillslopes increased because of the higher slope at the end of the plot, and so did the convergent hillslopes due to the concentration of surface runoff at the bottom of the plot. In addition, the examination of the simultaneous effect of profile and plan of CHs and soil RC on time of concentration and runoff peak of surface runoff showed that the percentage of the impact of profile and plan changes with soil RC changes. Finally, the analysis of hydrographs on CHs showed that start time of the runoff and time of concentration increased with increasing soil RC. However, in convex profiles, due to the high slope of the lower end of the plot, by increasing soil RC and consequently increasing its permeability, the runoff is discharged earlier and the time of concentration is reduced. As a result, the hydrographs of all three types of soil are almost overlapped. In order to obtain comprehensive results, it is recommended that experiments be carried out on non-uniform soils using different rainfall intensities and different slope steepness in CHs.

Author Contributions: Conceptualization and methodology, M.M., N.A., A.T., J.R.-C.; formal analysis and investigation, M.M., N.A., A.T. and M.K.-H.; data curation, M.K.-H.; writing—original draft preparation, M.M., N.A. and A.T.; writing—review and editing, M.K.-H. and J.R.-C.; supervision, N.A.

Funding: This research received no external funding.

Acknowledgments: We would like to thank Mahta Amanian, a senior undergraduate student at the University of British Columbia, Canada, for editing this paper.

Conflicts of Interest: The authors declare no conflict of interest.

References

1. Bosco, C.; De Rigo, D.; Dewitte, O.; Poesen, J.; Panagos, P. Modelling soil erosion at European scale: Towards harmonization and reproducibility. *Nat. Hazard. Earth Sys. Sci.* **2015**, *15*, 225–245. [CrossRef]
2. Garcia-Ruiz, J.M.; Beguería, S.; Nadal-Romero, E.; Gonzalez-Hidalgo, J.C.; Lana-Renault, N.; Sanjuán, Y. A meta-analysis of soil erosion rates across the world. *Geomorphology* **2015**, *239*, 160–173. [CrossRef]

3. Assouline, S.; Govers, G.; Nearing, M.A. Erosion and lateral surface processes. *Vadose Zone J.* **2017**, *16*, 1–4. [CrossRef]

4. Borrelli, P.; Robinson, D.A.; Fleischer, L.R.; Lugato, E.; Ballabio, C.; Alewell, C.; Meusburger, K.; Modugno, S.; Schütt, B.; Ferro, V.; et al. An assessment of the global impact of 21st century land use change on soil erosion. *Nat. Commun.* **2017**, *8*, 2013. [CrossRef]

5. Sadeghi, S.H.R.; Singh, V.P.; Kiani-Harchegani, M.; Asadi, H. Analysis of sediment rating loops and particle size distributions to characterize sediment source at mid-sized plot scale. *Catena.* **2018**, *167*, 221–227. [CrossRef]

6. Kinnell, P.I.A. Raindrop-impact-induced erosion processes and prediction: A review. *Hydrol. Process.* **2005**, *19*, 2815–2844. [CrossRef]

7. Marzen, M.; Iserloh, T.; Casper, M.C.; Ries, J.B. Quantification of particle detachment by rain splash and wind-driven rain splash. *Catena* **2015**, *127*, 135–141. [CrossRef]

8. Fernández-Raga, M.; Campo, J.; Rodrigo-Comino, J.; Keesstra, S.D. Comparative Analysis of Splash Erosion Devices for Rainfall Simulation Experiments: A Laboratory Study. *Water* **2019**, *11*, 1228. [CrossRef]

9. Sabzevari, T.; Saghafian, B.; Talebi, A.; Ardakanian, R. Time of concentration of surface flow in complex hillslopes. *J. Hydrol. Hydromech.* **2013**, *61*, 269–277. [CrossRef]

10. Kiani-Harchegani, M.; Sadeghi, S.H.R.; Asadi, H. Comparing grain size distribution of sediment and original soil under raindrop detachment and raindrop-induced and flow transport mechanism. *Hydrol. Sci. J.* **2018**, *63*, 312–323. [CrossRef]

11. Minea, G.; Ioana-Toroimac, G.; Moroşanu, G. The dominant runoff processes on grassland versus bare soil hillslopes in a temperate environment—An experimental study. *J. Hydrol. Hydromech.* **2019**, *67*, 1–8. [CrossRef]

12. Lavee, H.; Poesen, J.W.A. Overland flow generation and continuity on stone-covered soil surfaces. *Hydrol. Process.* **1991**, *5*, 345–360. [CrossRef]

13. Sabzevari, T.; Noroozpour, S. Effects of hillslope geometry on surface and subsurface flows. *Hydrogeol. J.* **2014**, *22*, 1593–1604. [CrossRef]

14. Troch, P.; Van Loon, E.; Hilberts, A. Analytical solutions to a hillslope-storage kinematic wave equation for subsurface flow. *Adv. Water Resour.* **2002**, *25*, 637–649. [CrossRef]

15. Hilberts, A.G.J.; Van Loon, E.E.; Troch, P.A.; Paniconi, C. The hillslope-storage Boussinesq model for non-constant bedrock slope. *J. Hydrol.* **2004**, *291*, 160–173. [CrossRef]

16. Chaplot, V.; Poesen, J. Sediment, soil organic carbon and runoff delivery at various spatial scales. *Catena* **2012**, *88*, 46–56. [CrossRef]

17. Orchard, C.M.; Lorentz, S.A.; Jewitt, G.P.W.; Chaplot, V.A.M. Spatial and temporal variations of overland flow during rainfall events and in relation to catchment conditions. *Hydrol. Process.* **2013**, *27*, 2325–2338. [CrossRef]

18. Talebi, A.; Hajiabolghasemi, R.; Hadian, M.R.; Amanian, N. Physically based modelling of sheet erosion (detachment and deposition processes) in complex hillslopes. *Hydrol. Process.* **2016**, *30*, 1968–1977. [CrossRef]

19. Danacova, M.; Vyleta, R.; Valent, P.; Hlavcova, K. The impact of slope gradients on the generation of surface runoff in laboratory conditions. In Proceedings of the 17th International Multidisciplinary Scientific GeoConference SGEM, Vienna, Austria, 27–29 November 2017; Volume 2017, pp. 677–684.

20. Eshghizadeh, M.; Talebi, A.; Dastorani, M. Thresholds of land cover to control runoff and soil loss. *Hydrol. Sci. J.* **2018**, *63*, 1424–1434. [CrossRef]

21. Zhao, L.; Hou, R.; Wu, F.; Keesstra, S. Effect of soil surface roughness on infiltration water, ponding and runoff on tilled soils under rainfall simulation experiments. *Soil. Till. Res.* **2018**, *179*, 47–53. [CrossRef]

22. Cerdà, A.; Rodrigo-Comino, J. Is the hillslope position relevant for runoff and soil loss activation under high rainfall conditions in vineyards? *Ecohydrol. Hydrobiol.* **2019**, *5*, 6. [CrossRef]

23. Kiani-Harchegani, M.; Sadeghi, S.H.R.; Ghahramani, A. Intra-storm Variability of Coefficient of Variation of Runoff and Soil Loss in Consecutive Storms at Experimental Plot Scale. In *Climate Change Impacts on Hydrological Processes and Sediment Dynamics: Measurement, Modelling and Management*; Springer: Berlin/Heidelberg, Germany, 2019; pp. 98–103.

24. Vermang, J.; Norton, L.D.; Huang, C.; Cornelis, W.M.; Da Silva, A.M.; Gabriels, D. Characterization of soil surface roughness effects on runoff and soil erosion rates under simulated rainfall. *Soil Sci. Soc. Am. J.* **2015**, *79*, 903–916. [CrossRef]

25. Ding, W.; Huang, C. Effects of soil surface roughness on interrill erosion processes and sediment particle size distribution. *Geomorphology* **2017**, *295*, 801–810. [CrossRef]

26. Vaezi, A.R.; Ebadi, M. Particle size distribution of surface-eroded soil in different rainfall intensities and slope gradients. *J. Water. Soil.* **2017**, *31*, 216–229. (In Persian)

27. Ferreira, C.S.S.; Walsh, R.P.D.; Steenhuis, T.S.; Shakesby, R.A.; Nunes, J.P.N.; Coelho, C.O.A.; Ferreira, A.J.D. Spatiotemporal variability of hydrologic soil properties and the implications for overland flow and land management in a peri-urban Mediterranean catchment. *J. Hydrol.* **2015**, *525*, 249–263. [CrossRef]

28. Kalantari, Z.; Ferreira, C.S.S.; Koutsouris, A.J.; Ahmer, A.K.; Cerdà, A.; Destouni, G. Assessing flood probability for transportation infrastructure based on catchment characteristics, sediment connectivity and remotely sensed soil moisture. *Sci. Total Environ.* **2019**, *661*, 393–406. [CrossRef]

29. Troch, P.A.; Paniconi, C.; Emiel van Loon, E. Hillslope-storage Boussinesq model for subsurface flow and variable source areas along complex hillslopes: 1. Formulation and characteristic response. *Water Resour. Res.* **2003**, *39*, 1316. [CrossRef]

30. Agnese, C.; Baiamonte, G.; Corrao, C. Overland flow generation on hillslopes of complex topography: Analytical solutions. *Hydrol. Process.* **2007**, *21*, 1308–1317. [CrossRef]

31. Talebi, A.; Troch, P.A.; Uijlenhoet, R. A steady-state analytical slope stability model for complex hillslopes. *Hydrol Process.* **2008**, *22*, 546–553. [CrossRef]

32. Geranian, M.; Amanian, N.; Taleb, A.; Hadian, M.R.; Zeini, M. Laboratorial Investigation of Effect of Plan Shape and Profile Curvature on Variations of Surface Flow In Complex Hillslopes. *Water Resour. Res.* **2013**, *9*, 64–72. (In Persian)

33. Sabzevari, T.; Noroozpour, S.; Pishvaei, M.H. Effects of geometry on runoff time characteristics and time-area histogram of hillslopes. *J. Hydrol.* **2015**, *531*, 638–648. [CrossRef]

34. Fariborzi, H.; Sabzevari, T.; Noroozpour, S.; Mohammadpour, R. Prediction of the subsurface flow of hillslopes using a subsurface time-area model. *Hydrogeol. J.* **2019**, *27*, 1401–1417. [CrossRef]

35. Evans, I. An integrated system of terrain analysis and slope mapping. *Z. Geomorphol.* **1980**, *36*, 274–295.

36. Sabzevari, T.; Talebi, A.; Ardakanian, R.; Shamsai, A. A steady-state saturation model to determine the subsurface travel time (STT) in complex hillslopes. *Hydrol. Earth Sys. Sci.* **2010**, *14*, 891–900. [CrossRef]

37. Rodríguez-Caballero, E.; Cantón, Y.; Chamizo, S.; Afana, A.; Solé-Benet, A. Effects of biological soil crusts on surface roughness and implications for runoff and erosion. *Geomorphology* **2012**, *145*, 81–89. [CrossRef]

38. Tavakoli, B. *Engineering Geology*; Payame Noor University Press: Tehran, Iran, 2011; p. 420.

39. Yong, C.T. *Sediment Transport: Theory and Practice*; McGraw-Hill: New York, NY, USA, 1996; p. 396.

40. Kiani-Harchegani, M.; Sadeghi, S.H.R.; Singh, V.P.; Asadi, H.; Abedi, M. Effect of rainfall intensity and slope on sediment particle size distribution during erosion using partial eta squared. *Catena* **2019**, *176*, 65–72. [CrossRef]

41. Cavalli, M.; Vericat, D.; Pereira, P. Mapping water and sediment connectivity. *Sci. Total Environ.* **2019**, *673*, 763–767. [CrossRef]

42. López-Vicente, M.; Ben-Salem, N. Computing structural and functional flow and sediment connectivity with a new aggregated index: A case study in a large Mediterranean catchment. *Sci. Total Environ.* **2019**, *651*, 179–191. [CrossRef]

43. Amanian, N.; Geranian, M.; Talebi, A.; Hadian, M.R. The effect of plan and slope profile on runoff initiation threshold. *Watershed Manag. Sci. Eng.* **2018**, *11*, 105–108. (In Persian)

44. Lin, Q.; Xu, Q.; Wu, F.; Li, T. Effects of wheat in regulating runoff and sediment on different slope gradients and under different rainfall intensities. *Catena* **2019**, *183*, 104196. [CrossRef]

45. Rodrigo-Comino, J.; Sinoga, J.R.; González, J.S.; Guerra-Merchán, A.; Seeger, M.; Ries, J.B. High variability of soil erosion and hydrological processes in Mediterranean hillslope vineyards (Montes de Málaga, Spain). *Catena* **2016**, *145*, 274–284. [CrossRef]

46. Fan, C.C.; Wang, H.Z. The behavior of wetting front on slopes with different slope morphologies during rainfall. *J. Hydro-Environ. Res.* **2019**, *25*, 48–60. [CrossRef]

47. Tucker, G.E.; Bras, R.L. Hillslope processes, drainage density, and landscape morphology. *Water Resour. Res.* **1998**, *34*, 2751–2764. [CrossRef]

48. Bonetti, S.; Richter, D.D.; Porporato, A. The effect of accelerated soil erosion on hillslope morphology. *Earth Surf. Process. Landf.* **2019**. Available online: https://doi.org/10.1002/esp.4694 (accessed on 3 December 2019). [CrossRef]

49. Baiamonte, G.; Singh, V.P. Overland flow times of concentration for hillslopes of complex topography. *J. Irrig. Drain. Eng.* **2015**, *142*, 04015059. [CrossRef]

50. Cossart, E.; Fressard, M. Assessment of structural sediment connectivity within catchments: Insights from graph theory. *Earth Surf. Dyn.* **2017**, *5*, 253–268. [CrossRef]

51. Cossart, E.; Viel, V.; Lissak, C.; Reulier, R.; Fressard, M.; Delahaye, D. How might sediment connectivity change in space and time? *Land Degrade. Dev.* **2018**, *29*, 2595–2613. [CrossRef]

52. Novara, A.; Gristina, L.; Guaitoli, F.; Santoro, A.; Cerdà, A. Managing soil nitrate with cover crops and buffer strips in Sicilian vineyards. *Solid Earth.* **2013**, *4*, 255–262. [CrossRef]

53. Kavian, A.; Saleh, I.; Habibnejad, M.; Brevik, E.C.; Jafarian, Z.; Rodrigo-Comino, J. Effectiveness of vegetative buffer strips at reducing runoff, soil erosion, and nitrate transport during degraded hillslope restoration in northern Iran. *Land Degrad. Dev.* **2018**, *29*, 3194–3203. [CrossRef]

54. Panagos, P.; Borrelli, P.; Meusburger, K.; van der Zanden, E.H.; Poesen, J.; Alewell, C. Modelling the effect of support practices (P-factor) on the reduction of soil erosion by water at European scale. *Environ. Sci. Policy.* **2015**, *51*, 23–34. [CrossRef]

55. Rodrigo-Comino, J.; Seeger, M.; Iserloh, T.; González, J.M.S.; Ruiz-Sinoga, J.D.; Ries, J.B. Rainfall-simulated quantification of initial soil erosion processes in sloping and poorly maintained terraced vineyards-Key issues for sustainable management systems. *Sci. Total Environ.* **2019**, *660*, 1047–1057. [CrossRef]

56. Ben-Salem, N.; Álvarez, S.; López-Vicente, M. Soil and water conservation in rainfed vineyards with common sainfoin and spontaneous vegetation under different ground conditions. *Water.* **2018**, *10*, 1058. [CrossRef]

57. Marzen, M.; Iserloh, T.; de Lima, J.L.; Fister, W.; Ries, J.B. Impact of severe rain storms on soil erosion: Experimental evaluation of wind-driven rain and its implications for natural hazard management. *Sci. Total Environ.* **2017**, *590*, 502–513. [CrossRef] [PubMed]

58. Jomaa, S.; Barry, D.A.; Heng, B.C.P.; Brovelli, A.; Sander, G.C.; Parlange, J.Y. Effect of antecedent conditions and fixed rock fragment coverage on soil erosion dynamics through multiple rainfall events. *J. Hydrol.* **2013**, *484*, 115–127. [CrossRef]

59. Zuazo, V.H.D.; Pleguezuelo, C.R.R. Soil-erosion and runoff prevention by plant covers: A review. *Agron. Sustain. Dev.* **2009**, *28*, 65–86. [CrossRef]

60. Feng, T.; Wei, W.; Chen, L.; Rodrigo-Comino, J.; Die, C.; Feng, X.; Ren, K.; Brevik, E.C.; Yu, Y. Assessment of the impact of different vegetation patterns on soil erosion processes on semiarid loess slopes. *Earth Surf. Process. Landf.* **2018**, *43*, 1860–1870. [CrossRef]

61. Rodrigo-Comino, J.; Novara, A.; Gyasi-Agyei, Y.; Terol, E.; Cerdà, A. Effects of parent material on soil erosion within Mediterranean new vineyard plantations. *Eng. Geol.* **2018**, *246*, 255–261. [CrossRef]

Article

Impact of Land-Use Changes on Spatiotemporal Suspended Sediment Dynamics within a Peri-Urban Catchment

C.S.S. Ferreira [1,*], R.P.D. Walsh [2], Z. Kalantari [3] and A.J.D. Ferreira [1]

[1] Research Centre for Natural Resources, Environment and Society (CERNAS), Polytechnic Institute of Coimbra, Higher Agricultural School, Coimbra 3045-601, Portugal; aferreira@esac.pt

[2] Department of Geography, College of Science, Swansea University, Swansea SA2 8PP, UK; R.P.D.Walsh@swansea.ac.uk

[3] Department of Physical Geography and Bolin Centre for Climate Research, Stockholm University, Stockholm SE-10691, Sweden; zahra.kalantari@natgeo.su.se

* Correspondence: cferreira@esac.pt; Tel.: +35-12-3980-2940

Received: 30 December 2019; Accepted: 25 February 2020; Published: 1 March 2020

Abstract: Understanding sediment dynamics in peri-urban catchments constitutes a research challenge because of the spatiotemporal complexity and variability of land-uses involved. This study investigates differences in the concentration of total sediments (TSC) and suspended sediments (SSC) in the small peri-urban Mediterranean Ribeira dos Covões catchment (40% urban area) in central Portugal. Suspended sediment responses at the catchment outlet (E) and in three upstream sub-catchments, during periods of urbanization (2011–2013) and stabilizing land-use (2017–2018) are compared for storm-event datasets encompassing similar ranges of rainstorm sizes and antecedent rainfall condition. The Quinta sub-catchment, with the lowest urban area (22%) but subject to major construction activities affecting 17% of its area, led to highest TSC and SSC during urbanization (attaining 4320 mg/L and 4184 mg/L, respectively), and a median reduction of 38% and 69%, respectively, during stabilization. Espírito Santo sub-catchment, with highest urban area (49%) and minor construction activities, displayed similar median TSC in both periods (258–240 mg/L) but highest SSC reduction (76%), highlighting the impact of the anthropogenic disturbance mainly on fine-particle sediments and a good connectivity with the stream. Porto Bordalo sub-catchment, with 39% urban area and subject to the construction of a four-line road covering 1.5% of its area, showed the lowest TSC and SSC concentrations and the lowest median reductions in both periods (31% and 64%, correspondingly), mainly because of the impact of an unplanned retention basin established with soil from the construction site. Overall, median TSC and SSC reduced 14% and 59% at E, from urbanization to stabilization. Information about sediment dynamics should guide stakeholders in establishing strategies to reduce sediment loads and mitigate the impacts on urban aquatic ecosystems.

Keywords: urbanization; land-use; suspended sediment concentration; spatiotemporal variation

---

1. Introduction

Society is facing increasing environmental challenges, such as land degradation, particularly in urban areas where most of the world's population reside. Erosion is one of the main soil threats identified by the European Commission [1] and a primary source of sediment to surface waters [2]. Although organic and inorganic materials eroded from the land surface can play an essential role in surface waters, by transporting nutrients to aquatic food webs, creating ecological habitats and structuring channel morphology [3], changes in sediment quantity and quality driven by human activities tend to have negative effects on (i) drainage capacity of rivers and hydraulic infrastructures, increasing flood hazard [2]; and (ii) ecological status of aquatic communities [4]. Suspended sediment, the dominant type of sediment generated within catchments [2], has been recognized as the most pervasive water quality problem globally [5], affecting products and services provided to society, such as water supply and recreation activities [6].

Excessive amounts of fluvial sediments harm the physical, chemical and biological processes in aquatic ecosystems. Suspended sediment increases turbidity and lead to reduced amount of light penetration, with detrimental impacts on photosynthesis, which affect dissolved oxygen concentration and primary productivity [7]. Suspended sediments are a major source of non-point pollution, since they can transport (i) nutrients, including nitrogen and phosphorus, contributing to eutrophication of rivers and lakes [8]; and (ii) contaminants, such as heavy metals [9,10], polycyclic aromatic hydrocarbons [11] and polychlorinated biphenyls [12]. Sediments have adverse impacts on development of fish [4], filter feeders such as mussels [6] and benthic invertebrate communities [13]. In Europe, only 38% of surface waters are in good chemical status [13]. In USA, pollution associated with sediment load is the most prevalent cause of impairment of aquatic systems [14]. In North America, the annual cost of suspended sediment influx to riverine ecosystems is estimated to range from $20 to $50 billion [15].

Understanding sediment dynamics is necessary to develop management strategies required to mitigate the impacts on the ecological status of the aquatic systems [4,16]. It is well-established that fluvial sediment loads and dynamics are affected by a combination of geological, lithological, topographical, climatic and land cover features of the catchments [6,17,18], as well as human activities, such as urbanization, mining and river regulation [19,20]. The complex spatiotemporal nature of these variables affects sediment-discharge relationships and determines their non-normality [21], making it difficult to understand and predict sediment concentrations and loads [4]. In Mediterranean areas, sediment concentrations vary by several orders of magnitude at a given discharge, depending on the spatial and temporal variability of intense rainstorms [22]. There have been many syntheses of knowledge of fluvial sediment dynamics in particular environments, including mountainous [4,23] and agricultural areas [24], as well as urban catchments [25,26]. In urban catchments, however, limited knowledge is available regarding the impact of distinct urban patterns (e.g., design, location within the catchment and stormwater design arrangements) on sediment yields [6]. Furthermore, limited research has focussed on peri-urban catchments [5], characterized by mixed land-use landscapes and dynamic and radical land-use change [27]. In these catchments, sediments from sealed surfaces (e.g., concrete, gravel roads) and house yards, can provide a relatively unlimited supply of sediments since they tend to be repaired by adding material from external catchments [6]. This input of particles coupled with increased runoff from urban surfaces and its effectiveness in eroding available sediment sources, lead to higher sediment loads than background values (e.g., in forest), even after the substantial increases recorded during construction works [2]. Furthermore, peri-urban catchments tend to comprise distinct mosaics of runoff sources and sinks which affect sediment connectivity between hillslope and the stream network [28,29] and lead to highly variable catchment sediment yields. Finally, high urbanization pressure in peri-urban catchments tends to lead to fast development of the built environment and dynamic changes in landscape patterns [30], making it more complex to assess temporal changes in sediment loads. Peri-urban expansion has been recorded even in countries with decreasing population, such as Portugal [31], highlighting the need to understand sediment dynamics and mitigate land degradation.

This study investigates spatial and temporal dynamics of fluvial suspended sediments in a Mediterranean peri-urban catchment, over a period of active urbanization and a subsequent period of relatively stable land-use. The specific objective is to assess the spatiotemporal variability of suspended sediment concentrations at the Ribeira dos Covões catchment outlet and in three upstream sub-catchments, located in central Portugal, characterized by different urban patterns, during storm events during two distinct periods of active urbanization and relatively stable land-use. This knowledge is relevant to guide urban planning and develop best management practices to reduce the anthropogenic impacts on water resources and aquatic ecosystems within peri-urban catchments.

2. Material and Methods

2.1. Study Area

The study focussed on the peri-urbanizing Ribeira dos Covões catchment, on the outskirts of the city of Coimbra, in central mainland Portugal (40°13′ N, 8°27′ W). The catchment, which drains into the much larger Mondego river, has a surface area of 6.2 km^2, with elevation ranging from 205 m in the headwaters to 30 m at the outlet. Catchment lithology comprises Jurassic dolomitic and marl-limestone units on the eastern side and Cretaceous and Tertiary sandstones conglomerates and mudstones in the west, with Plio-quaternary sandy-conglomerates and alluvial deposits in downstream areas [30]. The climate is humid Mediterranean, with a dry summer from June to August, and an autumn/winter period characterized by sequences of much wetter weather. Mean annual temperature is 15 °C and mean annual rainfall is 980 mm, with a highest daily fall since 1958 of 102 mm [30]. Storms of particularly high intensity over short periods tend to occur at the end of the summer dry season.

The catchment has perennial streamflow at the outlet, supplied by several springs, mainly located in the eastern sandstone area. Baseflow represents 33%–37% of streamflow. Mean annual runoff is 135 mm. This represents 17% of mean annual rainfall, but ranging from 14% in driest to 21% in wettest years [29].

From 1950 to 2011, population increased from 14,315 to 26,632 inhabitants, and is currently mainly devoted to the tertiary sector [32]. The increase of population over the last few decades led to significant land use changes, mainly rural up to 1972 to around 40% urban, 56% forest and 4% agricultural by 2012. In 2012, major disturbance activities included the construction of (i) an Enterprise Park in the extreme southwest part of the catchment, covering about 5% of the catchment area, which started in late 2008 with deforestation; and (ii) a four-lane highway crossing the northwest part of the catchment (Figure 1). Some detached houses were also built in SE and NW urban areas and some deforestation in the middle part of the catchment were recorded (Figure 1). After 2013, land-use stabilized because of a national economic crisis. Since 2017, despite a recovery of the national economy, little further urbanization (and catchment land-use change) has occurred. In 2018, clear-felling of small areas of forest was carried out for fire protection purposes, as a consequence of new and stricter legislation regarding the control of biomass in forest areas, following severe wildfires recorded in 2017.

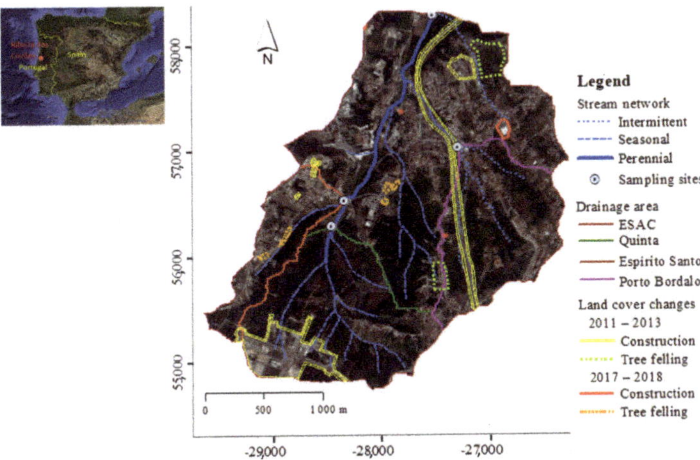

Figure 1. Location of the Ribeira dos Covões catchment in Portugal and the Iberian Peninsula (**left**), and the sampling sites (**right**) at the outlet (ESAC) and in three upstream sub-catchments (Quinta, Espírito Santo and Porto Bordalo). Additionally, shown are sub-catchment boundaries and the location of the main land use/cover interventions during the urbanization (2011–2013) and stabilization (2017–2018) periods. (Adapted from [33]).

In general, urban development is dispersed over the catchment, comprising distinct urban cores of varying design and population density, and with differing storm drainage systems. In the Enterprise Park, a detention basin was constructed to collect runoff from upslope paved surfaces and delay its discharge into the stream during large storms. This detention basin also led sediment retention especially during the construction phase. Nevertheless, sediment fingerprinting evidence showed that bare soil surfaces within the Enterprise Park area were the main source of sediments at the catchment outlet in 2012, though less dominant in 2015 due to increasing vegetation cover [27].

2.2. Catchment and Sub-Catchment Sampling Strategy

In order to investigate the spatial variability of suspended sediments, four sites within the Ribeira dos Covões catchment were selected for this study (Figure 1): the catchment outlet, ESAC, and three upstream sub-catchment sites draining areas of contrasting urban pattern. The three sub-catchments are (1) Quinta, with a relatively low urban area (22%) but including the Enterprise Park; (2) Espírito Santo, the most urbanized (49% urban), comprising mostly detached single-family houses, with relatively low surface sealing downslope and high imperviousness only in some upslope areas, with storm runoff infiltrating either in nearby or (after routing) downslope pervious areas (e.g., forest); and (3) Porto Bordalo, 39% urban, and comprising either rows of attached-houses or more closely packed detached houses with greater surface sealing, as well as part of the highway, with storm runoff from downslope impervious areas being piped directly to the stream. Quinta and Espírito Santo are located on sandstone whereas Porto Bordalo is in the limestone area. ESAC has a perennial flow regime, whereas Porto Bordalo has an ephemeral stream and Espírito Santo and Quinta have seasonal flows. The main characteristics of the four selected sites are summarised in Table 1.

Table 1. Main topographical, lithological, land-use (in 2016) and hydrological characteristics of the catchment and sub-catchments investigated in Ribeira dos Covões (adapted from [34]).

	ESAC–(Outlet)	Porto Bordalo	Espírito Santo	Quinta
Area (ha)	620	113	56	150
Mean slope (°)	10	12	8	4
Sandstone (%)	56	2	98	100
Limestone (%)	41	98	0	0
Alluvial (%)	3	0	2	0
Urban (%)	40	39	49	22
Woodland (%)	56	57	46	73
Agriculture (%)	4	4	5	5
Stream classification	Perennial	Ephemeral	Seasonal	Seasonal
Mean annual runoff (%)	14–21	11–12	26–31	13–14
Mean annual surface runoff (%)	9–13	8–9	20–21	8–9
Baseflow index (%)	36–39	1–2	23–29	33–39
Gauge elevation (m)	32	73	81	83

The four selected sites form part of a hydrological monitoring network installed in Ribeira dos Covões catchment in 2010 [34]. Chemical aspects of stream water quality (notably nutrients and heavy metals) at the sites between October 2011 and March 2013 were reported in Ferreira et al. [35]. In another previous study [27], geochemical properties of fluvial bed sediments collected in 2012 and 2015 were used to explore spatial and temporal changes in sediment sources in the catchment, using the sediment fingerprinting approach.

Water samples were collected at the four sites (ESAC, Quinta, Espírito Santo and Porto Bordalo) during discrete storm events through (i) a period of active urbanization, from April 2011 to March 2013 (12 storms), and (ii) a period of relatively stable land-use, between November 2017 and December 2018 (11 storms). Sampling started immediately before the commencement of rainfall (whenever possible) in order to include pre-storm baseflow sediment content, and then covered rising limb, peak and falling limb sections of the hydrograph. In PB it was not possible to collect water samples prior to individual storms due to the ephemeral regime of the stream. The number of samples taken varied between storms and sampling sites, depending on rainstorm magnitude and discharge response.

The selection of storms was based on weather forecasts and deigned to include (i) the first rainstorms after the summer dry season, and (ii) winter wet season storms of differing magnitude and following contrasting antecedent weather. Water samples were collected manually from the middle of stream sections in 250-mL polyethylene bottles, and stored in a dark chilled cooler ($\sim$4 °C) until reaching the laboratory. In total, 89 and 91 samples were collected at the catchment outlet (ESAC), 69 and 76 in Quinta, 67 and 87 in Espírito Santo and 83 and 90 in Porto Bordalo, during the urbanization and land-use stabilization periods, respectively.

Water level data recorded on dataloggers at 5-min intervals at each site were converted to discharge data using previously derived stage-discharge rating curves [34]. Then, 5-min rainfall data were provided by two rain gauges.

2.3. Laboratory Analysis

Total sediment (TS), including suspended sediment (SS, the portion of sediment retained by a filter) and dissolved sediment (TDS, the portion that passes through the filter), was analysed by the gravimetric method, after water evaporation (Method 2540-B) [36]. Electrical conductivity was measured in all the samples through potentiometric method (using a COND 51+ sensor). All samples were analysed in duplicate for each parameter, and mean values were used for data analyses.

For the samples collected during the land-use stabilization period, suspended sediment concentrations (SSC) were also quantified by filtering samples through pre-weighed 0.45 μm membranes, drying at 105 °C and reweighing (Method 2540-D) [36]. Suspended sediment concentrations for samples collected during the urbanization stage were estimated by subtracting TDS from TSC, where TDS was derived using site-specific linear regressions between TDS and specific conductance.

2.4. Data Analysis

Differences in TSC and SSC at the four sites (ESAC, Porto Bordalo, Espírito Santo and Quinta) were assessed for the urbanization and stabilization periods separately using the non-parametric Kruskal–Wallis H test, given the non-normal distribution of the data. For each site, temporal differences in TSC and SSC between the two periods were assessed using the Mann–Whitney U test. Statistically significant spatial and temporal differences were then further explored using the Least Significant Difference (LSD) post-hoc multiple comparison test. Statistical differences were evaluated at the 0.05 significance level. Linear correlations between sediment concentration variables and peak discharge, storm rainfall, maximum 15-min rainfall intensity (I_{15}) and 7- and 14-day antecedent rainfall (API_7 and API_{14}, respectively) were assessed using Pearson r coefficients. Statistical analyses were performed in IBM SPSS Statistics 25 software.

3. Results

3.1. Rainfall Characteristics and Runoff Response During Monitored Storms

Storm rainfall characteristics and storm runoff coefficients in the active urbanization (2011–2013) and relatively stable land-use (2017–2018) periods at the four monitored sites are shown in Table 2. Storm rainfalls of analysed events ranged from 2.4 to 46.7 mm during urbanization and from 4.0 to 29.2 mm in the stabilization period, covering events with return periods of up to 2 years. Although the largest storm was recorded during urbanization (storm 12), a greater number of relatively large storms (> 10 mm) were monitored during the stabilization period. Nevertheless, differences in the rainfall amount and antecedent rainfall prior to each storm event (API_7 and API_{14}) recorded during urbanization and stabilization periods were not statistically significant ($p > 0.05$). In addition, a broadly similar range of rainfall intensities (I_{15}) were experienced in both periods ($p > 0.05$).

Table 2. Rainfall characteristics and runoff response during the 12 and 11 storms monitored during the urbanization and stabilization periods ([a]: storms recorded after the summer; I_{15}: maximum rainfall intensity during 15 min; API7: 7-day antecedent rainfall; API_{14}: 14-day antecedent rainfall; E: catchment outlet; PB: Porto Bordalo; ES: Espírito Santo; Q: Quinta).

	Storm	Date	Rainfall				Runoff Coefficient (%)			
			Depth (mm)	I_{15} (mm/h)	API_7 (mm)	API_{14} (mm)	E	PB	ES	Q
Urbanization	1	19 April 2011	4.2	2.4	20.7	20.7	5.6	8.2	15.0	5.4
	2	29 April 2011	3.4	0.6	4.8	39.1	4.3	6.6	5.3	2.6
	3 [a]	23–24 October 2011	7.6	1.6	1.8	1.9	3.5	5.5	0.0	0.0
	4 [a]	26 October 2011	3.8	0.9	28.8	28.9	5.0	4.9	17.8	5.0
	5	02 November 2011	24.0	6.0	20.7	48.8	8.6	9.4	7.8	9.3
	6	14 November 2011	5.3	2.7	35.5	97.9	19.0	14.4	24.9	11.1
	7	16 December 2011	3.6	1.1	19.7	22.0	5.3	2.8	7.2	4.2
	8	04 May 2012	2.4	0.9	47.7	85.4	17.1	5.8	13.7	11.9
	9 [a]	25–26 September 2012	6.4	1.7	15.6	15.6	20.3	30.2	62.7	18.2
	10	08–10 January 2013	9.0	1.0	0.0	7.5	5.6	5.1	20.5	9.7
	11	15–17 January 2013	20.2	1.5	25.3	25.3	11.3	8.5	23.7	13.1
	12	25–29 March 2013	46.7	2.8	51.8	75.2	29.4	26.2	31.7	23.0
		Mean values	*11.4*	*1.9*	*22.7*	*39.0*	*11.3*	*10.6*	*19.2*	*9.5*
Stabilization	13 [a]	02 November 2017	4.0	1.0	0.4	8.0	2.0	15.9	3.4	0.0
	14	05 March 2018	9.0	2.4	113.8	114.2	19.6	2.6	23.3	16.0
	15	09 March 2018	7.2	1.2	64.4	134.8	17.4	2.8	25.0	11.1
	16	13 March 2018	14.0	5.2	46.6	155.0	20.4	6.5	22.1	27.4
	17	24 May 2018	29.2	5.8	7.2	2.4	7.3	3.7	9.6	15.1
	18	08 June 2018	20.8	1.6	13.6	18.8	16.3	5.6	14.2	14.7
	19 [a]	15 October 2018	13.8	4.4	14.8	15.0	8.4	37.8	15.4	5.1
	20	29 October 2018	16.8	2.2	0.2	11.0	3.8	11.1	6.4	6.2
	21	08 November 2018	6.2	0.8	47.0	64.6	10.9	20.8	11.6	6.5
	22	29 November 2018	16.3	1.6	46.4	66.6	17.2	10.5	32.1	22.1
	23	18 December 2018	7.8	1.6	32.0	33.6	15.8	5.6	36.8	8.3
		Mean values	*13.2*	*2.5*	*35.1*	*56.7*	*12.6*	*11.2*	*18.2*	*12.0*

Mean storm runoff coefficients were not significantly different between urbanization and stabilization periods for the four sampling sites ($p > 0.05$), although slightly lower in the latter period (Table 2). In general, storm runoff coefficients were greatest in Espírito Santo (3%–63%) ($p < 0.05$), with the largest urbanized contributing area, and similar between the other three sites ($p > 0.05$), although slightly lower in Quinta (0%–27%), with the smallest urban area. During the urbanization stage, the maximum storm runoff coefficient in Espírito Santo was twice as high as in the other sites, despite the active construction of the Enterprise Park in sub-catchment Quinta. Monitored storms included some of the first storms recorded after the summer, and no flow response was recorded in the sandstone Espírito Santo and Quinta sub-catchments, which become dry in summer, in storm 3 of the urbanization period and (in Quinta) in storm 13 of the stabilization period.

In all the four sites, peak discharge per storm was significantly correlated with event rainfall (r ranging from 0.89 for Porto Bordalo, $p < 0.01$, to 0.63 for Espírito Santo, $p < 0.05$) and I_{15} (r ranging from 0.88 for Quinta to 0.80 for Espírito Santo, $p < 0.01$) during the urbanization period. During the stabilization period, however, peak discharge was only correlated with I_{15} (r = 0.85–0.90, $p < 0.01$), except in the Quinta sub-catchment ($p > 0.05$).

3.2. Spatiotemporal Variation of Sediments

3.2.1. TSC and SSC Variations in the Urbanization and Stabilization Periods

TSC showed significant differences between the four sites during both periods. In general, TSC was lower in the Porto Bordalo sub-catchment in both periods (median values of 153 mg/L and 105 mg/L, respectively) than in Quinta, Espírito Santo and ESAC (medians 250–300 mg/L during urbanization and 154–258 mg/L during stabilization) (Figure 2a). The Quinta sub-catchment undergoing major land-use changes provided similar TSC during urbanization to those recorded at the catchment outlet (E) ($p > 0.05$). During stabilization, however, TSC was significantly lower in Quinta than at ESAC (median values of 154 mg/L and 258 mg/L, respectively) ($p < 0.05$).

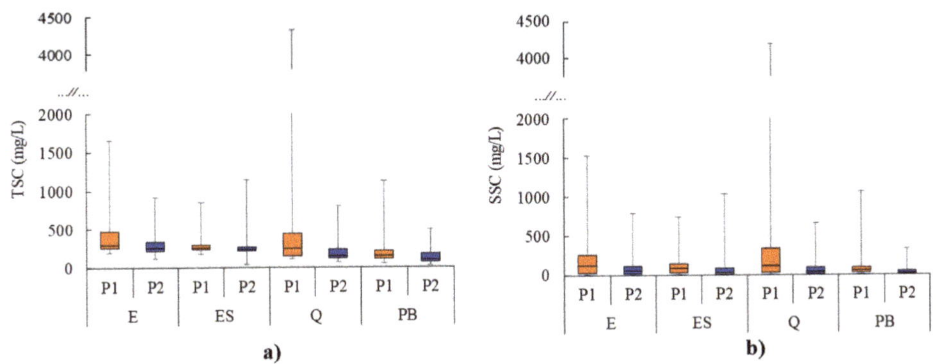

Figure 2. Box plots (median, upper (75) and lower (25) quartiles, maximum and minimum) showing the concentration of total sediment (TSC) (**a**) and suspended sediment (SSC) (**b**) in the four study sites (E: ESAC, ES: Espírito Santo, Q: Quinta, PB: Porto Bordalo), during urbanization (P1) and stabilization (P2) periods.

At all sites except Espírito Santo, there were significant differences in TSC between urbanization and stabilization periods ($p < 0.05$), with median TSC values in the stable period being 38%, 31% and 14% lower in Quinta, Porto Bordalo and ESAC, correspondingly, than during urbanization (Figure 3). The greater decrease occurred in the Quinta sub-catchment undergoing major land-use change. In contrast, the Espírito Santo sub-catchment displayed similar TSC in both periods (258 mg/L vs. 240 mg/L), since it was less affected by urbanization ($p > 0.05$).

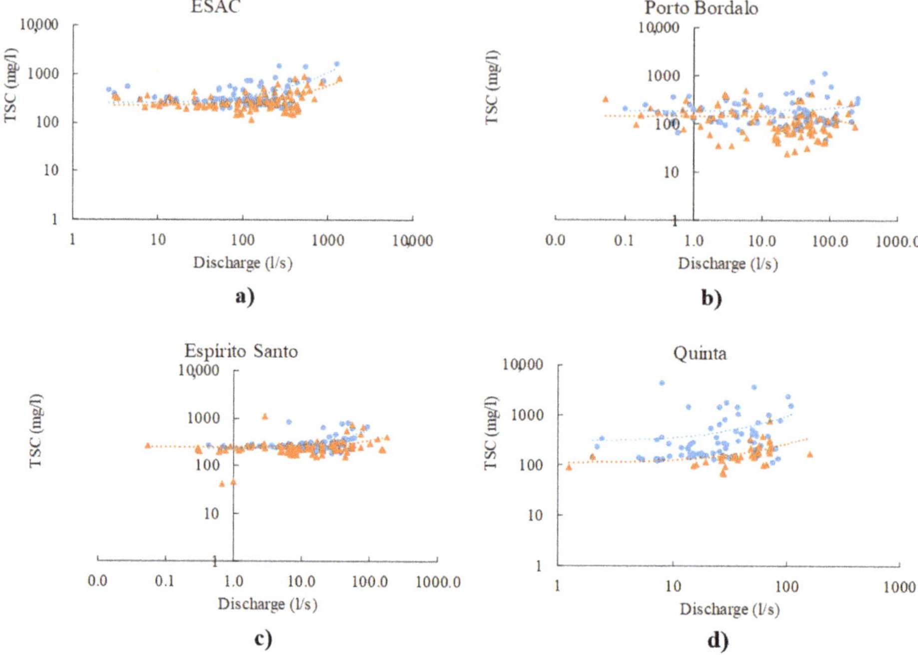

Figure 3. Relation between discharge and total sediment concentration (TSC) in the ESAC (**a**) Porto Bordalo; (**b**) Espírito Santo; (**c**) and Quinta; (**d**) sampling sites during urbanization and stabilization periods.

Similar spatiotemporal variations were recorded in SSC, with lower values in Porto Bordalo during urbanization (52 mg/L) and stabilization (19 mg/L) than at the other three sites (20–120 mg/L) (Figure 2b). Despite similar median SSCs, the greater urbanization activity in Quinta led to higher peak SSC there than at the catchment outlet (ESAC) and Espírito Santo ($p < 0.05$). At Q, 75% of measurements were higher than at ESAC (339 mg/L in Quinta vs. 254 mg/L in ESAC), and the maximum concentrations were almost 3 times higher (4184 mg/L vs. 1525 mg/L). During the stabilization period, however, similar SSC were recorded at Quinta, ESAC and Espírito Santo ($p < 0.05$).

Falls in median SSC (as percentages of urbanization values) between the urbanization and stabilization periods attained 76% in Espírito Santo, 69% in Quinta, 64% in Porto Bordalo and 59% in ESAC. The greatest percentage decrease recorded at Espírito Santo (median falling from 84 mg/L to 20 mg/L) occurred despite only minor land-use change during the urbanization period (Figure 1). In Espírito Santo, SSC decreases were greater for higher then smaller discharges (Figure 4). However, Espírito Santo recorded a higher maximum SSC value during the 'stabilization' period (1033 mg/L) than the previous urbanization period (743 mg/L), probably because of some forest clear-felling close to the stream (Figure 2b).

The catchment outlet (ESAC) recorded the lowest percentage fall in median SSC from urbanization to stabilization (120 mg/L to 49 mg/L). This decrease was clearer under smaller than larger discharges, contrary to Espírito Santo (Figure 4). Similar findings were noticed also in ESAC and Quinta, although the decrease in the curve relating SSC and the discharge is not so sharp from urbanization to stabilization in Quinta regarding higher values. In Porto Bordalo the decrease in SSC from urbanization to stabilization periods were affected similarly under low and higher discharges. An indication of baseflow SSC is given by the values for the first sample of each storm event, e.g., before the rise in streamflow, and the medians of SSC values were much lower than the storm event values, but also showed a fall between urbanization and stabilization periods from 65 to 5 mg/L in ESAC, 75 to 22 mg/L in Espírito

Santo and 45 to 32 mg/L in Quinta, and in Porto Bordalo from 33 to 17 mg/L. Suspended sediment represented a higher fraction of TSC during the urbanization than the stabilization period (Figure 2). In Q, affected by the construction of the large Enterprise Park, SSC represented in median 50% of TSC during urbanization, but only 22% during the stabilization period. In Porto Bordalo, SSC comprised 43% of the median TSC during the construction of the major road, covering 1.5% of the sub-catchment area, but only 19% of the median TSC during the stabilization period. A decrease in the proportion of SSC in TSC was also recorded at Espírito Santo (34% to 9%), despite little land-use change in either period. At the catchment outlet (ESAC), SSC represented 43% and 19% of TSC during the urbanization and stabilization periods, respectively.

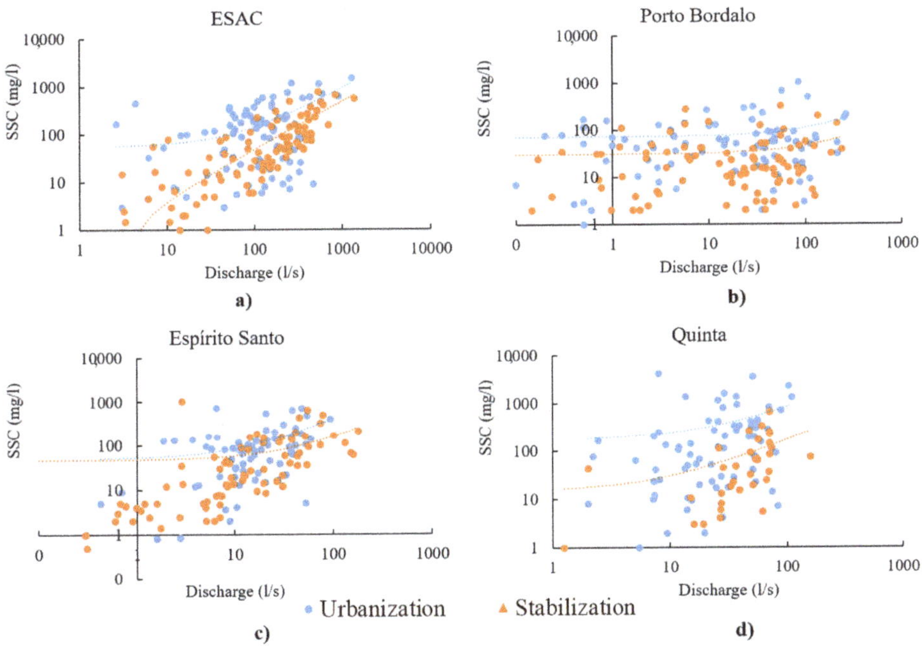

Figure 4. Relation between suspended sediment concentration (SSC) in the ESAC (**a**) Porto Bordalo; (**b**) Espírito Santo; (**c**) Quinta; (**d**) sampling sites during urbanization and stabilization periods.

3.2.2. Inter- and Intra-Storm Variations in SSC

Unsurprisingly, greatest SSC changes during storm events were recorded in the sub-catchment with greater land-use changes (Quinta), followed by the catchment outlet (ESAC), whereas Porto Bordalo (affected by the construction of the major highway) and Espírito Santo showed lower variability (Figure 5). During stabilization, however, highest SSCs and inter-storm variations were recorded in Espírito Santo, subject to clear-felling of trees close to the stream, followed by Quinta, ESAC and Porto Bordalo.

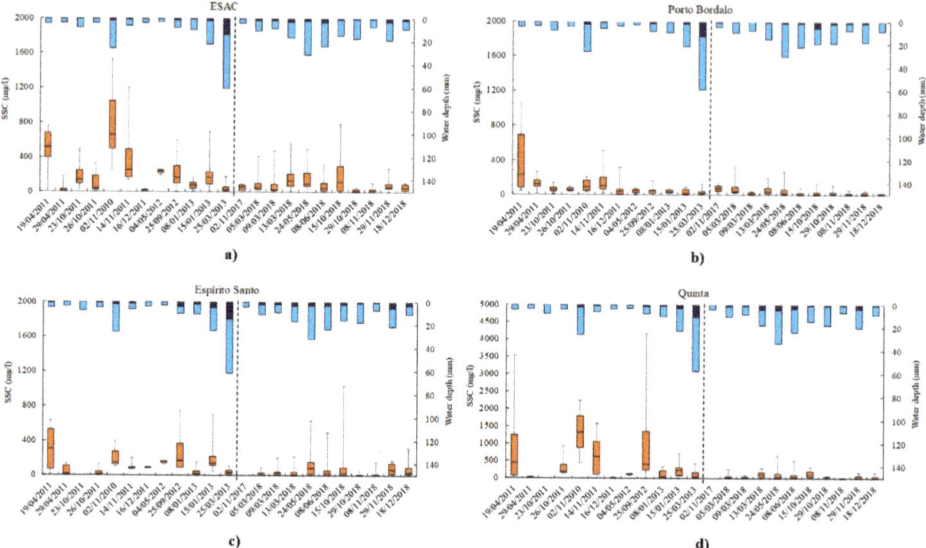

Figure 5. Temporal variability of suspended sediment concentration (SSC) and runoff depth between the four study sites. Dashed lines divide urbanization (2011–2013) and relatively stable land-use (2017–2018) periods. Note scale difference in Quinta regarding SSC.

In general, largest intra-storm variation was perceived in the first storm events monitored after the dry summer, both in the urbanization (e.g., 2 November 2010, 14 November 2011, 25 September 2012) and stabilization periods (15 October 2018, 29 November 2018) (Figure 5). Maximum SSC values of 4184 mg/L and 743 mg/L were recorded at Quinta and Espírito Santo, respectively, on 25/09/2012 during urbanization. This rainfall event triggered the beginning of streamflow in both streams, leading to peak concentrations recorded during the initial stage of discharge and with no baseflow contribution. Nevertheless, median SSC under baseflow conditions were 156 mg/L in Quinta and 249 mg/L in Espírito Santo, during urbanization, highlighting the considerable increase to peak SSC. Nevertheless, in most other storms peak SSC occurred either immediately before (e.g., 19 April 2011) or after (e.g., 08–10 January 2013) peak discharge (Figure 6). During the stabilization period, Espírito Santo displayed highest SSC concentrations in 2018 from late spring (769 mg/L in 24 May 2018) to the first storm after the summer dry period (1144 mg/L in 15 October 2018) (Figure 5) in storm events that occurred after the forest clear-felling episode of early spring that year. Again, in these storms, peak SSC occurred at the beginning of runoff, in contrast to peak discharge timing typical in other events (Figure 6). As a result of this within-period variation, peak SSC in Espírito Santo was not significantly correlated with rainfall amount and intensity, in both urbanization and stabilization periods ($p > 0.05$).

For Quinta, peak SSC also tended to occur at the beginning of discharge for the initial storms after the summer, but was associated with peak discharge for other storms (Figure 6). In Quinta, peak SSC was significantly correlated with I_{15} during urbanization ($r = 0.59$, $p < 0.05$) and with both rainfall amount and maximum intensity during stabilization ($r = 0.84$, $p < 0.01$ and $r = 0.70$, $p < 0.05$, respectively).

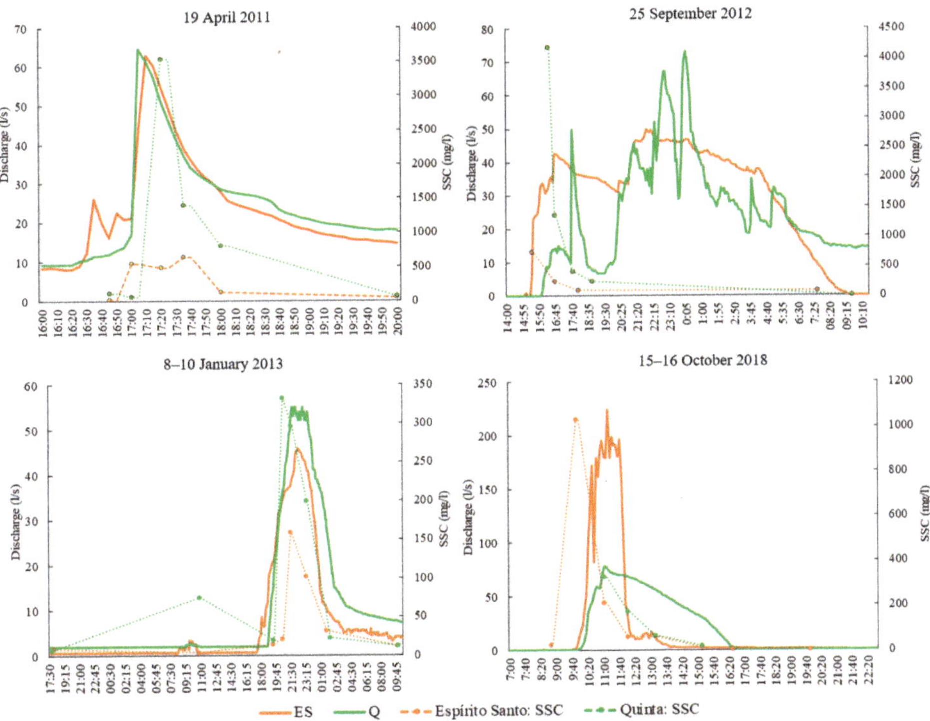

Figure 6. Variation of SSC and discharge in Espírito Santo (ES) and Quinta (Q) sub-catchments during contrasting storms, including under wettest season (19 April 2011 and 8–10 January 2013) and at the end of summer (25 September 2012 and 15–16 October 2018).

At the catchment outlet (ESAC), higher SSC values were also in part associated with the first storms after the summer (Figure 7), but also increasing from 23 October 2011 (729 mg/L) until 02 November 2011 (1656 mg/L) with increasing storm runoff coefficient (Table 2) during the urbanization period. In the stabilization period, SSC roses from 6 mg/L at baseflow to a peak of only 918 mg/L in the 15 October 2018 storm (the highest SSC recorded in 2017–2018) for a runoff coefficient (8%) similar to that recorded in the 02 November 2011 event. In most storm events, peak SSC was linked with peak discharge (e.g., 19 April 2011 and 02 November 2011), even in storms after the summer (e.g., 15–16 October 2018) (Figure 7). In ESAC, significant correlations were found between peak SSC and peak discharge and I_{15} during both the urbanization ($r = 0.63$, $p < 0.05$ and $r = 0.82$, $p < 0.01$, respectively) and stabilization ($r = 0.78$, $p < 0.01$ and $r = 0.69$, $p < 0.05$, respectively) periods.

In contrast to Quinta, Espírito Santo and ESAC, SSC variation in Porto Bordalo, which as an ephemeral flow regime, did not seem to be affected by the first storms after the summer (Figure 5). During urbanization, highest SSC was recorded in the 14 November 2011 event (598 mg/L), largely due to construction of a ditch just upstream of the gauging station. Additionally, the greatest intra-storm variation occurred in the wettest soil conditions of late winter (e.g., 19 April 2011 and 24 May 2018). Although SSC in Porto Bordalo generally increased with discharge (Figure 7), peak SSC per storm did not correlate significantly with peak discharge, storm rainfall or I_{15} ($p > 0.05$).

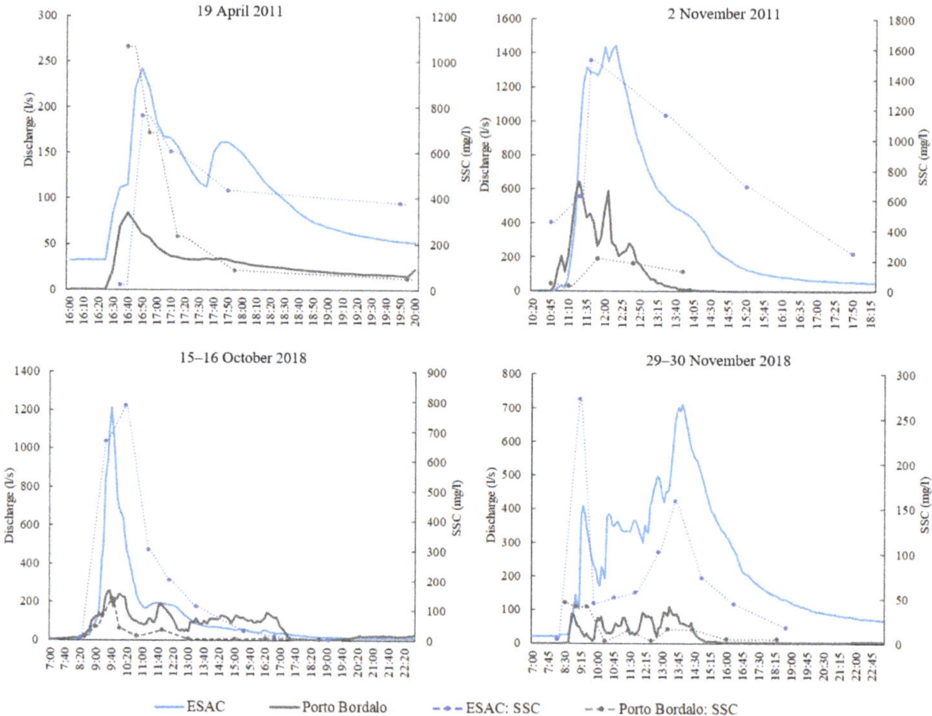

Figure 7. Variation of SSC and discharge at the catchment outlet (ESAC) and in the ephemeral stream Porto Bordalo in distinct storms.

4. Discussion

4.1. Impact of Urbanization on Stream Sediments

As found elsewhere [5,37], suspended sediment (TSC and SSC) in the Ribeira dos Covões catchment are affected considerably by urbanization (2011–2013). The Quinta sub-catchment, which underwent the most intense disturbance in the first period despite having the smallest (22 %) urban area (Table 1), displayed highest TSC and SSC than the other sub-catchments ($p < 0.05$). During the urbanization period, deforestation and construction affected 17% of Quinta sub-catchment, and led to median TSC and SSC being about twice has high as those recorded in Porto Bordalo (Figure 2), despite the latter having twice the urban area (39%) and the active construction of a highway covering 1.5% of the area. In addition, 75% of the TSC and SSC measurements in Quinta were 1.5- and 2.4- times, respectively, the values in Espírito Santo, despite the much higher (but pre-existing) 49% urban area of the latter.

The maximum SSC of 4184 mg/L recorded in Quinta is twice as high as that recorded in a mixed land-use catchment with 30% urban area (2388 mg/L) in Missouri, USA [5]. In this Missouri catchment, median SSC increased by 98%, with a 22% increase in the urban land-use, over four-year period of study [5]. The relatively high impact on SSC from the construction site in Quinta is noticed, despite being reduced by part of the sediments was trapped within a retention basin. This infrastructure slows discharge of runoff from the Enterprise Park area into the stream network, and induces sedimentation, which field observation confirms is added to by the growth of dense vegetation within the retention basin. Since this type of structures is designed to retain water, although they favour sedimentation, additional measures to mitigate sediment loads are required, such as grade control structures [38].

During the urbanization period, maximum TSC and SSC in Porto Bordalo were 1.3- and 1.4- times greater than those recorded in Espírito Santo (Figure 2), may be a consequence of the construction of the highway road. Deposition of excavated soil during construction, however, created an unplanned retention basin which retained part of the runoff and sediments from this source. Field observation during events show that this unplanned basin decreases considerably both flow and sediment connectivity between the construction site and the stream network, despite their close proximity (Figure 1). This has mitigated the impact of urbanization on sediment dynamics as measured at the gauging station downstream. Specialized grading to increase surface roughness and reduce runoff velocity, however, provides an erosion protection measure used in previous projects [38].

At the catchment outlet (ESAC), TSC during urbanization was slightly higher than at Quinta and Espírito Santo ($p > 0.05$) (Figure 2), and twice as high as at Porto Bordalo. Similar results were also recorded for SSC, except that Quinta displayed higher concentrations than at ESAC ($p < 0.05$). Although a tendency for SSC to fall with downstream distance has been recorded in previous studies due to increasing dilution by additional runoff [5], sediment measurements from Ribeira dos Covões do tend to indicate that Quinta and Espírito Santo are the main sources of sediment to the catchment outlet. This also accords with previous findings in Ribeira dos Covões, based on geochemical sediment fingerprinting properties of deposited sediments [27].

Construction sites tend to provide significant sources of sediments when bare soil is exposed, leading to high SSC in the affected streams [2,6]. Thus, previous studies have reported that under active urbanization, sediment yields can be anything from 3–420 times than sediment yields in agricultural and forest areas and that fine-grained sediment yields can reach 21–12,000 times these background rates [6].

Finer sediments from construction sites are more easily available then coarser sediments and can be washed off quicker during the subsequent storms. The greatest percentage decrease in median SSC between the two monitored periods occurred at Espírito Santo (Figure 2). This may indicate that, despite the minor nature of urbanization period construction activities in Espírito Santo (Figure 1), there is a relatively high connectivity between the sites and the stream network, favoured by their downslope location and quick runoff and sediment transport via roads to nearby the outlet (confirmed by field observations). Nevertheless, during the stabilization stage, in Espírito Santo, forest clear-felling in spring 2018 close to the stream constituted a visible new source of sediment. The loose soil was easily washed off in the subsequent storms, leading to very high SSC maxima in storms recorded between May and October 2018 (Figure 5). Several previous studies have shown that deforestation can enhance erosion and suspended sediment by up to several orders of magnitude [36]. These findings also agree with previous authors reporting changes in sediment sources over short timescales [2].

Local measures to mitigate erosion should be implemented to reduce the impacts of disruptive land-use changes, such as construction works and forest clear-felling. Examples of soil erosion mitigation strategies include stabilization measures, such as seeding and mulching, and the installation of physical barriers, e.g., gravel bags, slit fences and straw rolls [38].

4.2. Spatiotemporal Dynamics of Sediments between Urbanization and Stabilization Periods

Human activities involving landscape disturbance, such as construction works and forest clear-felling, even if in small parts of the catchment, have impacts on sediment dynamics. Under relatively stable land-use conditions (2011–2013), the four monitored sites displayed lower TSC and SSC than during the urbanization period (Figures 3 and 4), with similar arrays of rainstorm conditions ($p > 0.05$). Nevertheless, SSC decreases are more discernible at relatively low discharges in ESAC and Quinta, but at higher discharges in Espírito Santo. It is logical that anthropogenic disturbance sites and/or sources close to the stream network will lower the storm discharge threshold at which significant suspended sediment and runoff will be generated, as recorded in Quinta. It also would explain the greater percentage decrease in SSC from urbanization to stabilization phases at lower discharges (Figure 4). During larger storms, and consequent higher discharges, sediment sources more distant from the stream network may become relevant sources of sediments, such as may be the case of sediments from construction sites in Espírito Santo. Furthermore, other sediment sources over the catchment can become active only with increased soil saturation and flow connectivity in larger storms (especially in wet periods), thus explaining the smaller decreases in SSC during high discharges in ESAC (Figure 4).

In general, really high SSCs were experienced in just a few storms, particularly after the summer drought (Figure 5), associated with shorter but more intense rainstorms (Table 1), but also in some winter storms, notably those of 02 November 2011, 13 March 2018 and 24 May 2018 (Figure 5), with highest I_{15} ($\approx$6 mm/h, Table 2). Rainfall intensity determines the available raindrop energy to erode surfaces, and runoff rate transport sediment [39]. Studies elsewhere have also reported the impact of few extreme events on very high SSC and total suspended sediment loads in relatively small rivers [17].

Storms recorded after the summer are characterized by peak SSCs that vary by several orders of magnitude depending on antecedent peak discharge (Figures 6 and 7). During these storms, rainfall rapidly mobilizes sediments from active sources and/or dried, loose material in the dry sections of the drainage network. In addition, sediments transported from the hillslopes to low-order channels by relatively small early summer storms may remain stored until subsequent high flows, as described elsewhere by Gellis [37]. Peak discharge has been considered one of the key factors governing suspended sediment yields in small catchments [23].

Some studies have also reported sediment flushing from hillslopes and the channel during the first rainfall events after long dry periods [40], particularly from paved surfaces [41]. This may indicate that sediments are often supply limited instead of transport limited [17]. Thus, the meteorological and hydrologic characteristics of previous storms may influence sediment characteristics of the current storm [37]. The seasonal variation of sediment yields must be considered when planning land-use activities, such as urbanization and clear-felling. Some authors argue that scheduling activities that disturb established soil during the drier seasons, avoiding ground disturbance when water and wind are more likely, should be favoured [38]. This type of planning, however, must be considered carefully in climate regions such as the Mediterranean, recording short but high intensity storms immediately after the summer, since it may not be enough to mitigate erosion from construction sites effectively.

In the Ribeira dos Covões catchment, however, sediment dynamics differed between storms and sites. In Porto Bordalo, rises in SSC after the summer are less pronounced than at the other sites (Figure 5), possibly due to its ephemeral as rather than seasonal flow regime. At this site, high SSC values were recorded later in the wet season (e.g., 19 April 2011 and 13 March 2018, Figures 6 and 7), possibly due to increased connectivity of runoff and sediments eroded from more distant sources. This type of sediment behaviour, associated with increasing connectivity over the wet season, has also been reported in studies elsewhere [37].

Differences in available sediment sources affected by land-use changes and in antecedent rainfall are relevant parameters influencing intra-storm sediment variations, including suspended sediment hysteresis [26]. Complex sediment patterns can be characterized by (i) clockwise hysteresis, when sediments increase during the rising limb and decrease during the falling limb of the hydrograph, as a result, for example, of rapid sediment flushing and depletion in the stream network; (ii) anti-clockwise hysteresis, when sediment increase is delayed, driven by, for example, sediments dilution with stormwater runoff on the rising limb and arrival of sediments from more distant sources; and (iii) other complex hysteresis associated with figure-eight loops, as a result of exhaustion of a particular sediment source during severe storms, differences in the arrival time of sediments from distinct sources which become connected with the river over the rainstorm, as well as storage and resuspension of sediments within the channel during storms [2,23,37]. Hysteresis analysis was not possible in the current study, because the frequency of sampling was insufficient during some of the events.

4.3. Impact of Different Urban Patterns on Sediment Dynamics

In the stabilization period, the previous major construction activities are still to some extent affecting stream sediment dynamics. Thus in 2017–2018, Quinta remains the sub-catchment with highest TSC and SSC (Figure 2), despite the percentage reductions of 38% and 69%, respectively, in median values compared with during the urbanization period. This is due to the extent of remaining erodible bare surface at these sites. Previous studies have also showed that most human interventions affect SSC over an extended period of time [2]. The high concentrations of sediments recorded in 2017-18 in Quinta stream, however, indicate also that the retention basin is not being very effective in trapping the fine suspended sediment fraction, such that SSCs at Quinta are now similar to those measured at the catchment outlet at ESAC ($p > 0.05$).

In contrast, in Porto Bordalo, containing the four-lane road construction site, SSC values in the stabilization period (median 19 mg/L) were much lower than in Quinta (Figure 2). The continued low SSC values in Espírito Santo in the second period, despite the sub-catchment having the highest (49 %) urban area and a forest-felling episode in Spring 2018 can be linked to the upslope location of the urban area, the detached house pattern and a lack of connectivity of runoff with the stream network. Studies elsewhere reported higher sediment concentrations from catchments with larger urban land-use [37]. Suspended sediment yields from urban catchments tend to be 2–70 times higher than background levels (in agriculture and forest) [6]. This is because stream channel erosion can persist longer after urban development, and because in some peri-urban areas gravel roads and parking areas can provide continuing sources of sediments to waterways [6]. Sealed surfaces, such as concrete and asphalt, can also be a relevant source of anthropogenic particles to streams, supplementing urban streams with coarser sediment loads [18,41].

The type of urban pattern and its location within the landscape can also affect overland flow and stream sediment supply and dynamics [29,40]. In Espírito Santo, runoff from upslope detached housing is mostly dissipated in surrounding pervious soil, enhancing water retention and infiltration opportunities. In contrast, in Porto Bordalo, runoff from its urban surfaces is partially piped to the stream network, thereby enhancing flow and sediment connectivity between the urban areas and the stream network [34]. This may explain the similar TSC and SSC of Porto Bordalo and Espírito Santo during the urbanization period, considering the unplanned retention basin downslope the road construction area in Porto BordaloB. Urban drainage systems are known to have a high sediment transport efficiency [2].

Lower sediment concentrations in Porto Bordalo may be also partially linked with its marly limestone soils, compared with the sandstone soils in the other upstream catchments. Analyses of fluvial sediments in Ribeira dos Covões showed that the limestone area (covering 40% of the catchment) only provided 10% of the sediments deposited at the catchment outlet [27]. Some authors have reported suspended sediment yields to be primarily dependent on geology and soil type, together with climate, land-use and position in the catchment [6,37].

At the catchment outlet (ESAC), median SSC during stabilization (49 mg/L) is lower than the median concentrations recorded during both wet (139-300 mg/L) and drier seasons (7–102 mg/L) in Hinkson Creek Watershed, Missouri, USA [5]. This watershed has similar annual rainfall (1036 mm) than the Ribeira dos Covões catchment, it is less urbanized (30% vs. 40%) and has an area 17 times higher (230 km^2). In Missouri catchment, however, variations in estimated annual suspended sediment yields in individual years from 16 to 313 t/km^2 were mainly attributed to differences in annual total precipitation, rather than urbanization [5]. This comparison between sites indicates a relatively acceptable suspended sediment situation during the stabilization period, but episodically rather high sediment levels during the urbanization phase (median SSC of 120 mg/L). However, sediments from urban areas in Ribeira dos Covões are associated with some pollutants, such as heavy metals, which may impair aquatic ecosystems [27]. In this study area, concentrations of Pb, Cu and Zn attained 155 mg/kg, 188 mg/kg and 659 mg/kg, respectively, in fluvial sediments derived from urban areas [27]. In 2018, a preliminary survey also identified and quantified some types of UV filters (e.g., octocrylene and ethylhexyltriazone) and parabens (e.g., methyl) in fluvial sediments over the Ribeira dos Covões catchment [42]. Thus, although sediment loads decreased during the stabilization period, the chemical load may be higher than during urbanization period.

In Portugal, as in other countries, no regulatory framework or legal guidelines for erosion control measures are established, thus no erosion control is performed in construction sites and/or clear-felled forest stands, and no specific measures to protect water bodies are implemented. Measurements to mitigate erosion are mandatory only for farmers receiving financial support to implement the Common Agricultural Policy. In the last years, however, Portuguese government is providing guidelines no mitigate soil erosion but only in wildfire affected areas. This lack of regulatory framework leads to high erosion rates over the landscape, leading, for example to siltation. In 2018, Coimbra municipality spent 4 million euros to remove ~700,000 m^3 of sediments from a 3.2 km section of the Mondego river (just upstream of the confluence with Ribeira dos Covões catchment), where a dam was constructed in 1981.

A more complete understanding of the impacts of human disturbance, and the mosaic of urban types on sediment dynamics, and their chemical impact requires further investigation, namely in Ribeira dos Covões. Such information is relevant to mitigate degradation of aquatic ecosystems, and to establish efficient strategies to control erosion and reduce sediment loads and flood risk to the stream system. Simoni et al. [38] propose a structured conceptual planning approach, based on knowledge of hydrologic and sediment availability and sediment connectivity to channels, to design effective erosion and sediment flux mitigation measures. In some cases, such as those where sediment loads may impair water bodies used for human consumption, restrictions and prohibitions on land-use changes may be recommended, and used to integrate delimited safeguard zones into land-use planning [43].

5. Conclusions

Urbanization and distinct urban patterns affect sediment levels and fluxes in streams inside the peri-urban Ribeira dos Covões catchment. The construction of an Enterprise Park covering 17% of the Quinta sub-catchment led to 1.4-2.2 higher SSC than in the other two sub-catchments despite their much greater percentage urban area. Although the impact of the Enterprise Park construction was reduced by 69% (in terms of median storm SSC) in the 2017-18 stabilization period, the extent of bare soil surface in the Enterprise Park meant that the site (and the Quinta sub-catchment) remained the most important sediment source in the catchment. This was despite partial mitigation by the construction of a retention basin at the site.

The study shows that landscape disturbance, even in small parts of the catchment, are of great importance to sediment dynamics. In the Espírito Santo sub-catchment, the construction of a few houses also had an impact on stream suspended sediments, favoured by the effectiveness of roads in transporting runoff and sediments to downslope areas. In the same sub-catchment, a small area of forest clear-felling close to the stream also enhanced sediment concentrations in 2018, in storms over the subsequent six months. In Porto Bordalo sub-catchment, the construction of a four-lane highway covering 1.5% of the area also played a relevant role in TSC and SSC, despite significant mitigation due to an unintentional retention basin created by piles of spoil downslope of the construction site, which reduced connectivity with the stream network.

Apart from anthropogenic activities and land-use patterns, sediment dynamics are also affected by rainfall and discharge patterns. During both the urbanization and stabilization periods, suspended sediment concentrations were usually highest in storms recorded after the summer dry season associated with shorter but more intense rainstorms and increased sediment availability. Storm events later in the winter wet season, however, also led to high sediment fluxes, due to the role of increasing wetness in enhancing the connectivity between less active sediment sources and the stream network. This is more noticed in the limestone Porto Bordalo sub-catchment, than in the sandstone sub-catchments.

Mixed land-use mosaics provide distinct sources and sinks of runoff and sediments, reducing the connectivity of their fluxes downslope. Planners need to incorporate such spatial mosaic strategies, with urban areas located further from the stream network, to reduce downstream sediment problems. Furthermore, planning the best time of land-use changes, based on seasonal differences in peak sediment concentrations, together with the implementation of specific strategies to mitigate and retain sediments in source sites (e.g., construction sites and clear-felling forest stands), such as seedling and sediment traps, must be considered to develop a conceptual approach to mitigate sediment flux control and natural hazard risk mitigation.

Author Contributions: C.S.S.F. performed the field and laboratorial work and wrote the manuscript; R.P.D.W. supervised the research and supported the writing of the manuscript; Z.K. reviewed and edited the manuscript; A.J.D.F. supervised the research. All authors have read and agreed to the published version of the manuscript.

Funding: This research was funded by the Portuguese Science and Technology Foundation through the post-doctoral research grant of Carla Ferreira (SFRH/BPD/120093/2016).

Acknowledgments: The authors are grateful to Jaydeth Nascimento and Adélcia Veiga for technical support in the laboratory analysis.

References

1. Communication from the Commission to the Council, the European Parliament, the European Economic and Social Committee and the Committee of Regions. In *Thematic Strategy for Soil Protection (COM 2006.231)*; Commission of the European Communities: Brussels, Belgium, 2006.

2. Vercruysse, K.; Grabowski, R.C.; Rickson, R.J. Suspended sediment transport dynamics in rivers: Multi-scale drivers of temporal variation. *Earth-Sci. Rev.* **2017**, *166*, 38–52. [CrossRef]

3. Gray, A.B. The impact of persistent dynamics on suspended sediment load estimation. *Geomorphology* **2018**, *322*, 132–147. [CrossRef]

4. Duan, W.; Takara, K.; He, B.; Luo, P.; Nover, D.; Yamashiki, Y. Spatial and temporal trends in estimates of nutrient and suspended sediment loads in the Ishikari River, Japan, 1985 to 2010. *Sci. Total Environ.* **2013**, *461–462*, 499–508. [CrossRef] [PubMed]

5. Zeiger, S.; Hubbart, J.A. Quantifying suspended sediment flux in a mixed-land-use urbanizing watershed using a nested-scale study design. *Sci. Total Environ.* **2016**, *542*, 315–323. [CrossRef]

6. Russell, K.L.; Vietz, G.J.; Fletcher, T.D. Global sediment yields from urban and urbanizing watersheds. *Earth-Sci. Rev.* **2017**, *168*, 73–80. [CrossRef]

7. Atasoy, M.; Palmquist, R.B.; Phaneuf, D.J. Estimating the effects of urban residential development on water quality using microdata. *J. Environ. Manag.* **2006**, *79*, 399–408. [CrossRef]

8. Shojaeezadeh, S.A.; Nikoo, M.R.; McNamara, J.P.; AghaKouchak, A.; Sadegh, M. Stochastic modeling of suspended sediment load in alluvial rivers. *Adv. Water Resour.* **2018**, *119*, 188–196. [CrossRef]

9. Mukherjee, D.P. Dynamics of metal ions in suspended sediments in Hugli estuary, India and its importance towards sustainable monitoring program. *J. Hydrol.* **2014**, *517*, 762–776. [CrossRef]

10. Wijesiri, B.; Liu, A.; Deilami, K.; He, B.; Hong, N.; Yang, B.; Zhao, X.; Ayoko, G.; Goonetilleke, A. Nutrients and metals interactions between water and sediment phases: An urban river case study. *Environ. Pollut.* **2019**, *251*, 354–362. [CrossRef]

11. Yu, S.; Wu, Q.; Li, Q.; Gao, J.; Lin, Q.; Ma, J.; Xu, Q.; Wu, S. Anthropogenic land uses elevate metal levels in stream water in an urbanizing watershed. *Sci. Total Environ.* **2014**, *488–489*, 61–69. [CrossRef]

12. Dias-Ferreira, C.; Pato, R.L.; Varejão, J.B.; Tavares, A.O.; Ferreira, A.J.D. Heavy metal and PCB spatial distribution pattern in sediments within an urban catchment—Contribution of historical pollution sources. *J. Soils Sediments* **2016**, *11*, 1–12. [CrossRef]

13. European Environmental Agency (EEA). *European Waters—Assessment of Status and Pressures 2018*; EEA Report No 7/2018; Environmental Protection Agency: Luxembourg, 2018. Available online: https://www.eea.europa.eu/publications/state-of-water (accessed on 1 December 2019).

14. United States Environmental Protection Agency (USEPA). Water Quality Assessment and TMDL Information. Available online: https://iaspub.epa.gov/waters10/attains_index.home (accessed on 2 September 2019).

15. Miller, J.R.; Ferri, K.; Grow, D.; Villarroel, L. Hydrologic, geomorphic, and stratigraphic controls on suspended sediment transport dynamics, Big Harris Creek restoration site, North Carolina, USA. *Anthropocene* **2019**, *25*, 100188. [CrossRef]

16. Ehteram, M.; Ghotbi, S.; Kisi, O.; Ahmed, A.N.; Salih, G.H.A.; Fai, C.M.; Krishnan, M.; EL-Shfie, A. River Suspended Sediment Prediction Using Improved ANFIS and ANN Models: Comparative Evaluation of the Soft Computing Models. *Water* **2019**, *11*. in press.

17. Tramblay, Y.; Saint-Hilaire, A.; Ouarda, T.B.M.J.; Moatar, F.; Hecht, B. Estimation of local extreme suspended sediment concentrations in California Rivers. *Sci. Total Environ.* **2010**, *408*, 4221–4229. [CrossRef]

18. Ferreira, C.S.S.; Moruzzi, R.; Isidoro, J.M.G.P.; Tudor, M.; Vargas, M.; Ferreira, A.J.D.; de Lima, J.L.M.P. Impacts of distinct spatial arrangements of impervious surfaces on runoff and sediment fluxes from laboratory experiments. *Anthropocene* **2019**, *28*, 100219. [CrossRef]

19. Mao, L.; Carrillo, R. Temporal dynamics of suspended sediment transport in a glacierized Andean basin. *Geomorphology* **2017**, *287*, 116–125. [CrossRef]

20. Oliveira, K.S.S.; Quaresma, V.S. Temporal variability in the suspended sediment load and streamflow of the Doce River. *J. South Am. Earth Sci.* **2017**, *78*, 101–115. [CrossRef]

21. Gray, A.B.; Pasternack, G.B.; Watson, E.B.; Warrick, J.A.; Goñi, M.A. Effects of antecedent hydrologic conditions, time dependence, and climate cycles on the suspended sediment load of the Salinas River, California. *J. Hydrol.* **2015**, *525*, 632–649. [CrossRef]

22. Peña-Angulo, D.; Nadal-Romero, E.; Gonzalez-Hidalgo, J.C.; Albaladejo, J.; Andreu, V.; Bagarello, V.; Barhi, H.; Batalla, R.J.; Bernal, S.; Gienes, R.; et al. Spatial variability of the relationships of runoff and sediment yield with weather types throughout the Mediterranean basin. *J. Hydrol.* **2019**, *571*, 390–405.

23. Esteves, M.; Legout, G.; Navratil, O.; Evrard, O. Medium term high frequency observation of discharges and suspended sediment in a Mediterranean mountainous catchment. *J. Hydrol.* **2019**, *568*, 562–574. [CrossRef]

24. Walling, D.E. Tracing suspended sediment sources in catchments and river systems. *Sci. Total Environ.* **2005**, *344*, 159–184. [CrossRef] [PubMed]

25. Hammer, T.R. Stream channel enlargement due to urbanization. *Water Resour. Res.* **1972**, *8*, 1530–1540. [CrossRef]

26. Kemper, J.T.; Miller, A.J.; Welty, C. Spatial and temporal patterns of suspended sediment transport in nested urban watersheds. *Geomorphology* **2019**, *336*, 95–106. [CrossRef]

27. Ferreira, C.S.S.; Walsh, R.P.D.; Blake, W.H.; Kikuchi, R.; Ferreira, A.J.D. Temporal dynamics of sediment sources in an urbanizing Mediterranean catchment. *Land Degrad. Dev.* **2017**, *28*, 2354–2369. [CrossRef]

28. Ferreira, C.S.S.; Walsh, R.P.D.; Steenhuis, T.S.; Shakesby, R.A.; Nunes, J.P.N.; Coelho, C.O.A.; Ferreira, A.J.D. Spatiotemporal variability of hydrologic soil properties and the implications for overland flow and land management in a peri-urban Mediterranean catchment. *J. Hydrol.* **2015**, *525*, 249–263. [CrossRef]

29. Ferreira, C.S.S.; Walsh, R.P.D.; Nunes, J.P.C.; Steenhuis, T.S.; Nunes, M.; de Lima, J.L.M.P.; Coelho, C.O.A.; Ferreira, A.J.D. Impact of urban development on streamflow regime of a Portuguese peri-urban Mediterranean catchment. *J. Soils Sediments* **2016**, *16*, 2580–2593. [CrossRef]

30. Kalantari, Z.; Ferreira, C.S.S.; Walsh, R.P.D.; Ferreira, A.J.D.; Destouni, G. Urbanization development under climate change: Hydrological responses in a peri-urban Mediterranean catchment. *Land Degrad. Dev.* **2017**, *28*, 2207–2221. [CrossRef]

31. European Environment Agency. *The European Environment—State and Outlook 2015*; Synthesis Report; European Environment Agency: Copenhagen, Denmark, 2015. Available online: https://www.eea.europa.eu/ /soer-2015/synthesis/report (accessed on 28 December 2019).

32. Instituto Nacional de Estatística. *IX Recenseamento Geral da População*; Instituto Nacional de Estatística: Lisboa, Portugal, 2011. (In Portuguese)

33. Google Earth. *Image Landsat/Copernicus*; US Department of State Geographer: New York, WA, USA, 2019. Available online: https://www.eea.europa.eu/publications/state-of-water (accessed on 30 November 2019).

34. Ferreira, C.S.S.; Walsh, R.P.D.; Steenhuis, T.S.; Ferreira, A.J.D. Effect of Peri-urban Development and Lithology on Streamflow in a Mediterranean Catchment. *Land Degrad. Dev.* **2018**, *29*, 1141–1153. [CrossRef]

35. Ferreira, C.S.S.; Walsh, R.P.D.; Costa, M.L.; Coelho, C.O.A.; Ferreira, A.J.D. Dynamics of surface water quality driven by distinct urbanization patterns and storms in a Portuguese peri-urban catchment. *J. Soils Sediments* **2016**, *16*, 2606–2621. [CrossRef]

36. Baird, R.B.; Eaton, A.D.; Rice, E.W. *Standard Methods for the Examination of Water and Wastewater*, 20th ed.; American Public Health Association, American Water Works Association and Water Environment Federation: Washington, DC, USA, 1998.

37. Gellis, A.C. Factors influencing storm-generated suspended-sediment concentrations and loads in four basins of contrasting land use; humid-tropical Puerto Rico. *Catena* **2013**, *104*, 39–57. [CrossRef]

38. Simoni, S.; Vignoli, G.; Mazzorana, B. Enhancing sediment flux control and natural hazard risk mitigation through a structured conceptual planning approach. *Geomorphology* **2017**, *291*, 159–173. [CrossRef]

39. Barco, J.; Hogue, T.S.; Curto, V.; Rademacher, L. Linking hydrology and stream geochemistry in urban fringe watersheds. *J. Hydrol.* **2008**, *360*, 31–47. [CrossRef]

40. Helsel, D.R. *Land Use Influences on Heavy Metals in an Urban Reservoir System*; NTIS PB-296 724; Department of Commerce: New York, WA, USA, 1978. Available online: https://books.google.pt/books?id=Hx_ApCXk9LMC&pg=PA1206&lpg=PA1206&dq=Land+Use+Influences+on+Heavy+Metals+in+an+Urban+Reservoir+System (accessed on 2 January 2019).

41. Ferreira, A.J.D.; Soares, D.; Serrano, L.M.V.; Walsh, R.P.D.; Ferreira, C.M.D.; Ferreira, C.S.S. Roads as sources of heavy metals in urban areas. The Covões Catchment experiment; Coimbra; Portugal. *J. Soils Sediments* **2016**, *16*, 2622–2639. [CrossRef]

42. Ferreira, C.S.S.; Apel, C.; Bento, C.; Koetke, D.; Ferreira, A.J.D.; Ebinghaus, R. Assessment of UV Filters and Parabens in a small Portuguese Peri-urban catchment. *Proceedings* **2019**, *30*, 30. [CrossRef]

43. Restrepo, J.D.; Kettner, A.J.; Syvitski, J.P.M. Recent deforestation causes rapid increase in river sediment load in the Colombian Andes. *Anthropocene* **2015**, *10*, 13–28. [CrossRef]

Article

Effects of Tractor Passes on Hydrological and Soil Erosion Processes in Tilled and Grassed Vineyards

Giorgio Capello [1], Marcella Biddoccu [1,*], Stefano Ferraris [2] and Eugenio Cavallo [1]

[1] Institute for Agricultural and Earthmoving Machines (IMAMOTER), National Research Council of Italy, Strada delle Cacce, 73, 10135 Torino, Italy; g.capello@ima.to.cnr.it (G.C.); e.cavallo@imamoter.cnr.it (E.C.)

[2] Interuniversity Department of Regional and Urban Studies and Planning (DIST), Politecnico di Torino and Università di Torino, Castello del Valentino, Viale Mattioli, 39, 10125 Torino, Italy; stefano.ferraris@unito.it

* Correspondence: marcella.biddoccu@cnr.it; Tel.: +39-011-3977-723

Received: 31 August 2019; Accepted: 7 October 2019; Published: 12 October 2019

Abstract: Soil erosion is affected by rainfall temporal patterns and intensity variability. In vineyards, machine traffic is implemented with particular intensity from late spring to harvest, and it is responsible for soil compaction, which likely affects soil hydraulic properties, runoff, and soil erosion. Additionally, the hydraulic and physical properties of soil are highly influenced by vineyards' inter-rows soil management. The effects on soil compaction and both hydrological and erosional processes of machine traffic were investigated on a sloping vineyard with different inter-row soil managements (tillage and permanent grass cover) in the Alto Monferrato area (Piedmont, NW Italy). During the investigation (November 2016–October 2018), soil water content, rainfall, runoff, and soil erosion were continuously monitored. Field-saturated hydraulic conductivity, soil penetration resistance, and bulk density were recorded periodically in portions of inter-rows affected and not affected by the machine traffic. Very different yearly precipitation characterized the observed period, leading to higher bulk density and lower infiltration rates in the wetter year, especially in the tilled vineyard, whereas soil penetration resistance was generally higher in the grassed plot and in drier conditions. In the wet year, management with grass cover considerably reduced runoff (−76%) and soil loss (−83%) compared to tillage and in the dry season. Those results highlight the need to limit the tractor traffic, in order to reduce negative effects due to soil compaction, especially in tilled inter-rows.

Keywords: vineyards; soil management; tractor traffic; hydrological properties; erosion; runoff; hydraulic conductivity; soil water conservation

---

1. Introduction

Viticulture represents one of the most important agricultural activities worldwide, covering, at a global scale in 2018, 7.4 million ha [1]. In Europe, vineyards are mostly devoted to wine production, and, in 2016, they covered 3.3 million ha, and Italy, with 690.000 ha, was ranked third, after Spain and France [1]. The vineyard agro-ecosystem has relevant socio-economic impacts on the vine-growing regions, because of its interactions with the environment, landscape, the cultural and touristic features, and employment [2].

The vineyard agro-ecosystem needs to be carefully managed to preserve essential resources such as soil and water, and its overall socio-economic and environmental sustainability [3]. Since the last decade, a growing attention was paid to the impacts of agricultural activities on the ecosystem services, defined as "the direct and indirect contributions of ecosystems to human well-being" [4]. Such ecosystem services are negatively affected by soil degradation, namely soil compaction and soil erosion, offsite contamination, biodiversity reduction, and pressure on water resources [5–7]. Soil erosion and soil compaction were identified as two of the major threats that affect worldwide agricultural soils by the Soil Thematic Strategy from the European Union [8,9] and the FAO Status of the World's Soil Resources [10].

Vineyard operations are highly mechanized, so the traffic of tractors and other machines (e.g., harvesters) necessarily occurs along fixed paths. Traffic is particularly intense from late spring to harvest and has a relevant effect on soil compaction, on soil hydraulic properties, and, consequently, on runoff and soil erosion at field scale [11–13]. The soil management that is adopted in vineyards' inter-rows, as well as in other permanent crops, such as fruit and olive orchards, affects the hydrological response of the soil, the ecosystem services, and the cultural, landscape, and aesthetic values. Several studies report cover crops (or grass cover) in the inter-rows as a soil management practice adopted to reduce runoff and soil erosion in vineyards, with differences in its effectiveness depending on local conditions [14–21]. The use of vegetation cover in vineyards also has a relevant effect in improving biodiversity [22], soil organic matter and physical properties [23], and water availability and trafficability [24]. The vineyard inter-row soil management has a fundamental impact on the water balance at field scale, due to its effects on evapotranspiration, runoff formation, and several hydrological characteristics of the soil, such as hydraulic conductivity, soil water content, soil water retention, and ground-water recharge [25–29]. Depending on the pedoclimatic context, when not properly managed, cover crops may affect grapevine yield because of the competition for water and nutrients [30–32]. For this reason, in semi-arid environments, the soil is usually maintained bare, while, in regions with a less-dry climate, different soil managements to improve soil quality and ecosystem services are adopted [33]. For example, in the framework of the 2007–2013, the local regional Rural Development Program the Piedmont region (NW Italy) set up and supported measures for soil erosion prevention and the maintenance of soil organic carbon (SOC) levels based on the grass covering on more than 13,000 ha of orchards and vineyards [34].

Wide use of machinery during vineyard plantation and management in modern viticulture affects soil and water conservation. Deep ploughing and, occasionally, land levelling are carried out with heavy machinery before plantation [35,36]. Multiple tractor passes on fixed paths in the inter-rows are required every year for operations such as mechanical weeding, chemical spreading, green and winter pruning, and harvesting [37]. The repeated tractor traffic in inter-rows causes soil compaction on most of the vineyard surface [37], and when operations are performed on wet soil conditions, the risk of soil compaction worsens [38]. Soil compaction increases soil resistance to roots' exploration, reduces yields [39,40], and negatively affects soil physical fertility and soil organic carbon stock, resulting in the reduction of soil porosity, water infiltration capacity, and increased runoff, with a decrease of storage and supply of water in the soil [11–13]. Scaling-up spatially the effects of soil compaction, the increase of surface runoff at field-scale impacts the peak discharge at the catchments scale, and thus could have a relevant role in increasing flood risk [40].

In the sloping vineyards of the Alto Monferrato region, previous studies showed the effectiveness of inter-rows' permanent grass cover in reducing runoff and soil erosion, with respect to soil management with tillage [12,16,20,41]. Such behavior is likely related to soil compaction induced by tractor traffic, which also influences soil water infiltration and retention. This study aimed to investigate the blended effects of soil management and tractor traffic on the spatial and temporal variability of soil compaction and both hydrological and erosional processes. Soil water content, rainfall, runoff, and soil erosion were continuously monitored for two years in coupled field-scale plots, with inter-rows managed with permanent grass cover and tillage, respectively. In addition, periodic surveys were carried out during the growing seasons, to measure temporal variations in soil compaction and field-saturated hydraulic conductivity after the implementation of tractor passages.

2. Materials and Methods

2.1. Study Site

The study was carried out at the Tenuta Cannona Experimental Vine and Wine Centre of Agrion Foundation. It is located in the Alto Monferrato hilly area of Piedmont, North-West Italy, at an average elevation of 296 m above sea level (a.s.l.) (Figure 1). The study site lies on Pleistocenic fluvial terraces in the Tertiary Piedmont Basin, including highly altered gravel, sand, and silty-clay deposits, with red alteration products [42]. Soil has a clay-to-clay-loam texture, and it is classified as Typic Ustorthents, fine-loamy, mixed, calcareous, mesic [43], or Dystric Cambisols [44]. The climate is Csa (Hot-summer Mediterranean climate in the Köppen climate classification [45]). The average annual precipitation value recorded in the experimental site in the period 2000–2016 was 852 mm, ranging from a minimum of 539 mm (year 2007) to a maximum of 1336 mm (year 2002). The annual mean air temperature in the same period was 13 °C. Rainfall is mainly concentrated in October and November (about 40% of annual precipitation observed in autumn), when major runoff events usually occur [46], and, secondarily, in March. Summer, particularly July, is the driest season, with 12% of annual precipitation.

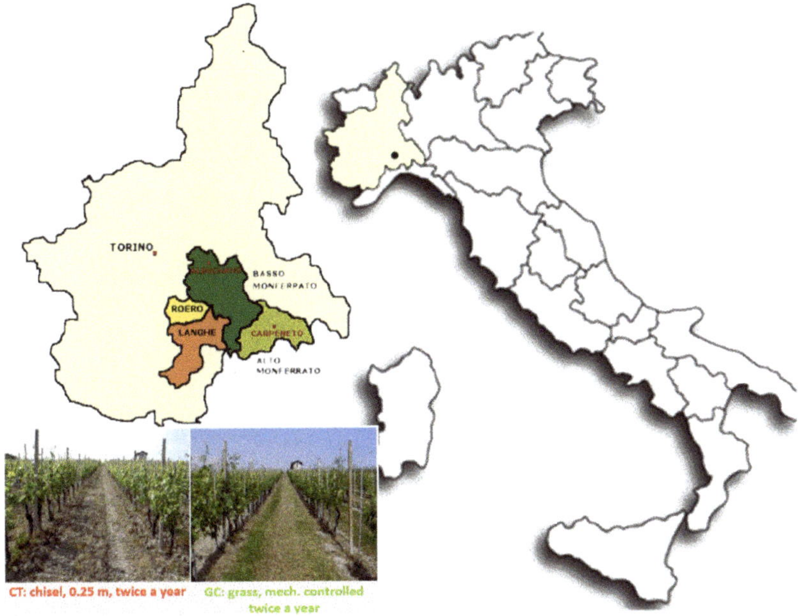

Figure 1. Location of the Tenuta Cannona Experimental Vine and Wine Centre of Agrion Foundation. Photos show the two soil managements.

The 2-years' experiment was carried out in two vineyard plots (1221 m^2, 6 rows aligned along the slope, spaced 2.75 m, where the vines are spaced 1.0 m along the row), located on a hillslope with average 15% slope and SE aspect. The vineyard was planted in 1988 with Barbera grape variety and managed according to conventional farming for wine production. The soil of the two plots was managed with different techniques since 2000. Twice a year, in spring and autumn, either cultivation with chisel at a depth of about 0.25 m or mulching of the spontaneous grass cover was carried out, in the conventional tillage plot (CT, hereafter) and in the controlled grass (GC), respectively. Most of the farming operations in the vineyard were carried out using tracked or tyred tractors (Table 1), carrying or towing implements, with passage intensification from spring to the grape harvest time.

The dates of tractor passages and field measurements are reported in Table 2, along with soil water content measured during surveys.

Table 1. Tractors' characteristics.

Model	New Holland TN95FA	New Holland TK80A
Engine power (CV/kW)	95/67.9	80/57.4
Front tyres	PIRELLI TM 700, 280/70 R18	Iron track
Rear tyres	PIRELLI TM 700, 420/70 R24	Iron track
Inflation pressure, front tyre (kPa)	150	-
Inflation pressure, rear tyre (kPa)	150	-
Total mass (kg)	2760	4280
Front mass (kg)	1000	-
Rear mass (kg)	1760	-
Max additional equipment mass (kg)	600	900

Table 2. Dates in bold indicate field measurements: values of soil water content (SWC, $m^3\ m^{-3}$) measured in the conventional tillage plot (CT) and in the controlled grass (GC) treatments in track (T) and no-track (NT) positions are indicated. Other dates indicate the passage of tractors in vineyard (the number of passages and if passages were tyred or tracked) and dates of execution of field operations (ripping and/or mulching).

Date	GC			CT		
dd/mm/yyyy	Passes N	SWC NT ($m^3\ m^{-3}$)	SWC T ($m^3\ m^{-3}$)	Passes N	SWC NT ($m^3\ m^{-3}$)	SWC T ($m^3\ m^{-3}$)
11/10/2016	-			ripped		
05/12/2016		0.374	0.387		0.300	0.357
05/12/2016	1 × tyred			1 × tyred		
07/12/2016		0.365	0.404		0.304	0.385
20/02/2017		0.349	0.381		0.308	0.369
21/02/2017	1 × tyred			1 × tyred		
23/02/2017		0.378	0.395		0.346	0.389
09/05/2017	1 × tyred			1 × tyred		
10/05/2017		0.337	0.380		0.286	0.374
11/05/2017	1 × tracked (mulched)			ripped		
25/05/2017	1 × tyred			1 × tyred		
31/05/2017		0.179	0.175		0.188	0.162
01/06/2017	1 × tracked			1 × tracked		
08/06/2017		0.133	0.149		0.082	0.167
09/06/2017	1 × tyres			1 × tyred		
16/06/2017	1 × tracked			1 × tracked		
23/06/2017	2 × tracked			2 × tracked		
26/06/2017	1 × tyred			1 × tyred		
04/07/2017	1 × tyred			1 × tyred		
12/07/2017	1 × tyred			1 × tyred		
25/07/2017	1 × tyred			1 × tyred		
26/07/2017		0.081	0.100		0.078	0.091
02/08/2017	3 × tyred			3 × tyred		
10/08/2017	1 × tyred			1 × tyred		
28/08/2017		0.060	0.119		0.048	0.042
12/09/2017	1 × tracked			1 × tracked		
13/09/2017	1 × tracked			1 × tracked		
27/09/2017		0.120	0.192		0.092	0.113
28/09/2017	-			ripped		
TOT YEAR 1	18			17		

Table 2. *Cont.*

Date	GC			CT		
dd/mm/yyyy	Passes N	SWC NT ($m^3\ m^{-3}$)	SWC T ($m^3\ m^{-3}$)	Passes N	SWC NT ($m^3\ m^{-3}$)	SWC T ($m^3\ m^{-3}$)
26/03/2018	1 × tracked			1 × tracked		
28/03/2018	1 × tyred			1 × tyred		
26/04/2018		0.237	0.219		0.278	0.356
27/04/2018	1 × tracked (mulched)			ripped		
08/05/2018	1 × tyres			1 × tyred		
16/05/2018	1 × tyres			1 × tyres		
17/05/2018		-	-	(no passes)	0.247	0.247
17/05/2018		0.322	0.377		0.266	0.342
25/05/2018	1 × tyres			1 × tyred		
28/05/2018	2 × tracked			2 × tracked		
01/06/2018	1 × tyred			1 × tyred		
10/06/2018	1 × tyred			1 × tyred		
11/06/2018		0.261	0.318		0.150	0.299
13/06/2018	2 × tracked			2 × tracked		
18/06/2018	1 × tracked			1 × tracked		
18/06/2018	1 × tyred			1 × tyred		
19/06/2018	1 × tyred			1 × tyred		
29/06/2018	1 × tyred			1 × tyred		
03/07/2018	1 × tyred			1 × tyred		
04/07/2018		-	0.257		0.258	0.293
09/07/2018	1 × tyred			1 × tyred		
13/07/2018	1 × tyred			1 × tyred		
19/07/2018	1 × tyred			1 × tyred		
27/07/2018	1 × tyred			1 × tyred		
30/07/2018		0.196	0.253		0.115	0.138
07/08/2018	1 × tyred			1 × tyred		
11/09/2018		0.167	0.256		0.156	0.170
26/09/2018	1 × tracked			1 × tracked		
27/09/2018	1 × tracked			1 × tracked		
10/10/2018		0.256	0.293		0.258	0.275
25/10/2018	1 × tyred			1 × tyred		
	-			ripped		
TOT YEAR 2	25			22		

2.2. Measurements

The experiment was conducted in the period November 2016–October 2018. Measurements were periodically carried out in the two plots, both in the track position (T), which is the portion of inter-row affected by the passage of tractor wheels or tracks, as this is where the compressive effects tend to concentrate [47], and in the middle of the inter-row, identified as the no-track position (NT), that is not affected by direct contact with tractor wheels or tracks. Thus, measurements were carried out in four positions: CT-T and CT-NT in the tilled plot, and GC-T and GC-NT in the grassed plot. Periodic measurements (Table 2) were carried out to obtain values of compaction, namely soil penetration resistance (PR), bulk density (BD), and the associated initial soil water content (SWC). They were performed monthly in the growing season (depending on weather conditions) and two times before and after winter passages. Infiltration tests were performed according to the simplified falling head (SFH) technique [48] to detect the temporal variability of the field-saturated soil hydraulic conductivity (K_{fs}) at the surface of the vineyard inter-rows. A weather station and a monitoring system measured and recorded rainfall, runoff, and soil loss in the experimental plots.

2.2.1. Rainfall, Runoff, and Soil Water Content

From November 2016 to October 2018, rainfall and runoff related to 93 events were recorded for the two plots. Rainfall was recorded at 10 min intervals by a rain-gauge station, with 0.2 mm resolution, placed near the plots. Based on the records, RIST (Rainfall Intensity Summarization Tool) [49] was used to obtain the rainfall amount (P), event duration (D), rainfall maximum intensity at 15, 30, and 60 min time intervals (MAX15, MAX 30, and MAX60), rainfall energy (E) (based on the equation proposed by Brown and Foster [50]), and the event erosivity index (EI30) [51] for each precipitation event. Rainfall events were defined as the time between the initiation and cessation of rainfall with a lack of rainfall for at least 12 h, in order to separate long-lasting events. Isolated events with less than 1 mm of rainfall were omitted from the analysis, because they were not significant for runoff initiation or for soil moisture changes. Runoff generated by rainfall was collected separately for each plot. Each vineyard portion was hydraulically bounded, and runoff was collected at the extremity by a channel, connected with a sedimentation trap, and then a tipping bucket device measured the hourly volumes of runoff (RO) and the runoff rates (RC) in CT and GC. Runoff samples were collected to obtain sediment yield (SY) for erosive events. Furthermore, if sedimentation occurred in the channels and sediment trap, then sediment yield was collected and weighted (see Biddoccu et al. [46], for details). Soil moisture was monitored by indirect method [52] by capacitance/frequency domain sensors (ECH$_2$O-5TM sensors, Decagon Devices Inc., Pullman, WA, USA), gravimetrically calibrated, and placed at 0.1, 0.2, and 0.3 m depth in each plot in NT and T positions. Soil water content was obtained every hour from the average of 1 min measurements and stored by a Decagon EM50 Datalogger.

2.2.2. Soil Compaction Measurements

Soil compaction was evaluated with two different methods: bulk density (BD) and penetration resistance (PR). The use of two separate methods was meant to offer a robust experimental setup, capable of internal corroboration and multiple detection capacity, where the effects that may elude one method are captured by the other [46,53].

For each of the four positions, 5 PR profiles and 9 BD were yielded, with a total number of 20 PR and 36 BD for each recurrent measurement. Measurements were repeated 10 times in Year 1 (200 PR and 360 soil cores), and 8 times in Year 2 (160 penetration resistance profiles and 288 soil cores). Dates of field sampling, along with soil water content measured in the same day, and dates of tractor passages in each plot are reported in Table 2. For reference values, the survey farther from tractor passages (20/2/2017) was chosen for the GC plot, whereas values recorded on 17/5/2018 were selected for the CT plot, that were obtained just after the tillage without tractor passages. The standard Proctor compaction test [54] was performed on soil samples taken in the two treatments for both years.

(1) Penetration resistance (PR)

Soil penetration resistance was measured using a dynamic penetrometer, built according to the design of Herrick and Jones [55]. The cone used for the tests had an ASAE-standard 30° cone angle (3.22 cm^2 base area). The penetrometer was placed vertically, with the cone tip inserted into the soil. The mass of the slide hammer was 2 kg, falling from a height of 0.3 or 0.4 m. The number of hits necessary to reach the penetration depths of 3.5 − 7.0 − 10.5 − 14.0 − 17.5 − 21.0 − 24.5 − 28.0 cm was recorded manually to evaluate changes in penetration resistance values down the soil profile. At each survey, 5 repetitions were performed in the T and NT positions, both in the GC and in the CT plot. The PR was then estimated as the work done by the soil to stop the movement of the penetrometer, divided by the cone travelled distance [55] according to the following equation:

$$R_s = \frac{W_s}{P_d} \tag{1}$$

where R_s is soil resistance (N), W_s is work done by the soil (J), and P_d is cone travel distance through the soil (m).

The work done by the soil was calculated as the change in the kinetic energy of the penetrometer, according to Equation (2) [56]. As the penetrometer was driven into the soil by the hammer, the kinetic energy of the hammer was transferred to the penetrometer cone. Therefore, the work done by the soil was equal to the kinetic energy (KE) transferred to the cone from the penetrometer when the hammer contacted the strike plate, which was calculated as follows:

$$KE = W_s = \frac{1}{2}mv^2 \tag{2}$$

where v is hammer velocity (m s^{-1}) and m is hammer mass (2 kg).

In turn, velocity (v) was calculated as:

$$v = \sqrt{v_0^2 + 2a(x)} \tag{3}$$

where x is falling height (0.3 m), a is gravity acceleration (9.81 m s^{-2}), and v_0 is initial velocity (assumed as null).

The previous calculations assume that all of the hammer's kinetic energy was transferred to the cone.

Finally, average PR was estimated for each depth interval of soil travelled by a given number of hammer strikes as:

$$R_s = \frac{a \cdot m \cdot x \cdot n}{A \cdot P_d} \tag{4}$$

where m is hammer mass (2 kg), n is number of strikes, and P_d is penetration depth (m).

PR is directly correlated with BD and shows an inverse relationship with SWC, but those relationships are not linear over a wide range of values of BD and SWC. Several studies suggest to operate cone penetrometer measurements at water contents close to a standardized matric potential to obtain comparable results [57]. To allow comparison of measurements taken at different SWC conditions, Busscher [58] introduced the practice of normalizing PR readings to a common SWC value. According to the procedure illustrated by Vaz et al. [59], the measured data were used to parameterize the exponential function proposed by Jakobsen and Dexter [60]:

$$PR = \exp(a + b \cdot BD + c \cdot SWC) \tag{5}$$

that expresses PR as function of SWC and BD. The PR mean values for 10-cm depth intervals were associated to corresponding BD and SWC values measured during each survey. Each of the four datasets obtained in different treatments and positions during the study period was used to obtain the best fitting equations, minimizing root mean square error. Then PR data for the 10 cm depth intervals were normalized using a common SWC$_{corr}$ value that was set at 0.300 m^3 m^{-3}.

(2) BD (and SWC)

BD is a dynamic soil property that varies according to its structure, which can be altered by flora and microorganisms, by agricultural practices, by trampling, or by heavy vehicle traffic, but also by the impact of precipitation. To determinate BD and SWC, core samples (V = 100 cm^3) were collected in the T and NT position both in the GC and in the CT plot, at the depth of 0–0.10, 0.10–0.20, and 0.20–0.30 m. Then, samples were weighed before and after oven-drying at 105 °C for 48 h. SWC was determined by gravimetric method from each sample. Three samples were collected at each position and depth; thus, BD and SWC were calculated as the average of values obtained from the three repetitions.

2.2.3. Field-Saturated Soil Hydraulic Conductivity (K$_{fs}$)

The K$_{fs}$ was measured in the four investigated positions at each periodic sampling by infiltration tests carried out according to the simplified falling head technique (SFH) proposed by Bagarello et al. [48].

With respect to the original test setup, a larger second ring concentric to the inner one was inserted to assure one-dimensional flow. The two PVC cylinders were 0.30 m high, and the diameters were 0.305 and 0.486 m, for the inner and the external ring, respectively. They were inserted in the soil to a minimum depth of 0.05 m. The applied water volumes were 7.0 L in the inner ring and 10.8 L in the external one. According to Bodhinayake et al. [61], the slope of the experimental plots does not affect the measurements significantly. In winter, the soil temperature was checked to be sure that the soil was not frozen. Next to the investigated area and, after the water infiltration, inside the inner ring, undisturbed soil cores ($V = 100$ cm^3) were collected at the depth of 0–0.07 m, in order to determine initial and saturated volumetric water content values (SWC_i and SWC_s). Overall, 38 infiltration tests were carried out in Year 1 and 36 in Year 2.

2.2.4. Statistical Analysis

The statistical frequency distributions of the data were assumed to be normal for the SWCi and the BD, and log-normal for the K_{fs}, as common for these variables [62,63]. Arithmetic means were used to represent the SWC, PR, and BD results. Geometric means were used to represent the K_{fs}. Mean values of BD, PR, and K_{fs} obtained for each position were analyzed statistically, by means of *t*-test [64], in order to find significant difference: (i) with respect to the undisturbed soil conditions, by comparing values obtained in each survey with the reference ones, for BD and PR; (ii) for BD topsoil values, in selected dates—between the two plots, within the same date and position (i.e., CT-T vs. GC-T) and (iii) between values measured in the same plot and same position in corresponding survey dates in the two monitored years; and (iv) for K_{fs}, between average annual values measured in different position and plot. The elected significance level for all tests was $\alpha < 0.05$.

3. Results

3.1. Rainfall

The first year observed during the present study (from November 2016 to October 2017, Year 1 hereafter) was characterized by cumulated precipitation lower than the Mean Annual Precipitation (MAP) (569 mm, 67% of MAP). In particular, the period June–October 2017 was very dry. The second year (November 2017–October 2018, Year 2 hereafter) was a relatively wet period, with cumulated precipitation of 1125 mm (132% MAP). Total erosivity (EI30) of rainfall felt in the entire period of observation was 4715 MJ mm ha^{-1} h^{-1}, and only 12% of the erosivity was due to events that occurred during Year 1.

Table 3 shows the classification of rainfall events following the cumulated rainfall depth. The maximum rainfall event depth was 127.2 mm in Year 1 (autumn 2016) and 217.4 mm in Year 2 (autumn 2018); the latter was the only event with more than 200 mm of precipitation in the observed period. More than 60% of the rainfall events cumulated less than 10 mm during each period of observation. Rainfall events with cumulated rainfall between 1 and 50 mm gathered most of the precipitation fallen during each year of observation, namely 76% and 53% in Year 1 and Year 2, respectively. The average 15 min rainfall intensity and erosivity of events belonging to the two first classes were lower than 17 mm h^{-1} and 63 MJ mm ha^{-1} h^{-1}. Only three events were observed for the class 50–100, with all of them occurring during Year 2, characterized by mean EI30 equal to 430 MJ mm ha^{-1} h^{-1}: the corresponding cumulated erosivity accounted for the 27% of the total. Only three rainfall events with more than 100 mm of precipitation were observed. These three events accounted for 38% of the total erosivity (average 595 MJ mm ha^{-1} h^{-1}). This class showed the highest rainfall erosivity (more than 1600 MJ mm ha^{-1} h^{-1}). Just two events showed very high erosivity (>1000 MJ mm ha^{-1} h^{-1}), and they occurred in autumn 2018 (27/10) and in summer 2018 (16/07). The latter also reached the maximum rainfall intensity (84.3 mm h^{-1}).

Table 3. Classification of rainfall events recorded in the two years of observation, with indication, for each class, of number of events, average rainfall depth (mm) and duration (h), average 15 min maximum intensity (mm h^{-1}), average rainfall erosivity (MJ ha^{-1}), cumulated (total for each class) rainfall (mm), erosivity (MJ mm ha^{-1} h^{-1}), and runoff (mm) in the two monitored plots.

Year 1	Rainfall Events Category	Number of Events	Av. Rainfall Depth (mm)	Av. Duration (h)	Av. Max 15	Av. Rainfall Erosivity EI30	Cumulated Rainfall	Cumulated Rainfall Erosivity EI30	cum RO CT	cum RO GC
	1–10	28	3.8	6.2	6.3	3.3	106.4	91.4	0.1	0.0
	10–50	12	27.1	27.1	9.7	31.0	325.6	371.5	5.1	2.9
	50–100	0								
	100–200	1	127.2	70.2	8.4	117.2	127.2	117.2	0.5	0.8
	>200	0								
		41					559.2	580.1		
Total P (mm)	569						76%			

Year 2	Rainfall Events Category	Number of Events	Av. Rainfall Depth (mm)	Av. Duration (h)	Av. Max 15	Av. Rainfall Erosivity EI30	Cumulated Rainfall	Cumulated Rainfall Erosivity EI30	cum RO CT	cum RO GC
	1–10	29	3.6	10.8	4.1	1.6	104.2	47.3	0.1	0.1
	10–50	18	27.1	25.6	16.9	62.8	488.2	1130.7	14.4	2.8
	50–100	3	65.1	56.0	36.5	430.3	195.4	1291.0	8.7	2.9
	100–200	1	101.4	105.3	7.2	64.2	101.4	64.2	27.1	2.6
	>200	1	217.4	57.2	45.8	1602.0	217.4	1602.0	40.5	11.2
		52					1106.6	4135.2		
Total P (mm)	1125						53%			

3.2. Soil Moisture

In general, SWC was higher in T than in NT and in GC than in CT (Figure 2). During winter and until May of Year 1 (Figure 2), the SWC in the CT soil was higher (over 0.250 m^3 m^{-3}) at depths of -30 and -40 cm than in more superficial layers. No data are available for that period for the GC plot. During the summer, after tillage and mowing operations, the soil was very dry in consequence of the absence of rain. Since July, SWC in CT was lower than 0.200 m^3 m^{-3} (except at -10 cm depth in CT-NT), and it reached values close to 0.150 m^3 m^{-3} after mid-September. In GC, the lowest values were measured at mid-August, with values lower than 0.200 m^3 m^{-3} for the most superficial layer. Nevertheless, after the following rainfall events, an increase in soil moisture, which was not evident in the CT plot, was recorded. Furthermore, from June to September, the SWC measured in CT positions both at -10 and -20 cm was higher than soil moisture at higher depth, whereas, before the tillage, the trend was the opposite. Monitoring also shows that, in winter, every rain event caused higher SWC increase in deepest soil layers of the CT plot than in the superficial ones. In Year 2 (Figure 2), winter and early-spring precipitation resulted in SWC higher than 0.250 and 0.300 m^3 m^{-3} along the entire soil profile, in CT-NT and GC, respectively (measurements for CT-T not available) until mid-April. Then, during summer, many rainfall events occurred and caused higher variability in soil moisture than in the previous year, with SWC increasing after precipitation events in the two treatments. In fact, SWC was higher than 0.200 m^3 m^{-3} during most of the summer, both in CT and in GC. In both years, the value at -30 cm in GC-T does not generally drop below the most superficial one.

3.3. Soil Compaction

3.3.1. Proctor Test

The volumetric soil water content at the maximum compaction was evaluated by the standard Proctor compaction test [54] for the soil of each plot (Figure 3), using samples collected in both years. For each treatment, the results were similar in the two Years: SWC at maximum compaction was always higher in GC than in CT, equal to 0.331 and 0.289 m^3 m^{-3}, respectively; and, in both tests, the corresponding BD was higher in CT than in GC and reached, respectively, 1.61 and 1.48 g m^{-3}.

3.3.2. Bulk Density

The reference bulk density ranged between 1.10 and 1.34 kg m^{-3} and between 1.17 and 1.32 kg m^{-3} in the GC and CT plots, respectively. During the study period, BD ranged from 1.18 and 1.50 kg m^{-3} and from 1.09 and 1.44 kg in the GC-T and GC-NT positions, respectively (Figure 4a). The highest BD values were observed in the most superficial layer on 26/07/2017 (after 14 passages) in GC-T and on 17/5/2018 in GC-NT (after 5 passages). In GC-T, BD was significantly higher than the reference values (t-test, $p < 0.05$) in the most superficial layer in almost all dates after five tractor passages, in both monitored seasons. The only exception were the values obtained on 28/8/2017. In the deepest layer of GC-T (20–30 cm), the BD significantly increased only after 18 passages in 2017 and after 26 passages in 2018. In the GC-NT, the BD in the most superficial layer was significantly higher with few exceptions from December 2016 to June 2017 in the first season, from late April to June and in the final survey in the second season. In the 10–20 cm layer, the BD increase was significant only on 26/7/2017. At the 20–30 cm depth, the BD was significantly higher just after the first winter passage (7/12/2016), and then, on some dates in the second season, namely after two and five passages in the springtime and in the final survey.

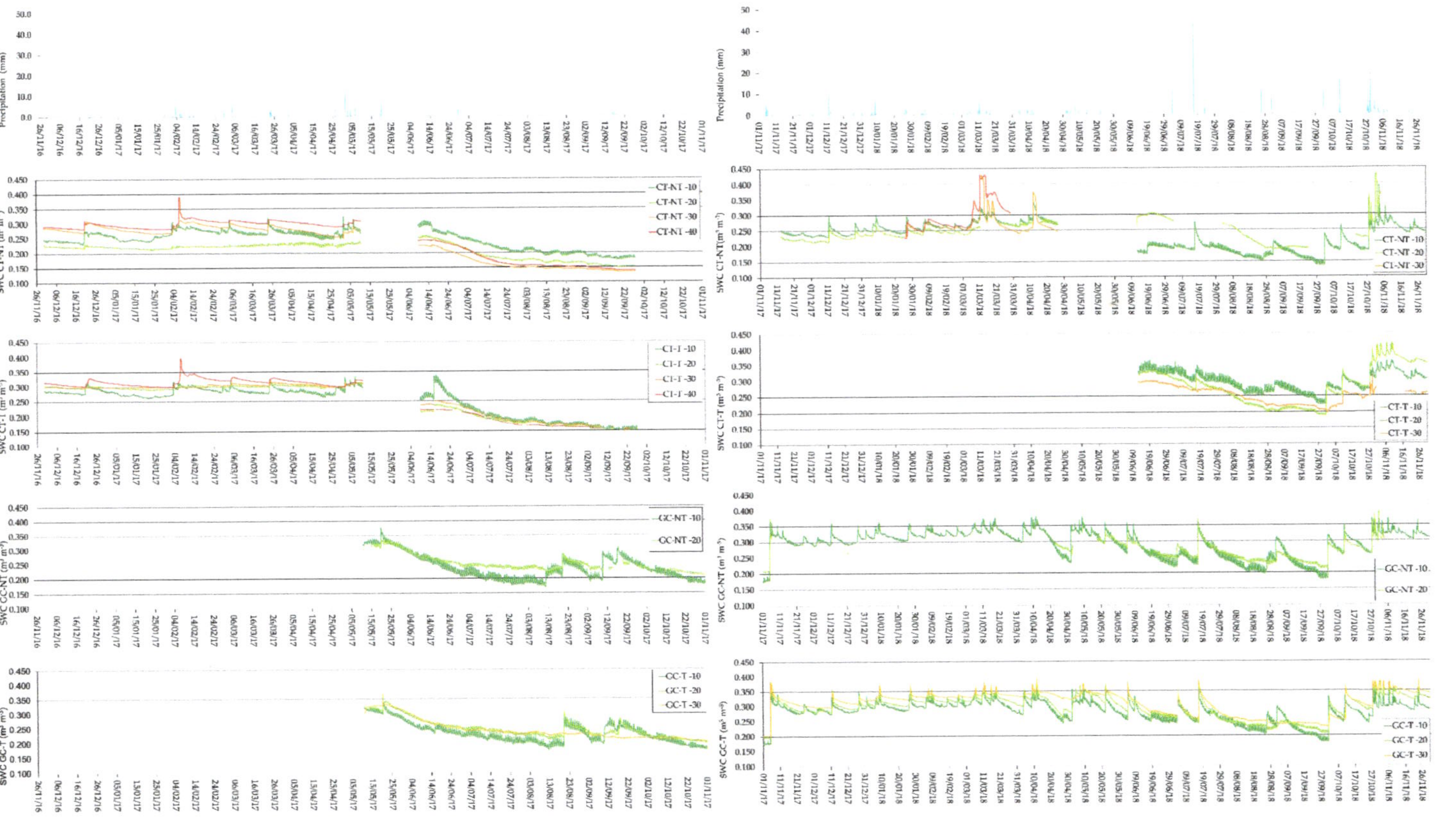

Figure 2. Precipitation (**top**), CT (**middle**) and GC (**bottom**) volumetric soil water content in Year 1 (left side) and Year 2 (right side).

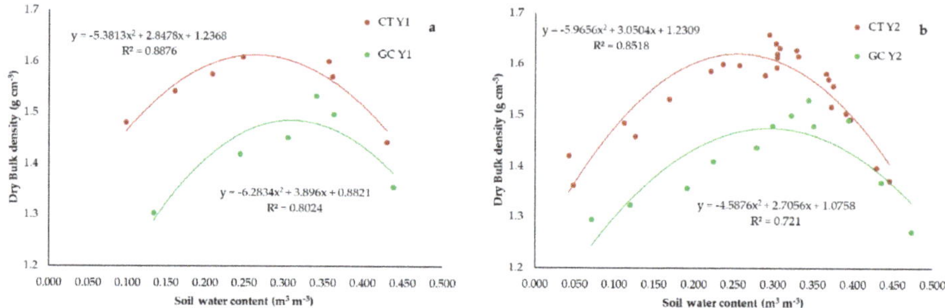

Figure 3. Standard Proctor compaction test: Year 1 (**a**) and Year 2 (**b**).

In the CT plot, BD assumed values ranging from 1.16 to 1.60 kg m^{-3} and from 1.05 to 1.43 kg m^{-3} in the CT-T and CT-NT positions, respectively (Figure 4b). The highest BD was obtained on 26/07/2017 (after 10 passages) in CT-T and on 30/07/2018 in CT-NT (after 19 passages). In CT-T, Year 1, at the depth 0–10 cm, BD was significantly higher than the reference in all dates before the spring tillage, on 26/07/2017, just after the first harvest, and in all dates during Year 2. At deeper layers, in Year 1, BD was significantly higher than reference at 26/07/2017 and on 27/09/2017 (in the latter date only at 10–20 cm depth), but it increased significantly during the entire Year 2 at 10–20 cm depth and, in the summertime, at 20–30 cm depth.

Table 4 shows the BD mean values measured in the topsoil in each plot and position in the two years in selected dates, which represent similar conditions with respect to the tillage operations and traffic with tractors. In Year 1, the values of BD in CT-T were significantly higher than in GC-T after one to three tractor passages occurred before spring tillage; afterward, BD was lower in CT-T, being the difference significant in July (after more than 10 passages) and at the end of the season. In the NT position, BD was higher in GC than in CT during most of the surveys, without significant difference. During Year 2, BD measured in the track position was always higher in CT than in GC, with a significant difference only before harvest. In the same date, BD in the middle of the inter-row was also significantly higher in CT than in GC. BD was higher in GC-NT than in CT-NT only during the survey carried out after one to two passages following spring tillage, and, in that case, the difference was statistically significant. The comparison between the two monitored seasons revealed significant differences during almost the entire season in the CT-T treatment, with highest BD values obtained in Year 2. In the NT position treatments, the BD was significantly higher in Year 2 in the first survey after spring tillage (after one to two tractor passages) in GC and in CT, after harvest.

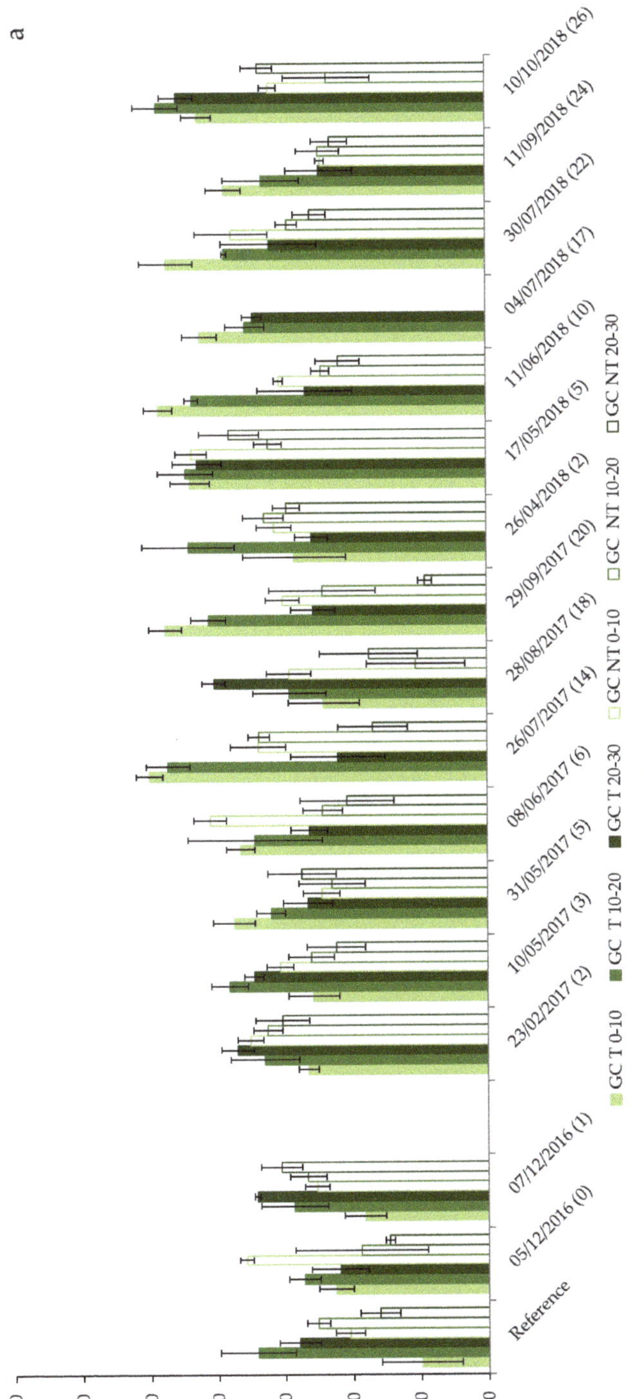

Figure 4. *Cont.*

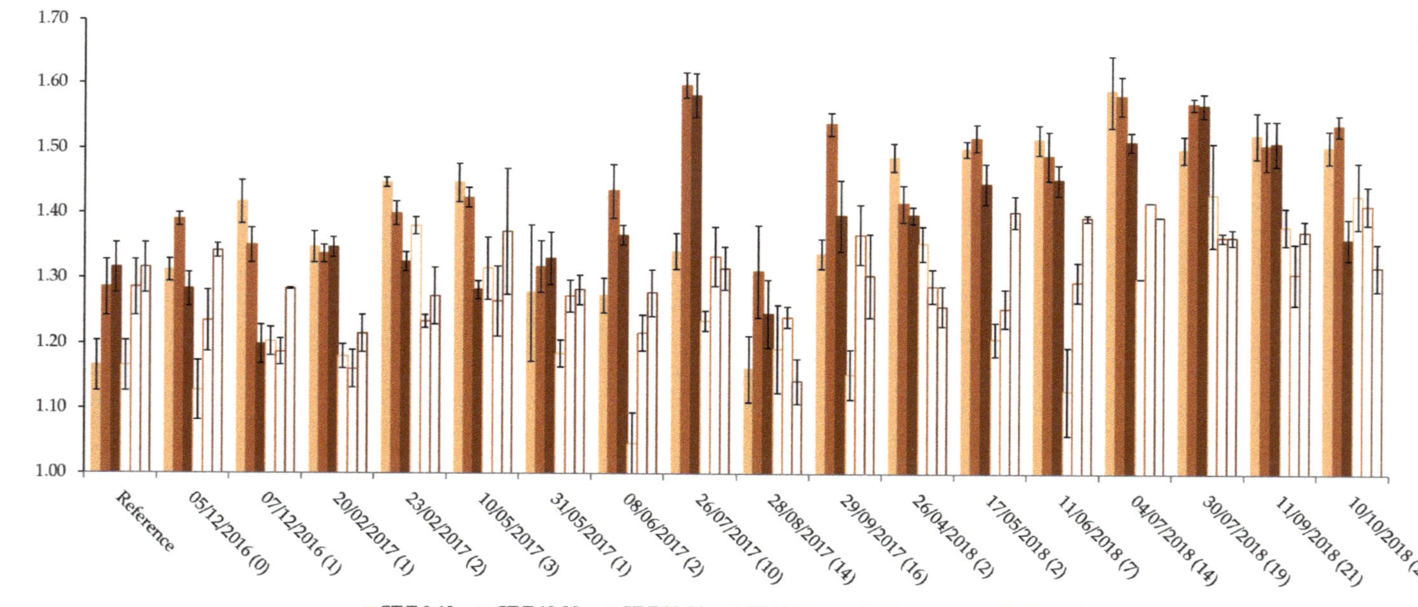

Figure 4. Mean bulk density (BD) (g m^{-3}) measured along the period of observation at three depths in track (T) and no-track (NT) positions in the two vineyard plots ((**a**) GC = with grassed inter-rows and (**b**) CT = with tilled inter-rows). The "b" letters indicate significant difference with the corresponding reference values ("a" are omitted), obtained with *t*-test (p = 0.05). Number in brackets after the date indicates the number of tractor passages occurred with respect to reference soil conditions (minimum BD for GC and after tillage for CT).

Table 4. Values of SWC, BD, and K_{fs} measured in the topsoil in each plot and position in the two years in selected dates, which represent similar conditions with respect to the tillage operations and traffic with tractors, and total values of runoff and soil losses measured in the two vineyard plots in the two monitored years. Bold values indicate significant differences (according to *t*-test, $p = 0.05$) between the values measured in the same date and same position, but different plot (i.e., CT-T vs. GC-T). Italic values indicate significant differences (according to *t*-test, $p = 0.05$) observed in the same plot and position between corresponding surveys in the two observed periods (i.e., differences in CT-T after 1–2 passages in Year 1 and in Year 2). Dates in *italic* indicates tillage and mowing operations.

Year 1 (2016–2017)			CT-T			CT-NT			CT All		GC-T			GC-NT			GC All	
	Date	NP	SWC_{10} m^3 m^{-3}	BD_{10} $g\,m^{-3}$	K_{fs} mm h^{-1}	SWC_{10} m^3 m^{-3}	BD_{10} $g\,m^{-3}$	K_{fs} mm h^{-1}	Runoff mm	SL kg ha^{-1}	SWC_{10} m^3 m^{-3}	BD_{10} $g\,m^{-3}$	K_{fs} mm h^{-1}	SWC_{10} m^3 m^{-3}	BD_{10} $g\,m^{-3}$	K_{fs} mm h^{-1}	Runoff mm	SL kg ha^{-1}
Autumn tillage/mowing	*11/10/2016*																	
After 1st passage	07/12/2016	1–1	0.357	**1.42**	1	0.300	1.20	1297			0.387	**1.18**	49	0.374	1.25	242		
After 2–3 passages (tyred)	10/05/2017	3–3	0.374	**1.45**	3	0.286	1.32	599			0.380	**1.26**	53	0.337	1.31	145		
Spring tillage/mowing	*11/05/2017*																	
After 1–2 passages (tyred)	31/05/2017	1–5	0.162	1.28	26	0.188	1.18	219			0.175	1.38	86	0.179	1.24	1484		
After more than 10 passages	26/07/2017	10–14	0.091	**1.34**	5	0.078	1.24	430			0.100	**1.50**	na	0.081	1.34	na		
Before harvest	28/08/2017	14–18	0.042	*1.16*	23	0.048	1.19	33			0.119	1.24	57	0.060	1.29	99		
End of season (after harvest)	27/09/2017	16–20	0.113	**1.34**	16	0.092	*1.15*	68			0.192	**1.48**	527	0.120	1.30	52		
Total									5.62	5.8							3.67	3.3

Year 2 (2017–2018)			CT-T			CT-NT			CT all		GC-T			GC-NT			GC all	
	Date	NP	SWC_{10} m^3 m^{-3}	BD_{10} $g\,m^{-3}$	K_{fs} mm h^{-1}	SWC_{10} m^3 m^{-3}	BD_{10} $g\,m^{-3}$	K_{fs} mm h^{-1}	Runoff mm	SL kg ha^{-1}	SWC_{10} m^3 m^{-3}	BD_{10} $g\,m^{-3}$	K_{fs} mm h^{-1}	SWC_{10} m^3 m^{-3}	BD_{10} $g\,m^{-3}$	K_{fs} mm h^{-1}	Runoff mm	SL kg ha^{-1}
Autumn tillage/mowing	*28/09/2017*																	
After 1st passage	na																	
After 2–3 passages (tyred)	26/04/2018	2–2	0.356	*1.49*	2	0.278	1.35	576			0.219	1.28	99	0.237	1.31	440		
Spring tillage/mowing	*27/04/2018*																	
After 1–2 passages (tyred)	17/05/2018	2–5	0.342	1.50	3	0.266	**1.21**	2642			0.377	1.44	45	0.322	**1.44**	173		
After more than 10 passages	04/07/2018	14–17	0.293	*1.59*	4	0.258	1.3	120			0.257	1.42	371	*na*	*na*	151		
Before harvest	11/09/2018	21–24	0.170	**1.52**	4	0.156	**1.38**	35			0.256	**1.39**	12	0.167	**1.24**	32		
End of season (after harvest)	10/10/2018	23–26	0.275	1.50	7	0.258	1.43	46			0.293	1.43	190	0.256	1.32	450		
Total									90.78	3089.3							19.55	524.0

3.3.3. Soil Penetration Resistance

Figure 5 shows the profiles of soil penetration resistance (PR) in each treatment in the two years. In GC, the mean values of reference PR along the profile varied between 1.5 and 3.7 MPa and from 2.5 to 4.3 MPa in NT and T positions, respectively. In GC-NT, during the 2016–2017 winter, PR did not significantly increase with respect to reference values after one and three tractor passages. After 5 passages, on 31/5/2017, the mean PR values increased significantly (up to 9.5 MPa) in the first 20 cm of soil profile, and later in the entire soil profile until the end of the season, with a maximum value of about 25 MPa at 7 cm depth at the end of July 2017. In 2018, mean PR was not higher than 7.4 MPa, values obtained at maximum depth before the harvest. Nevertheless, after only two tractor passages, the mean PR was significantly higher than the reference values up to 17.5 cm depth, and later it increased gradually at higher depth, until it was significantly different along the entire profile after the 2018 harvest. In Year 1, in the GC-T position, the PR significantly increased through the first centimetres of soil after three passages. Starting from the late spring, the increase of PR was significant (with mean values exceeding 10 MPa) up to 10.5 cm of depth on 31/05/2017 and then along the entire soil profile. In Year 2, the PR was significantly higher than the reference values up to 10.5 cm depth and at maximum depth after two tractor passages following the spring tillage. A decrease of PR was detected in the following survey (17/5/2018), and then it increased gradually during the summer, exceeding 15 MPa between 24 and 30 cm of depth on 4/7/2018. Finally, after harvest, PR generally decreased.

Reference values for mean PR in CT ranged between 0.3 and 2.5 MPa. In CT-NT, after three tractor passages, the mean PR measured in spring 2017 was about 1 MPa, with a significant increase with respect to reference values. Later, after a single passage following the spring tillage operation, the PR showed significant increase between 10 and 20 cm (0.9 and 1.7 MPa) and at maximum depth (3.5 MPa). At the end of July (after 10 passages), the soil PR was higher (3.1–8.3 MPa) than the reference below 10.5 cm, and then in the entire soil profile up to the end of season, it exceeded 9 MPa at maximum depth. In Year 2, the soil had high PR just after two early spring passages. Following spring tillage, the soil showed high PR at maximum depth again after the first two passages, and then the entire soil profile showed higher PR than the reference values, reaching highest mean values at maximum depth (7.6 MPa). In CT-T, after the first tractor passage following the autumn tillage operations, and until the spring tillage, the soil showed higher soil PR than the reference values up to 10.5 cm depth. After tillage in May 2017, a single tractor passage caused a significant increase of PR up to 25 cm in depth (mean values between 2.0 and 6.4 MPa), and, in the rest of the season, the entire profiles showed PR significantly higher than the reference (up to 24.8 MPa). During Year 2, after two tractor passages the PR was significantly higher, up to 10.5 cm, and up to 28 cm in following measurements (up to 11.1 MPa at maximum depth before harvest).

In Figure 5, it is evident, in both the treatments, how original PR values are higher when the SWC is lower, and the opposite. Such fluctuations were softened when normalized values were used. CT-NT is the only one position where most of normalized PR values are lower than 2.5 MPa (which is, according to Whalley et al. [65], the limit value generally accepted beyond which root elongation is significantly restricted); meanwhile, GC-T was the condition with higher values, above 6 MPa. The track positions presented normalized PR value always higher than NT positions, as well GC positions present always higher values than CT positions.

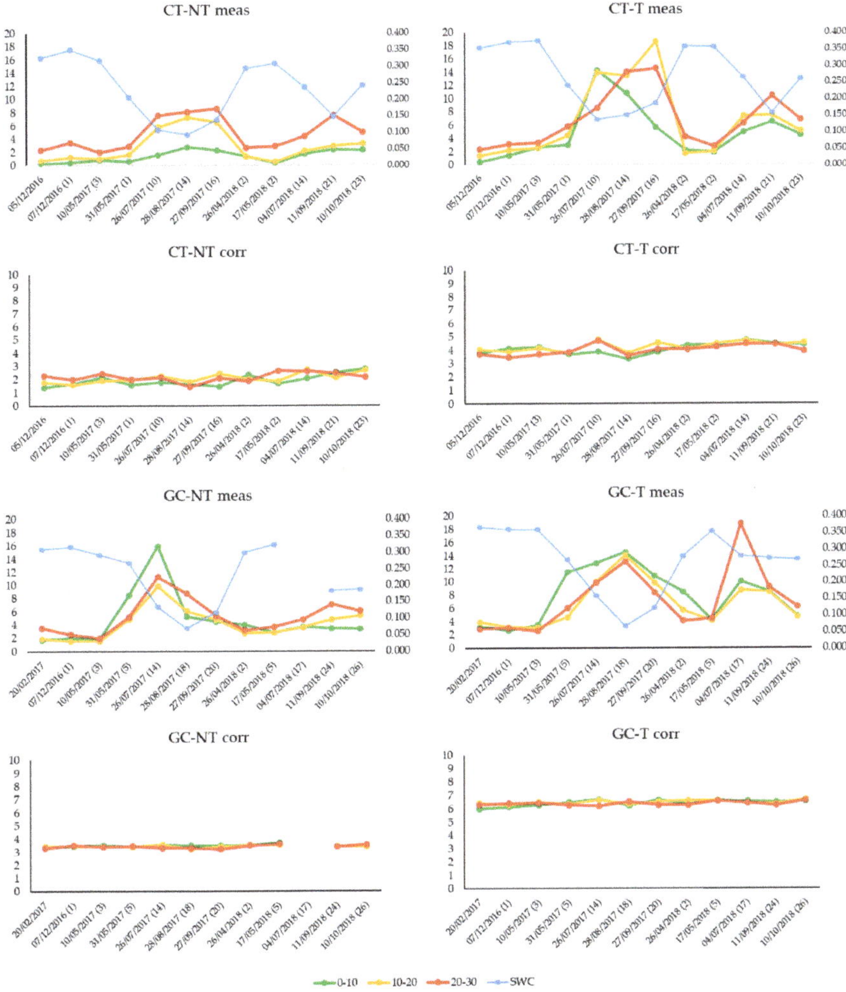

Figure 5. Values of soil penetration resistance (PR) in different positions of each treatment, for depth intervals 0–10, 10–20, and 20–30 cm, with corresponding mean SWC, and values of corresponding normalized penetration resistance, corrected for value set at SWCcorr = 0.300 m^3 m^{-3}.

3.4. Field-Saturated Hydraulic Conductivity

The mean values of the field-saturated hydraulic conductivity (K$_{fs}$) measured in the plots are shown in Figure 6. The mean annual values of K$_{fs}$ ranged between 4 and 402 mm h^{-1}, and they are significantly different when comparing NT and T measurements in CT in each season. Whilst the CT-NT mean values are not significantly different from GC-NT, the T and the overall mean values differ significantly between the two plots, both in each season and in the entire period of observation. Table 4 shows K$_{fs}$ values measured in each plot and position in the two years in selected dates. In CT-T, all K$_{fs}$ values were lower than 26 mm h^{-1}, and they were lower than 10 mm h^{-1} after only one tractor passage on wet soil following the tillage operations. In CT-NT, values of K$_{fs}$ ranged from 33 to 2642 mm h^{-1}, with the lowest values (<50 mm h^{-1}) measured before harvest. In the grassed plots, the K$_{fs}$ values ranged from 33 to 1484 mm h^{-1} and from 12 to 527 mm h^{-1} in GC-T and GC-NT, respectively. Most of the measured values were higher than 50 mm h^{-1}.

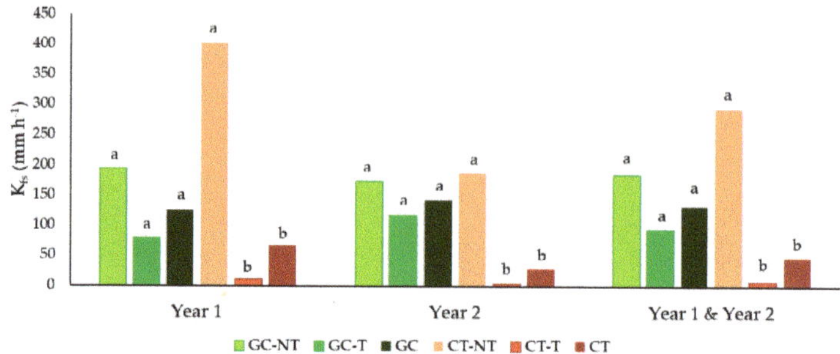

Figure 6. Mean values of the field-saturated hydraulic conductivity (K_{fs}) measured in the plots. Letters indicate significant difference between series, obtained with *t*-test ($p = 0.05$).

3.5. Runoff and Soil Losses

Figure 7 shows monthly runoff (RO) and soil loss (SL). As a consequence of 569 mm of precipitation (67% MAP) in Year 1, the runoff coefficient was lower than 1% and soil losses were only 5.8 and 3.3 kg ha^{-1} in CT and GC, respectively (Table 4). Sediment yield was 0.1% and 0.2% of the average annual soil loss measured in the 2000–2016 period (6.6 and 1.5 Mg ha^{-1} in CT and GC, respectively, according to Capello et al. [66]).

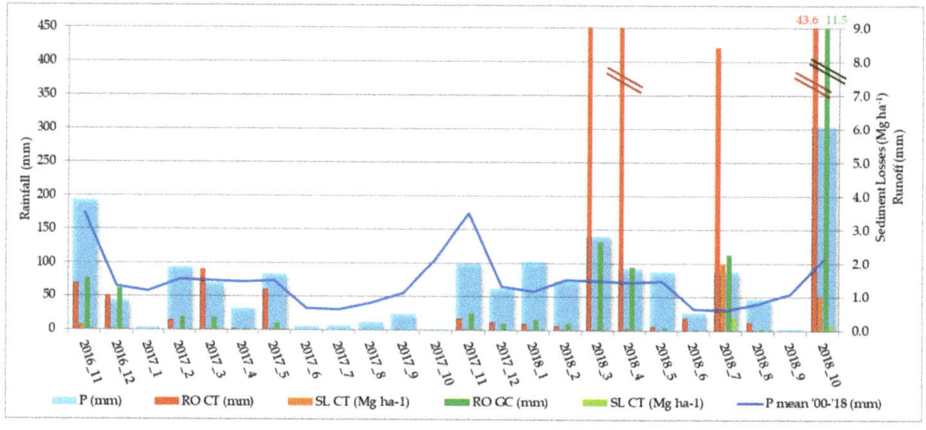

Figure 7. Monthly precipitation (P) and mean precipitation in 2000–2018 period (P mean), runoff (RO), and soil loss (SL) in CT and GC.

Year 2 was characterized by precipitation higher than MAP (149%), and, consequently, the runoff coefficient was 8% and 2% in CT and GC, respectively. In 2018, runoff was mainly concentrated during four rainfall events (10/3, 11/4, 16/7, and 27/10) that accumulated more than 94% and 91% of the annual runoff in CT and GC, respectively. The highest runoff coefficient was recorded during the 10/3 event in CT (26.7%) and the 27/10 event in GC (5.2%). Soil losses were 3.1 and 0.5 Mg ha^{-1} in CT and GC, respectively (that is 47% and 33% of the average 2000–2016 annual soil loss). They were concentrated during the rainfall occurred the 16/7/2018: 2 Mg ha^{-1} of soil loss in CT (64%) and 0.38 Mg ha^{-1} in GC (76%).

4. Discussion

During the study, Year 1 was not rainy, particularly in the period June–October 2017, when, after tillage and mowing operations, the soil was very dry. Summer rainfall increased the soil moisture in the topsoil in GC, without getting deeper, while no changes were recorded in CT. After a few sunny days, the soil dried rapidly, returning to a low SWC value. In addition, after spring tillage, the soil surface in CT maintained the "ripped" appearance, and the superficial crust that is usually observed in tilled inter-rows after first rainfall events [46,67], or shallow ruts due to the traffic, were not observed. The GC soil was less turfed than in other years, and there were many up to 2 cm size cracks. In Year 2, precipitations were higher than MAP, particularly abundant in spring months, and relevant also in summer. Consequently, soil moisture was higher than $0.200 \ cm^3 \ cm^{-3}$, even in summer months, when tractor traffic is frequent. In both years, SWC gravimetrically measured during surveys resulted generally higher in GC than in CT and in T than in NT. Increase in BD, PR, and SWC after traffic operations was measured by Barik et al. [68], where an increase in BD, at almost constant gravimetric water content, corresponds to an increase of the volumetric water content. Higher volumetric SWC and BD were also observed in grassed inter-rows, rather than tilled, by Bogunovic et al. [69] in a Croatian vineyard.

The inter-annual high variability of rainfall and soil moisture conditions were reflected by the different evolution of BD and PR values along the two growing seasons. As expected, during Year 1, BD generally increased already during winter and spring, also in the CT-T position, since the first tractor passage occurred in December 2016. On 08/06/2017, after six (GC) and two (CT) tractor passages, BD did not show significant differences with reference in both plots. On 26/07/2017, very high BD values were observed, that in CT-T below 10 cm of depth reached the maximum value, very close to the proctor value identified for the CT soil ($1.60 \ g \ cm^{-3}$). In the same date, the highest seasonal BD values were observed in the GC-T position up to 20 cm depth, exceeding the GC proctor values. Since the winter and the last tillage operation 14 and 10 tractor passages have occurred in GC and CT, respectively. The last passage was performed just the day before the survey (25/07/2017), and a minor storm (only 3 mm of rain) was recorded the previous evening (24/07/2017 between 18.00 and 19.00), but moisture sensors did not show increasements of SWC in any position. The high compaction level was likely due to the increasing number of passages, even if they were performed on soil with SWC lower than $0.200 \ m^3 \ m^{-3}$. Since May, and until the post-harvest survey, BD was higher in GC than in CT, especially in the most superficial layer, as Guzmán et al. [25] already observed in vineyards of the Montilla-Moriles region (with MAP of 604 mm). During the following survey (28/08/2017), BD decreased to lower values despite the additional tractor passages, whereas it was expected to increase, as usually observed in vineyards with various soil management [61]. This behaviour could be explained with a recovery of soil properties thanks to some wet/dry cycles consequent to rainfall events that occurred between the last tractor passage and the survey. An increase of the K_{fs} was also observed in CT-T. O'Keefe [70] showed that changes of the bulk density after three wetting/drying cycles can be significantly different. Also the measured PR values were very high in all positions at the end of July, up to 25 MPa at the 7 cm depth in GC-NT. Similarly, Bogunvic et al. [69] observed PR values significantly higher in a grassed inter-row than in tilled one. High values of PR were generally measured during the 2017 summer, with dry soil (mean SWC $< 0.200 \ m^3 \ m^{-3}$ at 0–30 cm depth, during the measurements). The PR profiles were similar to those obtained by Vaz et al. [59] in dry, sandy clay loam soil, with PR higher than 16 MPa. Such high values are not comparable with measurements obtained with moister soil. Looking at the corrected PR values (calculated for $SWC_{corr} = 0.300 \ m^3 \ m^{-3}$), the temporal trend of PR_{corr} reflects the one of BD, as expected due to correction equation, and the highest PR_{corr} values were associated with the GC treatment, both in T and in NT positions. The most limiting K_{fs} values in CT-T (lower than $20 \ mm \ h^{-1}$) were associated to BD higher than $1.42 \ g \ cm^{-3}$ up to 20 cm depth. The inverse relationship between BD and K_{fs} is especially evident in T position with both soil managements, as already observed in previous seasons [12,66]. During summer 2017, despite higher BD and PR values in GC-T than in CT-T, K_{fs} in GC-T was always higher than $50 \ mm \ h^{-1}$, favouring more water infiltration and avoiding the conditions for Hortonian runoff during rainfall events. An increase of surface cracks was observed during summer, especially in the GC plot, and the presence of preferential flow ways can explain the increase of hydraulic conductivity [70]. In the NT position of the two plots, the K_{fs} was higher than 100 mm h^{-1} until the end of August.

In Year 2, surveys were not carried out during the winter. Since the first survey, which occurred on 26/4/2018 after two tractor passages, the BD showed some values higher than the reference values. The frequent tractor passages on the moist soil during spring and summer resulted in BD approaching or overcoming the proctor value in the first 20 cm of soil both in GC-T, and in CT-T during almost all surveys until the end of August. Highest values were reached in T in GC-T after 10 passages, and in CT-T after 14 passages. In Year 2, the range of PR values was less wide than in the previous season, reflecting the lower variations of SWC. However, this parameter showed an increasing trend during spring and summer: the highest values were reached on the same dates as BD. Considering PR$_{corr}$ values, they resulted similar to Year 1, but generally higher in CT-T in the second observed season than in the first one, with most of values exceeding 4.1 MPa. Similar to Year 1, the lowest values of K_{fs} (lower than 10 mm h^{-1}) were observed in CT-T, where soil always showed a BD higher than 1.42 g cm^{-3}, at least in the first 20 cm of soil. On the contrary, the PR, both directly measured and corrected, seems to be less related with variations of K_{fs}. As an example, GC-T, for which PR$_{corr}$ always showed higher values than other treatments, did not show the lowest values of K_{fs}. The combination of weather conditions, soil, and traffic management solutions resulted in relevant runoff during some spring, summer, and autumn rainfall events, especially in the CT plot.

In both Years, K_{fs} in CT resulted significantly lower in T than NT and both positions of GC: in particular, after just one tractor passages on wet soil after tillage operation, K_{fs} in CT-T fell down from value over 1000 to near 1 mm h^{-1}. Those results highlight the need to limit the soil surface directly interested by tractor traffic, in order to reduce negative effects due to soil compaction. In their study on the spatial variability of soil compaction in the Languedoc vineyard region, Lagacherie et al. [30] highlighted that the use of tractor and trailed tools is the mainly responsible for compaction, representing this management practice as the cause of 42% of the compacted soil in their study.

In Year 1, runoff and soil erosion were very low, in consequence of the absence from December to October of rainfall events with $p > 50$ mm and intensity (MAX_15) >16 mm, which Bagagiolo et al. [12] demonstrated to be responsible for relevant runoff in sloping vineyards of Monferrato. In this case, soil management and tractor passages influenced hydrological and soil degradation processes to a lesser extent than usual. In fact, runoff and soil losses were reduced by 35% and 43% in the GC compared to the CT plot, whereas the average event reduction by grass cover in the study area resulted in 55% and 79% for runoff and soil losses, respectively [16]. Nevertheless, high-erosive events can occur also in semi-arid climates, and studies in Portugal and Spain [14,71,72] already showed how inter-row's vegetative cover can reduce soil losses in vineyards, compared to the management with tillage, especially during the most intense precipitations. More than 90% of runoff in Year 2 was concentrated during four rainfall events, which can be classified, according to Bagagiolo et al. [16], as "long lasting" (10/3, 11/4, 27/10), for which saturation-excess runoff occurred (driven by high SWC, as evident in the graph of Figure 2), and "intense" (16/7), for which infiltration-excess runoff occurred, as observed by Biddoccu et al. [34]. When analyzing the 16/7 event, it is clear how K_{fs} in CT-T was 20 times lower (4.2 mm h^{-1}) than the 15 min maximum intensity (84.3 mm h^{-1}). Consequently, it is almost certain that there was runoff along the track, notwithstanding, in the no-track position, the K_{fs} was higher (120.1 mm h^{-1}). In GC, K_{fs} was higher (371.2 and 150.7 mm h^{-1} in GC-T and GC-NT, respectively) than the intensity; this can explain why runoff was higher in CT (8.27 mm) than in GC (2.26 mm).

In these cases (both in saturation-excess and in infiltration-excess runoff) the effect of the soil management, as well as traffic over wet soil, was evident: runoff and soil loss were considerably higher in CT than in GC. In such a climate, with some possible intense or long lasting events, the soil management with permanent grass cover provides more soil hydraulic conductivity and soil water recharge, reducing runoff and soil losses. In particular, in autumn and winter, because the runoff generally occurs due to soil saturation in CT [12], rainfall events caused runoff higher than GC despite the autumn ripping and the usually higher K_{fs}, in respect to the other seasons and to the management with grassing [73]. Lateral flow under the soil surface could also occur when infiltrating water moves laterally and locally along an inclined hydraulically restrictive layer, such as the compacted plough pan, as observed by Jiang et al. [74]. The loss of water in the winter period can be the cause of the lower water recharge of the entire soil profile in the CT vineyard, compared to GC, especially in the less-rainy seasons, as directly measured by Gaudin et al. [29].

The mostly erosive events were characterized by high rainfall amount (peculiar of autumn and winter) or by their high intensity (spring and summer storms). In both the cases, GC erosion was reduced in respect to CT. Moreover, the considerable soil loss of the 16/07/2018 (3.1 and 0.5 Mg ha^{-1} in CT and GC, respectively) could be influenced by frequent tractors and people passages over dry soil, which can favor presence of detached soil particles that can be transported by runoff and also on occasions of a little precipitation, as observed by Kirchhoff et al. [17] in the German Mosella vineyards.

5. Conclusions

The effect of the soil management and tractor passes over wet soil was evident in the wet year (Year 2), where GC reduced considerably runoff (−76%) and soil loss (−83%) compared to CT. When weather conditions present possible intense or long-lasting events, the soil management with permanent grass cover provides greater soil hydraulic conductivity (most of the measured values were higher than 50 mm h^{-1}) and soil water recharge, reducing runoff and soil losses.

Soil bulk density and penetration resistance in CT-T increase, compared to GC-T, after only one to three tractor passages following tillage operation, especially in the topsoil (first 10 cm). Soil compaction affects water infiltration, especially in the wet year. In CT, one tractor passage on wet soil after tillage operation dramatically reduced K_{fs} from over 1000 to near 1 mm h^{-1}, while in GC, K_{fs} remained above the usual rain-intensity values, allowing water to infiltrate the soil. Consequently, runoff and soil erosion were higher in the tilled plot, even if lower than the long-period average values. Soil benefits from tillage only if there isn't any machine traffic; therefore, management in alternate rows, with traffic only in grassed rows, could be a valid alternative that is worth investigating in future studies. This study raises interest on issues such as the effects of traffic on the subsoil compaction and the role of the plow pan on the subsurface runoff, and their effects on water balance (water input), as well as on a spatial scale wider than on a single plot.

Author Contributions: Conceptualization, G.C. and M.B.; methodology, G.C. and S.F.; validation, M.B.; formal analysis, M.B.; data curation, G.C.; writing—original draft preparation, G.C. and M.B.; writing—review and editing, S.F. and E.C.; visualization, G.C. and M.B.; supervision, M.B. and S.F.; project administration, E.C.; funding acquisition, E.C.

Funding: This research was partially funded by the Fondazione Giovanni Goria and Fondazione CRT (Bando Talenti della Società Civile 2016). Part of this research was carried out within the framework of the WATER4EVER Project, which is funded under the WaterWorks2015 2 Call, with contribution of EU and MIUR.

Acknowledgments: The Authors thank Giuseppe Delle Cave, Giancarlo Padovan, Giorgia Bagagiolo, and Guarino Benvengù for the support in field operations and in laboratory analysis; the staff of the Tenuta Cannona Experimental Centre, which collaborated managing the vineyards and in sample collections. The content of this article reflects only the authors' views, and the WaterWorks2015 Consortium is not liable for any use that may be made of the information contained therein.

Conflicts of Interest: The authors declare no conflicts of interest. The funders had no role in the design of the study; in the collection, analyses, or interpretation of data; in the writing of the manuscript; or in the decision to publish the results.

References

1. OIV-Organizzazione Internazionale della Vigna e del Vino. Available online: http://www.oiv.int/ (accessed on 15 August 2019).
2. Martucci, O.; Arcese, G.; Montauti, C.; Acampora, A. Social Aspects in the Wine Sector: Comparison between Social Life Cycle Assessment and VIVA Sustainable Wine Project Indicators. *Resources* **2019**, *8*, 69. [CrossRef]

3. Vaudour, E.; Costantini, E.; Jones, G.V.; Mocali, S. An overview of the recent approaches to terroir functional modelling, footprinting and zoning. *Soil* **2015**, *1*, 287–312. [CrossRef]

4. Millennium Ecosystem Assessment. *Ecosystems and Human Well-Being: Synthesis*; Island Press: Washington, DC, USA, 2005; ISBN 1-59726-040-1.

5. Beaufoy, G. *EU Policies for Olive Farming; Unsustainable on All Counts*; BirdLife International-WWF: Brussels, Belgium, 2001.

6. Scheidel, A.; Krausmann, F. Diet, trade and land use: A socio-ecological analysis of the transformation of the olive oil system. *Land Use Policy* **2011**, *28*, 47–56. [CrossRef]

7. Fernandez Escobar, R.; de la Rosa, R.; Leon, L.; Gomez, J.A.; Testi, F.; Orgaz, M.; Gil-Ribes, J.A.; Quesada-Moraga, E.; Trapero, A. Evolution and sustainability of the olive production systems. In *Present and Future of the Mediterranean Olive Sector*; Arcas, N., Arroyo López, F.N., Caballero, J., D'Andria, R., Fernández, M., Fernandez, E.R., Garrido, A., López-Miranda, J., Msallem, M., Parras, M., et al., Eds.; CIHEAM/IOC: Zaragoza, Spain, 2013; pp. 11–42.

8. Communication from the Commission to the Council, the European Parliament, the European economic and social Committee and the Committee of the Regions. *Thematic Strategy for Soil Protection*; CEC: Brussels, Belgium, 2006.

9. European Parliament. *Proposal for a Directive of the European Parliament and of the Council Establishing a Framework for the Protection of Soil and Amending Directive 2004/35/EC*; CEC: Brussels, Belgium, 2006.

10. FAO & ITPS. *Status of the World's Soil Resources (Main Report)*; FAO: Rome, Italy, 2015; p. 608.

11. Hamza, M.A.; Anderson, W.K. Soil compaction in cropping systems: A review of the nature, causes and possible solutions. *Soil Tillage Res.* **2005**, *82*, 121–145. [CrossRef]

12. Biddoccu, M.; Ferraris, S.; Pitacco, A.; Cavallo, E. Temporal variability of soil management effects on soil hydrological properties, runoff and erosion at the field scale in a hillslope vineyard, North-West Italy. *Soil Tillage Res.* **2017**, *165*, 46–58. [CrossRef]

13. Ferrero, A.; Usowicz, B.; Lipiec, J. Effects of Tractor Traffic on Spatial Variability of Soil Strength and Water Content in Grass Covered and Cultivated Sloping Vineyard. *Soil Tillage Res.* **2005**, *84*, 127–138. [CrossRef]

14. Ferreira, C.S.S.; Keizer, J.J.; Santos, L.M.B.; Serpa, D.; Silva, V.; Cerqueira, M.; Ferreira, A.J.D.; Abrantes, N. Runoff, sediment and nutrient exports from a Mediterranean vineyard under integrated production: An experiment at plot scale. *Agric. Ecosyst. Environ.* **2018**, *256*, 184–193. [CrossRef]

15. Gómez, J.A. Sustainability using cover crops in Mediterranean tree crops, olives and vines—Challenges and current knowledge. *Hung. Geogr. Bull.* **2017**, *66*, 13–28. [CrossRef]

16. Bagagiolo, G.; Biddoccu, M.; Rabino, D.; Cavallo, E. Effects of rows arrangement, soil management, and rainfall characteristics on water and soil losses in Italian sloping vineyards. *Environ. Res.* **2018**, *166*, 690–704. [CrossRef]

17. Kirchhoff, M.; Rodrigo-Comino, J.; Seeger, M.; Ries, J.B. Soil erosion in sloping vineyards under conventional and organic land use managements (Saar-Mosel Valley, Germany). *Cuadernos de Investigación Geográfica* **2017**, *207*, 119–140. [CrossRef]

18. Comino, J.R.; Senciales, J.M.; Ramos, M.C.; Martínez-Casasnovas, J.A.; Lasanta, T.; Brevik, E.C.; Ries, J.B.; Sinoga, J.R. Understanding soil erosion processes in Mediterranean sloping vineyards (Montes de Málaga, Spain). *Geoderma* **2017**, *296*, 47–59. [CrossRef]

19. Novara, A.; Gristina, L.; Saladino, S.S.; Santoro, A.; Cerdà, A. Soil erosion assessment on tillage and alternative soil managements in a Sicilian Vineyard. *Soil Tillage Res.* **2011**, *117*, 140–147. [CrossRef]

20. Biddoccu, M.; Opsi, F.; Cavallo, E. Relationship between runoff and soil losses with rainfall characteristics and long-term soil management practices in a hilly vineyard (Piedmont, NW Italy). *Soil Sci. Plant Nutr.* **2014**, *60*, 92–99. [CrossRef]

21. Biddoccu, M.; Zecca, O.; Audisio, C.; Godone, F.; Barmaz, A.; Cavallo, E. Assessment of long-term soil erosion in a mountain vineyard, Aosta Valley (NW Italy). *Land Degrad. Dev.* **2018**, *29*, 617–629. [CrossRef]

22. Winter, S.; Bauer, T.; Strauss, P.; Kratschmer, S.; Paredes, D.; Popescu, D.; Landa, B.; Guzmán, G.; Gómez, J.A.; Guernion, M.; et al. Effects of vegetation management intensity on biodiversity and ecosystem services in vineyards: A meta-analysis. *J. Appl. Ecol.* **2018**, *55*, 2484–2495. [CrossRef] [PubMed]

23. Novara, A.; Minacapilli, M.; Santoro, A.; Rodrigo-Comino, J.; Carrubba, A.; Sarno, M.; Venezia, G.; Gristina, L. Real cover crops contribution to soil organic carbon sequestration in sloping vineyard. *Sci. Total Environ.* **2019**, *652*, 300–306. [CrossRef] [PubMed]

24. Garcia, L.; Celette, F.; Gary, C.; Ripoche, A.; Valdés-Gómez, H.; Metay, A. Management of service crops for the provision of ecosystem services in vineyards: Areview. *Agric. Ecosyst. Environ.* **2018**, *251*, 158–170. [CrossRef]

25. Guzmán, G.; Cabezas, J.M.; Sánchez-Cuesta, R.; Zaller, J.G.; Gómez, J.A. A field evaluation of the impact of temporary cover crops on soil properties and vegetation communities in southern Spain vineyards. *Agric. Ecosyst. Environ.* **2019**, *272*, 135–145. [CrossRef]

26. López-Vicente, M.; Álvarez, S. Stability and patterns of topsoil water content in rainfed vineyards, olive groves, and cereal fields under different soil and tillage conditions. *Agric. Water Manag.* **2018**, *201*, 167–176. [CrossRef]

27. Haruna, S.; Nkongolo, N.; Anderson, S.; Eivazi, F.; Zaibon, S. In situ infiltration as influenced by cover crop and tillage management. *J. Soil Water Conserv.* **2018**, *73*, 164–172. [CrossRef]

28. Capello, G.; Biddoccu, M.; Ferraris, S.; Pitacco, A.; Cavallo, E. Year-round variability of field-saturated hydraulic conductivity and runoff in tilled and grassed vineyards. *Chem. Eng. Trans.* **2017**, *58*, 739–744. [CrossRef]

29. Gaudin, R.; Celette, F.; Gary, C. Contribution of runoff to incomplete off season soil water refilling in a Mediterranean vineyard. *Agric. Water Manag.* **2010**, *97*, 1534–1540. [CrossRef]

30. Celette, F.; Gaudin, R.; Gary, C. Spatial and temporal changes to the water regime of a Mediterranean vineyard due to the adoption of cover cropping. *Eur. J. Agron.* **2008**, *29*, 153–162. [CrossRef]

31. Fourie, J.C. Soil Management in the Breede River Valley Wine Grape Region, South Africa. Organic matter and macro-nutrient content of a medium-textured soil. *S. Afr. J. Enol. Vitic.* **2012**, *33*, 105–114. [CrossRef]

32. Guilpart, N.; Metay, A.; Gary, C. Grapevine bud fertility and number of berries per bunch are determined by water and nitrogen stress around flowering in the previous year. *Eur. J. Agron.* **2014**, *54*, 9–20. [CrossRef]

33. Ruiz-Colmenero, M.; Bienes, R.; Marques, M.J. Soil and water conservation dilemmas associated with the use of green cover in steep vineyards. *Soil Tillage Res.* **2011**, *117*, 221–223. [CrossRef]

34. IPLA. Programma di Sviluppo Rurale 2007–2013—Monitoraggio Ambientale. Available online: http://www.regione.piemonte.it/agri/psr2007_13/dwd/monitoraggio/PSRvalutazioneFinale2016.pdf (accessed on 29 December 2018).

35. Cerdà, A.; Keesstra, S.D.; Rodrigo-Comino, J.; Novara, A.; Pereira, P.; Brevik, E.; Giménez-Morera, A.; Fernández-Raga, M.; Pulido, M.; di Prima, S.; et al. Runoff initiation, soil detachment and connectivity are enhanced as a consequence of vineyards plantations. *J. Environ. Manag.* **2017**, *202*, 268–275. [CrossRef] [PubMed]

36. Ramos, M.C.; Martínez-Casasnovas, J.A. Impact of land levelling on soil moisture and runoff variability in vineyards under different rainfall distributions in a Mediterranean climate and its influence on crop productivity. *J. Hydrol.* **2006**, *31*, 131–146. [CrossRef]

37. Lagacherie, P.; Coulouma, G.; Ariagno, P.; Virat, P.; Boizard, H.; Richard, G. Spatial variability of soil compaction over a vineyard region in relation with soils and cultivation operations. *Geoderma* **2006**, *134*, 207–216. [CrossRef]

38. Marinello, F.; Pezzuolo, A.; Cillis, D.; Chiumenti, A.; Sartori, L. Traffic effects on soil compaction and sugar beet (*Beta vulgaris* L.) taproot quality parameters. *Span. J. Agric. Res.* **2017**, *15*, 11. [CrossRef]

39. Spinelli, R.; Magagnotti, N.; Cavallo, E.; Capello, G.; Biddoccu, M. Reducing soil compaction after thinning work in agroforestry plantations. *Agrofor. Syst.* **2019**, *93*, 1765–1779. [CrossRef]

40. Alaoui, A.; Roggerb, M.; Pethc, S.; Blöschlb, G. Does soil compaction increase floods? A review. *J. Hydrol.* **2018**, *557*, 631–642. [CrossRef]

41. Tropeano, D. Rate of soil erosion processes on vineyards in Central Piedmont (NW Italy). *Earth Surf. Process. Landf.* **1984**, *9*, 253–266. [CrossRef]

42. Boni, A.; Casnedi, R. *Note illustrative della Carta Geologica d'Italia alla scala 1:100,000 Fogli 69 e 70 Asti-Alessandria*; Servizio Geologico d'Italia, Poligrafica & Cartevalori: Ercolano (Napoli), Italy, 1970; p. 64.

43. Soil Survey Staff. *Keys to Soil Taxonomy*, 11th ed.; USDA-Natural Resources Conservation Service: Washington, DC, USA, 2010.

44. FAO/ISRIC/ISSS. *World Reference Base for Soil Resources*; World Soil Resources Report, No. 84; FAO: Rome, Italy, 1998.

45. Kottek, M.; Grieser, J.; Beck, C.; Rudolf, B.; Rubel, F. World Map of the Köppen-Geiger climate classification updated. *Meteorol. Z.* **2006**, *15*, 259–263. [CrossRef]

46. Biddoccu, M.; Ferraris, S.; Opsi, F.; Cavallo, E. Long-term monitoring of soil management effects on runoff and soil erosion in sloping vineyards in Alto Monferrato (North-West Italy). *Soil Tillage Res.* **2016**, *155*, 176–189. [CrossRef]

47. Sohne, W. Druckverteilung im Boden und Boden-verformung unter Schlepperreifen. *Grundlagen der Landtechnik* **1953**, *5*, 49–63.

48. Bagarello, V.; Iovino, M.; Elrick, D. A simplified falling-head technique for rapid determination of fieldsaturated hydraulic conductivity. *Soil Sci. Soc. Am. J.* **2004**, *68*, 66–73. [CrossRef]

49. ARS-USDA. RIST Rainfall Intensity Summarization Tool. 2015. Available online: http://www.ars.usda.gov/Research/docs.htm?docid=3251 (accessed on 4 April 2019).

50. Brown, L.C.; Foster, G.R. Storm erosivity using idealized intensity distributions. *Trans. ASAE* **1987**, *30*, 379–386. [CrossRef]

51. Renard, K.G.; Foster, G.R.; Weesies, G.A.; McCool, D.K.; Yoder, D.C. *Predicting Soil Erosion by Water: A Guide to Conservation Planning with the Revised Universal Soil Loss Equation (RUSLE)*; US Department of Agriculture Agricultural Handbook No.703; USDA: Washington, DC, USA, 1997.

52. Raffelli, G.; Previati, M.; Canone, D.; Gisolo, G.; Bevilacqua, I.; Capello, G.; Biddoccu, M.; Cavallo, E.; Deiana, R.; Cassiani, G.; et al. Local- and Plot-Scale Measurements of Soil Moisture: Time and Spatially Resolved Field Techniques in Plain, Hill and Mountain Sites. *Water* **2017**, *9*, 706. [CrossRef]

53. Ampoorter, E.; Goris, R.; Cornelis, W.; Verheyen, K. Impact of mechanized logging on compaction status of sandy forest soils. *For. Ecol. Manag.* **2007**, *241*, 162–174. [CrossRef]

54. Proctor, R. Fundamental principles of soil compaction. *Eng. News Rec.* **1933**, *111*, 245–248, 286–289, 348–351.

55. Herrick, J.E.; Jones, T.L. A dynamic cone penetrometer for measuring soil penetration resistance. *Soil Sci. Soc. Am. J.* **2002**, *66*, 1320–1324. [CrossRef]

56. Halliday, D.; Resnick, R. *Physics, for Students of Science and Engineering*, combined edition; John Wiley & Sons: Hoboken, NJ, USA, 1963.

57. Busscher, W.J.; Bauer, P.J.; Camp, C.R.; Sojka, R.E. Correction of cone index for soil water content differences in a coastal plain soil. *Soil Tillage Res. Amst.* **1997**, *43*, 205–217. [CrossRef]

58. Busscher, W.J. Adjustment of flat-tipped penetrometer resistance data to a common water content. *Trans. ASAE* **1990**, *33*, 519–524. [CrossRef]

59. Vaz, C.M.P.; Manieri, J.M.; de Maria, I.C.; Tuller, M. Modeling and correction of soil penetration resistance for varying soil water content. *Geoderma* **2011**, *166*, 92–101. [CrossRef]

60. Jakobsen, B.F.; Dexter, A.R. Effect of soil structure on wheat root growth, water uptake and grain yield. A computer simulation model. *Soil Tillage Res.* **1987**, *10*, 331–345. [CrossRef]

61. Bodhinayake, W.; Si, B.C.; Noborio, K. Determination of hydraulic properties in sloping landscapes from tension and double-ring infiltrometers. *Vadose Zone J.* **2004**, *3*, 964–970. [CrossRef]

62. Warrick, A.W. *Environmental Soil Physics*; Hillel, D., Ed.; Academic Press: San Diego, CA, USA, 1998; pp. 665–675.

63. Bagarello, V.; Sgroi, A. Using the simplified falling head technique to detect temporal changes in field-saturated hydraulic conductivity at the surface of a sandy loam soil. *Soil Tillage Res.* **2007**, *94*, 283–294. [CrossRef]

64. Snedecor, G.W. *Statistical Methods Applied to Experiments in Agriculture and Biology*; Iowa State College Press: Ames, IA, USA, 1957.

65. Whalley, W.R.; To, J.; Kay, B.D.; Whitmore, A.P. Prediction of the penetrometer resistance of soils with models with few parameters. *Geoderma* **2007**, *137*, 370–377. [CrossRef]

66. Capello, G.; Biddoccu, M.; Ferraris, S.; Pitacco, A.; Cavallo, E. Effects of Soil Management and Traffic on Water and Soil Conservation in Agriculture: A Case Study in Piedmont. In Proceedings of the 15th Biennial Conference Euromediterranean Network of Experimental and Representative Basins, At Coimbra, Portugal, 9–13 September 2014.

67. Alagna, V.; Bagarello, V.; Di Prima, S.; Guaitoli, F.; Iovino, M.; Cerdà, A. Using Beerkan experiments to estimate hydraulic conductivity of a crusted loamy soil in a Mediterranean vineyard. *J. Hydrol. Hydromech.* **2019**, *67*, 191–200. [CrossRef]

68. Barik, K.; Aksakal, E.L.; Islam, K.R.; Sari, S.; Angin, I. Spatial variability in soil compaction properties associated with field traffic operations. *Catena* **2014**, *120*, 122–133. [CrossRef]

69. Bogunovic, I.; Bilandzija, D.; Andabaka, Z.; Stupic, D.; Comino, J.R.; Cacic, M.; Brezinscak, L.; Maletic, E.; Pereira, P. Soil compaction under different management practices in a Croatian vineyard. *Arab. J. Geosci.* **2017**, *10*, 340. [CrossRef]

70. O'Keefe, S. The Recovery of Soils after Compaction: A Laboratory Investigation into the Effect of Wet/Dry Cycles on Bulk Density and Soil Hydraulic Functions. Ph.D. Thesis, Cranfield University at Silsoe, National Soil Resources Institute, Cranfield, UK, 2009.

71. Gómez, J.A.; Llewellyn, C.; Basch, G.; Sutton, P.B.; Dyson, J.S.; Jones, C.A. The effects of cover crops and conventional tillage on soil and runoff loss in vineyards and olive groves in several Mediterranean countries. *Soil Use Manag.* **2011**, *27*, 502–514. [CrossRef]

72. Marques, M.J.; García-Muñoz, S.; Muñoz-Organero, G.; Bienes, R. Soil conservation beneath grass cover in hillside vineyards under mediterranean climatic conditions (Madrid, Spain). *Land Degrad. Dev.* **2010**, *21*, 122–131. [CrossRef]

73. Capello, G.; Biddoccu, M.; Cavallo, E. Seasonal Variability and Effect of Tractor Passes on Soil Compaction, Field-Saturated Hydraulic Conductivity, Runoff and Soil Erosion in Tilled and Grassed Vineyards. In Proceedings of the Sixth International Congress on Mountain and Steep Slope Viticulture, San Cristobal de la Laguna (Isla de Tenerife), Spain, 26–28 April 2018; ISBN 978-88-902330-5-0.

74. Jiang, X.; Liu, X.; Wang, E.; Li, X.G.; Sun, R.; Shi, W. Effects of tillage pan on soil water distribution in alfalfa-corn crop rotation systems using a dye tracer and geostatistical methods. *Soil Tillage Res.* **2015**, *150*, 68–77. [CrossRef]

Article

Impact of Soil Conservation and Eucalyptus on Hydrology and Soil Loss in the Ethiopian Highlands

Demesew A. Mhiret [1,2], Dessalegn C. Dagnew [2,3], Tilashwork C. Alemie [4], Christian D. Guzman [5], Seifu A. Tilahun [1], Benjamin F. Zaitchik [6] and Tammo S. Steenhuis [1,7,*]

1   Faculty of Civil and Water Resources Engineering, Bahir Dar Institute of Technology, Bahir Dar University, Bahir Dar, Ethiopia; demisalmaw@gmail.com (D.A.M.); satadm86@gmail.com (S.A.T.)
2   Blue Nile Water Institute, Bahir Dar University, Bahir Dar, Ethiopia; cdessalegn@yahoo.com
3   Institute of Disaster Risk Management and Food Security Studies, Bahir Dar University, Bahir Dar, Ethiopia
4   Amhara Regional Agricultural Research Institute, Bahir Dar, Ethiopia; tilashworkchanie@gmail.com
5   Department of Civil and Environmental Engineering, University of Massachusetts, Amherst, MA 01003, USA; cdguzman@umass.edu
6   Department of Earth and Planetary Sciences, Johns Hopkins University, Baltimore, MD 21218, USA; zaitchik@jhu.edu
7   Department of Biological and Environmental Engineering, Cornell University, Ithaca, NY 14853, USA
*   Correspondence: tss1@cornell.edu; Tel.: +1-607-255-2489

Received: 25 September 2019; Accepted: 30 October 2019; Published: 2 November 2019

Abstract: The Ethiopian highlands suffer from severe land degradation, including erosion. In response, the Ethiopian government has implemented soil and water conservation practices (SWCPs). At the same time, due to its economic value, the acreage of eucalyptus has expanded, with croplands and pastures converted to eucalyptus plantations. The impact of these changes on soil loss has not been investigated experimentally. The objective of this study, therefore, is to examine the impacts of these changes on stream discharge and sediment load in a sub-humid watershed. The study covers a nine-year period that included installation of SWCPs, a three-fold increase from 1.5 ha in 2010 to 5 ha in 2018 in eucalyptus, and the upgrading of an unpaved to the paved road. Precipitation, runoff, and sediment concentration were monitored by installing weirs at the outlets of the main and four nested watersheds. A total of 867 storm events were collected in the nine years. Runoff and sediment concentration decreased by more than half in nine years. In the main watershed W5, we estimated that evapotranspiration by eucalyptus during the dry phase (November to May) increased approximately from 30 mm a^{-1} in 2010 to 100 mm a^{-1} in 2018. In watershed W3 it increased from 2 mm a^{-1} to 400 mm a^{-1}, requiring more rainfall before saturation excess runoff began in the rain phase. The reduction in runoff led to a decreased sediment load from 70 Mg ha^{-1} a^{-1} in 2010 to 2.8 Mg ha^{-1} a^{-1} in 2018, though the reduction in discharge may have negative impacts on ecology and downstream water resources. SWCPs became sediment-filled and minimally effective by 2018. This indicates that these techniques are either inappropriate for this sub-humid watershed or require improved design and maintenance.

Keywords: Ethiopian highlands; eucalyptus; gully; soil loss; soil and water conservation practices

1. Introduction

Land degradation and associated soil loss is a major global ecological problem [1–4]. The Ethiopian highlands are especially degraded [4–6]. Land degradation is accelerated by anthropogenic factors, such as population growth, cultivation of steep slopes, clearing of vegetation, overgrazing, and increased soil erosion [5,7–9].

In response to the severe drought in the 1970s, the Ethiopian government started to implement soil and water conservation practices (SWCPs) to reverse the trend in soil degradation. In 2012, the government expanded its conservation efforts, requiring rural farmers to volunteer their labor in January and February each year to install centrally planned SWCPs. The effectiveness of these practices is being debated [10]. In the semi-arid highlands, SWCPs perform well in conserving moisture and generally increase crop yields [3]. In the sub-humid and humid highland regions, with rainfall in excess of potential evaporation during the rainy phase, conserving moisture is not a priority. Instead, preventing saturation of the root zone is a priority. Consequently, the purpose of SWCPs is to safely remove the excess water [11,12]. Practices that increase infiltration, such as Fanya- Juu ("throw uphill") bunds, infiltration furrows, and stone bunds have been shown to be effective in decreasing soil losses in the first five years after implementation [9,13] except in watersheds with gullies downstream [14,15]. Gullies have been identified as a critical factor in soil loss from catchments in the sub-humid Ethiopian highlands [12,14,16,17].

Gully erosion is largely a consequence of forests being replaced by agricultural lands that are cultivated year after year [12]. The continuous cultivation following deforestation decreases organic matter content, that causes soil degradation and hardpan formation [16]. The hard pan reduces deep percolation rates and increases the interflow and surface runoff [12,16,18]. This in turn results in soil saturation in valley bottoms, reducing the cohesive soil strength and enhancing gully formation.

One approach that has been applied to understand relationships between runoff (Q) and suspended sediment concentration (SSC) in the presence of SWCPs is the analysis of hysteresis loops [19–21]. Analysis of the Q–SSC relationship indicated five hysteresis loop patterns that can be reduced to three categories [20]: clockwise, counter-clockwise, and mixed loops. Clockwise loops occur when the concentration on the rising limb is greater than the concentration on the recession limb at equal discharge and are characterized by a sediment peak before the discharge peak. It indicates that sediment becomes more limited during the storm [22]. Mixed loops are those where either the concentration is relatively constant throughout the runoff event or small sediment peaks occur before and after the discharge peak. Mixed loops occur when sediment is equally available during the runoff event [23–25]. Counter-clockwise loops have a greater concentration during the recession than during the rising limb and indicate that a new sediment source is accessed during the recession. This can be related to gully banks failing during the storm [23]. The fact that gully bank erosion has a specific, process-relevant signal in hysteresis loops suggests that analysis of these loops may be useful in assessing the relative roles of upland soil erosion and gully erosion in the presence of SWCPs and other land-use change. We note that in some contexts, the travel time from a distant sediment source has been invoked as an explanatory variable for the type of hysteresis loop [21,26,27], but this explanation does not apply in small watersheds with sampling intervals less than the time of concentration. In the Birr watershed of the upper Blue Nile basin, gully rehabilitation through planting vegetation (e.g., *Sesbania sesban*) was proven effective [28].

The objective of this study is to investigate the impact of the implementation of SWCPs and the expansion of eucalyptus acreage on discharge and soil loss patterns. The study was carried out in the 95-ha Debre Mawi watershed in the sub-humid Ethiopian highlands. In the main watershed, four nested watersheds were defined. Discharge and sediment loss were monitored at each watershed by installing a weir at the outlets. Samples were collected at 10-min time step during runoff events. The hysteresis patterns of suspended sediment concentrations and discharge were identified and analyzed in each watershed. Discharge and sediment concentration and loads were compared and evaluated.

2. Materials and Methods

2.1. Description of the Study Area

The Debre Mawi watershed (Figure 1) is located in the northern Ethiopian highlands, Amhara Regional State, 30 km from Bahir Dar on the road to Adet. The outlet is located at 37°22′11″ E and 11°18′17″ N. The elevation ranges from 2195 m near the outlet to 2308 m in the southeast. Slopes vary from 1% to 30%. The climate is categorized as sub-humid with a mean annual rainfall of 1240 mm a^{-1} [29]. Seventy percent of the rainfall falls from June to September. The mean daily temperature is 20 °C. Four smaller watersheds were nested in the main 95 ha watershed (Figure 1). These watersheds were W1 (8.8 ha), W2 (11.0 ha), W3 (6.4 ha), and W4 (10.4 ha). The entire watershed was named W5. The weirs at the outlets were given the same number as the watershed. So W1 was monitored by weir 1 and so on.

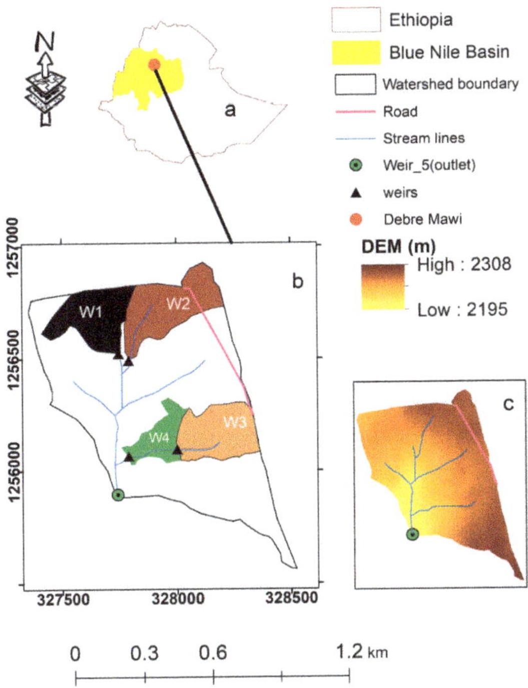

Figure 1. Location of Debre Mawi watershed and map of its four nested watersheds: (**a**) The location of the Debre Mawi watershed in the Upper Blue Nile Basin. (**b**) The Debre Mawi watershed consisting of the main watershed W5 together with the nested watersheds W1, W2, W3, and W4. The coordinates are in UTM. (**c**) The digital elevation map of the Debre Mawi watershed.

The lower part of the soil profile consists of shallow, highly weathered and fractured basalt overlain by dark-brown clay and light-brown, wet, sticky-clay. Intrusive dikes perpendicular to the flow direction blocks subsurface flow and results in the emergence of springs during the wet season [17,30,31]. The major types of soil in the watershed are Nitisols, Vertisols, and Vertic Nitisols [32]. Nitisols are fertile forest soils with high base saturation (>35%) and red, clay-loam soils covering the upper part of the watershed. These are very deep, well-drained, permeable soils and are suited for cereal cultivation. Vertisols, characterized by expanding montmorillonite clay, are found at the bottom part of the watershed [17]. They form shrinkage cracks on drying and swell during the rainy season. The mid-slope of the watershed is dominated by reddish-brown Vertic Nitisols with good permeability and high moisture retention capacity. These soils are particularly well suited for '*teff*' (*Eragrostis abyssinica*) production [31].

During the experimental period, farmers planted eucalyptus trees, and the acreage of trees increased from 1.5 ha to 5 ha (Figure 2). By 2018, 74% of the watershed was cropped, 15% grass and 5% eucalyptus trees, and 6% sparse vegetation. The expansion of eucalyptus was mainly on the cropland in Watershed W3 and on the erosive-prone land near the outlet of the main watershed. In addition, the small trees in 2010 were full-grown by 2018.

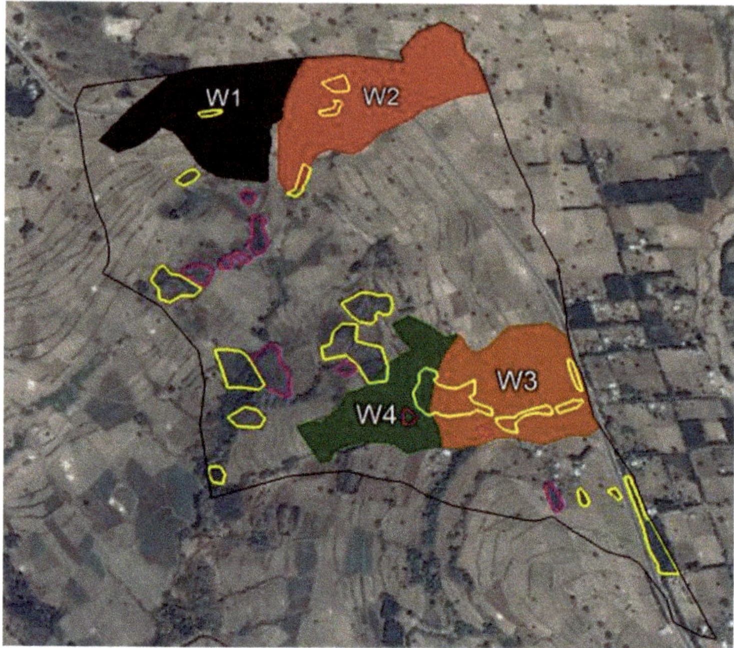

Figure 2. Eucalyptus tree expansion in the Debre Mawi Watershed (Google earth). Black is the boundary of the main watershed; purple indicates the eucalyptus area in 2010, and yellow indicates additional eucalyptus by 2018. W1, W2, W3, and W4 are the nested watersheds in the main watershed W5.

The grasslands are found in the valley bottomlands, which are too wet for cropping during the rain phase. Shrubs are found on the stony, steep, and shallow soils on the hillslopes. Crops grown are *teff*, maize (*Zea mays*), finger millet (*Eleusine coracana*) barley (*Hordeum vulgare*), and wheat (*Triticum aestivum*) [10,14,16]. The small nested watersheds W2 and W4 had a land use distribution that is comparable with the main watershed. Watershed W1 that had the largest portion of grassland and Watershed W3 had the largest eucalyptus acreage (Table 1).

Table 1. The land cover acreage (ha) in Debre Mawi and nested sub-watersheds in 2010 and 2018. Grassland did not change, so only one value is given.

Water-Shed	Cultivated		Grass Land	Shrub		Eucalyptus		Total	Existence of Gully
	2010	2018		2010	2018	2010	2018		
1	3.0	2.8	5.2	0.6	0.6	0	0.2	8.8	No gully
2	8.0	7.3	2.6	0.4	0.7	0	0.7	11	Upland gully formed in 2017
3	5.1	4.1	0.6	0.7	0.7	0.1	1.1	6.5	No gully
4	8.0	7.6	0.9	1.5	1.5	0	0.4	10.4	Gully became stable
5	69	67.5	14	10.5	8.5	1.5	5	95	15 small gullies; 1 gully of 1500 m².

Five-meter-deep and twenty-meter-wide gullies can be seen in the valley bottomlands (Figure 3a). Formation of the gullies in the Debre Mawi watershed started in the 1980s following the removal of indigenous forests, which in turn caused an increase in surface and subsurface runoff [16,23]. Gullies were initiated in the Debre Mawi watershed at the locations that were springs 40 years ago [16].

Active gullies are found in the periodically saturated bottomland portion of the main watershed, and a two-meter deep gully emerged in the saturated portion of watershed W2 in 2017.

Figure 3. Photos depicting features in the Debre Mawi watershed: (**a**) Five-meter deep gully in the saturated bottomland of the Debre Mawi watershed. (**b**) Bund with *Sesbania grandiflora* in 2017. The bund, together with the infiltration furrow, was constructed in 2012. The infiltration furrow was filled with sediment in 2018 and is not visible.

Starting early in 2012 and ending in 2014, the government mandated SWCPs to be implemented by farmers as part of a national campaign. The SWCPs consisted of bunds with infiltration furrows that were installed on the contour on cultivated and grazing lands with slopes ranging from 3% to 32% according to guidelines of the Ethiopian Ministry of Agriculture. The infiltration furrows were 0.5 m deep and 0.5 m wide. The bunds varied in height from 0.3 to 0.6 m. The horizontal spacing between the bunds was 32 m. The spacing was reduced on steep lands so that the maximum difference in elevation was 1.5 m. Sesbania grandiflora was planted on the bunds as animal fodder and for the strengthening of the sides (Figure 3b).

As we observed from Google Images at the end of the implementation period in 2014, 70% of the watershed was treated with SWCPs. However, they were not maintained. Consequently, in the saturated bottomland areas within a three-year period, most furrows were filled with sediment, and only the bunds were visibly covered with grass (Figure 4a). In one case, where the rate of filling was less than the transport capacity of the sediment out of the furrow, the furrow acted as a cutoff drain. The concentrated flow resulted in a gully (Figure 4b) that formed in 2013 and then cut out a path downstream in the remainder of the years. The infiltration furrows on the sloping lands in the remainder of the watershed filled with sediment more slowly. By 2018, all were filled with sediment, and only the bunds remained visible as green strips (Figure 4a). In addition, without government support, farmers voluntarily installed off-contour traditional furrows (or fesses in Amharic) after the plowing was finished. These served to remove excess rainfall. The dimensions were determined by the size of the local ox-driven plows (*Marsha*) and were approximately 20–30 cm wide and 10–15 cm deep. Local farmers report that fesses were preferred because, unlike deep furrows, they do not hinder tillage operation.

Figure 4. Soil and water conservation practices. (**a**) Fesses (plowed off-contour furrows; 20–30 cm wide and 10–15 cm deep) and 20–30 cm high bunds (green strips) in the Debre Mawi watershed in July 2018. Water is ponded at the surface, and the fesses carry off the excess water. Infiltration furrows uphill of the bunds have been filled with sediment and are not visible. (**b**) Infiltration furrow installed in 2012 in the saturated bottomland carries off interflow from the upland and caused the 2.5 m deep gully in the foreground. The photo was taken in August 2015. (**c**) Fifty to sixty centimeter wide rills formed by runoff concentrated by a bund on 18 July 2017.

Several other activities took place in the watershed, such as upgrading the unpaved road to a paved highway with a stone drainage ditch. The construction started in 2014 and lasted until 2016. The road crosses the watershed in the northern part of the eastern boundary (Figure 2). The unpaved road drainage ditch discharged into watershed W2. Some of the runoff came from the watershed across the road. The center of the paved road was distinctly higher than the unpaved road and prevented the water from crossing the road [15]. The exact amount of discharge is not known and varied during construction. Watershed W3 did not receive road drainage. Some drainage entered watershed W5 just north of watershed W3 (Figure 2). The weirs 1, 2, and 3 at the outlet of watersheds W1, W2, and W3, respectively, were located above a volcanic dike that interrupted the interflow, and the water table came to the surface. The outlet of watershed W4 is below watershed W3 on the same drainage path. The land between weir 3 and weir 4 is relatively flat and, therefore, subject to saturation during the rain phase [33]. Hence, all watersheds had saturated areas above the weir during the rain phase, which was covered by grass that tolerates water tables at shallow depths (Table 1). The acreage of grassland indicated the periodically saturated area in watershed W1 above weir 1 was the largest, and watershed W3 had the smallest. Watershed 2 received an unknown amount of storm runoff from the main road.

2.2. Data Collection

Precipitation: The five-minute rainfall amounts were recorded with an automatic tipping bucket rain gauge in the center of the watershed W5. The rain gauge was located at 37°25′21″ E and 11°21′31″ N. Precipitation was measured during the rain phase (June to October). Data from the Adet Agricultural Research Center 7 km south of the watershed were used to fill the 4 days missing in 2015 data and the 3 days in 2017.

Stream Flow: Runoff was measured during rain events from June to September for each of the five weirs. Plot scale measurements were not considered. Flow depth and velocity were measured at 10-min intervals from the time that the water became turbid until the water was clear. The surface velocity was determined by measuring the velocity of a float inserted 5 m upstream of the weirs, by recording the time taken by the float to reach the weir. The discharge was estimated as the product of a cross-sectional area and 2/3 of the surface velocity [34].

Suspended Sediment Concentration: One-liter water samples for sediment analysis were collected directly after the stage height measurement. The sediment concentration was determined by filtering the sample using filter papers with a pore size of 2.5 µm. The weight of the sediment was determined after drying for 24 h at 105 °C;

Eucalyptus expansion: Google Earth Image and field observation were used to map the expansion of the eucalyptus trees in the watershed. ArcGIS 10.3 was used for spatial data analysis.

2.3. Data Analysis

The procedure by [20] was used to analyze hysteresis in the sediment–discharge relationship for all storm events with five or more observations. First, discharge and sediment concentration are plotted over time. The loops can then be identified by comparing the timing of the sediment and discharge peaks. Mixed loops had two or more sediment peaks with at least one before and after the discharge peak. For clockwise loops, the only sediment peak occurred before the discharge peak, and for counter-clockwise loops, the sediment peak was after the discharge peak.

Other excel based statistical tools used were the Mann–Kendall trend test [35] to examine the trends in precipitation, sediment concentration, and runoff ratios, and the ANOVA (F test) for temporal and spatial changes in sediment concentrations between weirs.

3. Results and Discussion

3.1. Precipitation and Discharge

Precipitation during the rain phase (June-September) varied from a minimum of 832 mm in 2012 to a maximum of 1040 mm in 2017 (Figure 5). The maximum daily precipitation was 152 mm on 11 June 2014. The precipitation did not have a significant trend from 2010 to 2018, as confirmed by a Mann–Kendall [35] trend test.

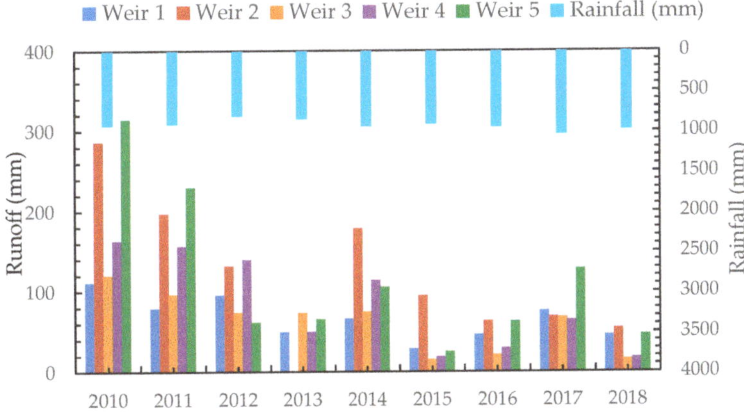

Figure 5. Precipitation and discharge in Debre Mawi watershed from 2010 to 2018.

The annual direct runoff for the main and nested watersheds was greatest in 2010 and 2011 before practices were installed, and then it decreased (Figure 5). The decrease in direct runoff was statistically significant (Mann–Kendall trend test at 5% level) [35]. In 2010, the maximum annual runoff depth was 314 mm during the rain phase for watershed W5 and 275 mm in watershed W2. The smallest annual direct runoff was in 2015.

The runoff coefficient is a good measure for assessing changes in hydrology. The ratio is defined as the discharge divided by the precipitation during the rain phase. Unlike discharge, it is minimally sensitive to precipitation amounts. Figure 6 shows that all runoff coefficients were decreasing from 2010 to 2015 and then remained the same or inclined slightly from 2016 through 2018. The minimum according to the exponential fit of the average runoff in all watersheds was in June 2016. The decrease in runoff means that more of the rainfall infiltrated and less became runoff. Evaporation is at the potential rate during the rainy season and will not affect the runoff coefficient [36].

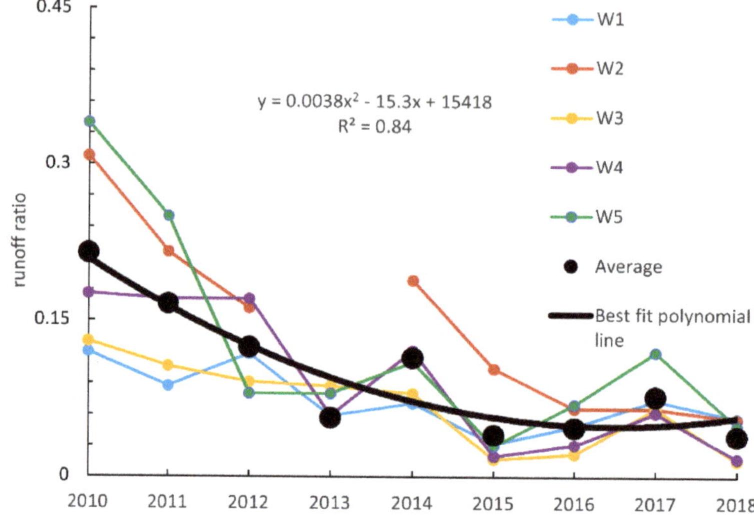

Figure 6. Runoff coefficients during the rain phase from 2010 to 2018 for the Debre Mawi watershed. Note watershed W2 was not monitored in 2013. The black line is the best polynomial fit line of the average runoff ratio of the five watersheds.

The runoff pattern of watersheds W2 and W5 was different than the other three nested watersheds W1, W3, and W4 (Figures 5 and 6). However, only before 2016 in watershed W2 was the discharge and runoff ratio significantly greater than the other nested watersheds because the drainage water from the unpaved road drained through watershed W2 [15]. After the road construction ended in 2016, the annual discharge of W2 was statistically the same as the other nested watersheds.

Both the runoff depth per unit area of watershed and the runoff ratio for the main watershed W5 was generally greater than the nested watersheds without the road drainage W1, W3, and W4 during high rainfall years (for example, 2010 and 2017; Figures 5 and 6). During moderate rainfall years (2015, for example), the discharge per unit area of W5 was nearly equal to the nested watersheds W1, W3, and W4. In 2012, the driest year, discharge per unit area of the main watershed W5 was less than any of the nested watersheds (Figure 5). This is likely due to the hydrological behavior of the grassed valley bottom area above weir 5 that becomes saturated during the rainy season. This area is a source of runoff during the wetter years [33], but we hypothesize that in a dry year, such as 2012, it acted as a sink for the runoff of the upper watershed rather than a source for runoff.

Another good measure to assess the changes in the watersheds is the amount of rain needed after the dry phase for the first runoff to occur. Figure 7 shows for all the watersheds the effective cumulative rain needed since the beginning of the rain phase to obtain 3 mm or more cumulative runoff. Effective rainfall is defined as rainfall minus potential evapotranspiration. Cumulative rainfall to generate runoff increased throughout the duration of the experiment, but the increase is not the same for all watersheds. The slope of the regression lines indicated that for watershed W1 and W2, the rainfall before runoff occurs increased by 30–35 mm a^{-1} and for the other three watersheds between 50–60 mm a^{-1} (Figure 7). Before the expansion of eucalyptus tree in watershed W3, the cumulative rainfall to generate 3 mm cumulative runoff (Figure 7) was less than W2 and W1, but after the great expansion of the trees in watershed W3 (Figure 2), it required higher amount of cumulative runoff to generate 3 mm cumulative runoff (Figure 7) than the two nested watersheds W1 and W2.

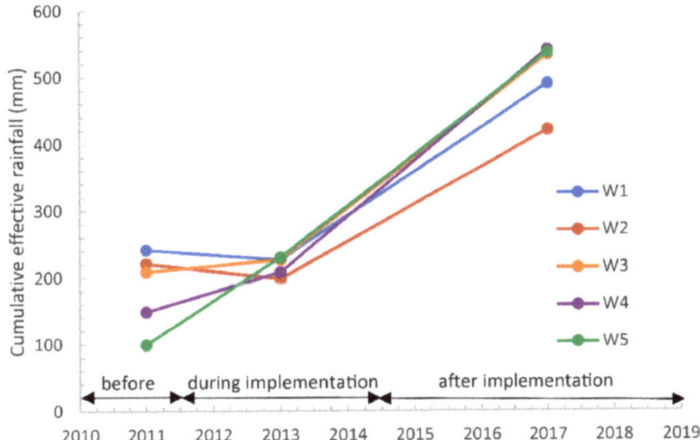

Figure 7. Average cumulative rainfall required to generate greater or equal to 3 mm runoff before (2010–2011), during (2012–2014), and after (2015–2018) SWCPs and eucalyptus tree expansion.

3.2. Effect of Changes in the Watershed on Discharge

To explain these changes in the runoff patterns, there are three likely causes, as discussed above, implementation of the SWCPs, increase in eucalyptus tree acreage, change in road drainage from a paved road, and precipitation. Since changes in runoff at the outlet are integral to all processes that take place within the watershed, it is difficult to sort out the exact cause with certainty. However, comparing the differential in runoff ratio and the amount of storage after the dry phase together with our observations, some tentative interpretations are possible. We will first discuss the SWCPs, followed by the eucalyptus trees and finally the road.

3.2.1. Soil and Water Conservation Practices

The government imposed SWCPs were installed in the period from 2012–2014. However, maintenance was poor, and consequently, during the five-year period after SWCP implementation, all infiltration furrows filled up, and only the bunds covered with grass remained visible (Figure 4a). In the periodically saturated bottomlands, SWCPs initiated a gully (Figure 4b) and in the uplands flow was concentrated by the bunds and caused erosion in some locations (Figure 4c) The long-term decrease in runoff at the watershed scale, as shown in Figures 5 and 6, was not expected because of the poor maintenance of the SWCPs and especially based on findings in the 99 ha Anjeni watershed where the implementation of off-contour infiltration furrows had a minimal impact on runoff [37]. In Debre Mawi watershed, the infiltration furrows were on the contour, and the initial short-term decrease in runoff in 2015 could be explained. However, because the infiltration furrows started to be filled with sediment (Figure 4a), the capacity to store water in the infiltration furrows decreased. This should lead to a gradual increase in runoff to levels before the installation, such as observed in the Anjeni watershed. However, that was not the case in the Debre Mawi watershed: The runoff ratio and runoff stayed below the 2010/2011 levels. The reason is the increasing amount of water removed during the dry phase by evapotranspiration by the expanding acreage of the eucalyptus trees, as we will argue below.

3.2.2. Eucalyptus Trees

Eucalyptus acreage increased three-fold over the nine years that the streamflow was measured. Before our observations started in 2010, 1.5 ha of trees were located on the most erodible lands in watershed W5 on the main stem below watershed W1 where crop production was not judged profitable by the farmers (Table 1, Figure 2). All were young seedlings in 2010 (refer to Figure 2). Plantings after

2010 were mostly on cropped lands, as the increasing need for charcoal and building material in the cities made eucalyptus cultivation more profitable than annual crops. In 2018, eucalyptus tree coverage was a total of 5 ha with 1.13 ha in watershed W3, 0.74 ha in watershed W2, 0.4 ha in watershed W4, and minimal in watershed W1 (Table 1, Figure 2).

Research in the Ethiopia highlands has shown that the increase in eucalyptus will increase water removal by evapotranspiration from the watershed during the dry phase [38–40]. A study in the Fogera Plain near Lake Tana found that the evapotranspiration of eucalyptus during the dry season was twice the potential evaporation of 4–5 mm/day due to additional energy of the dry wind [38]. Similarly, experiments south of Lake Tana found that during the dry phase, eucalyptus decreased the water content of the soil faster than native trees in a similar watershed south of Lake Tana [39]. Finally, in a watershed study in the central Ethiopian highlands, the shift from agricultural fields to eucalyptus reduced surface runoff by 21% [40].

The water that is removed from the watershed during the dry season needs to be filled up before runoff occurs. We proposed that the delay in runoff is caused by this phenomenon. The conceptual argument is as follows: The infiltration rates are high in the Debre Mawi watershed [41], and therefore, runoff occurs only when the soil becomes saturated [41,42]. Saturation of the soil occurs when the amount of rainfall during the rain phase equals the amount of water removed during the dry phase. Thus, a later start to runoff during the wet season indicates a larger amount of water removed during the dry season.

Thus, assuming that the dry season lasts for 200 days and the enhanced evapotranspiration of an area of eucalyptus trees is 10 mm d^{-1}, this area of eucalyptus trees can potentially evaporate 2 m a^{-1} of soil water. Since 5% of the area in watershed W5 consists of the eucalyptus trees, the 2 m of evaporation of water over 5% areas average to 100 mm over 100% of the watershed. A similar calculation shows that in watershed W3, where the eucalyptus trees take up 20% of the area, the average evaporation by the eucalyptus accounts for approximately 400 mm over watershed W3. Thus, the decrease in storage noted in Figure 7 can be only partially ascribed to the eucalyptus trees except for W3, where the 50 mm a^{-1} over nine years is approximately equal to the 450 mm. The decrease in runoff, as shown in Figure 5, is in the same order as the amount of water removed by the eucalyptus trees.

The runoff ratio of weir 4 and weir 3 decreased in time. (Figure 8). Weir 4 is located below weir 3 on the same stream. A flat area between the two weirs saturated in the rain monsoon phase. The explanation of the decreasing runoff ratio in Figure 8 is as follows: The discharge from watershed W3 decreased with the expansion of the eucalyptus trees. The reduced discharge decreased the extent of the saturated area in watershed W4. This, in turn, decreased the saturation excess runoff from watershed W4 through weir 4 further than the incoming recharge from watershed W3. Supporting this argument is that in 2013, a year with low rainfall and discharge, the runoff ratio of weir 4 and 3 was only 0.6 indicating that in that year the extent of the saturated area was the smallest when the runoff of watershed W3 was the smallest (Figure 8)

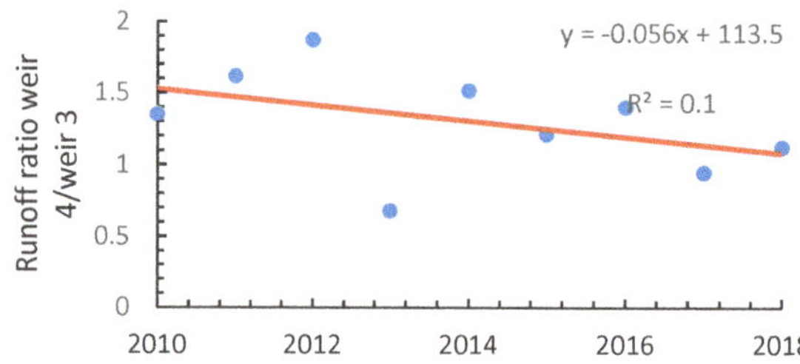

Figure 8. Runoff ratio of weir 4/weir 3 over time for the Debre Mawi watershed.

3.2.3. Road Construction

Finally, road construction changed the hydrology of watersheds W1 and W2 after 2014. The runoff coefficient of watershed W2 was much greater than the nested watersheds from 2010 to 2015 (Figures 5 and 6), and it was similar from 2016 to 2018. The runoff coefficient for watershed W1 decreased less than that of other watersheds over the nine years (Figures 5 and 6), indicating that another source of water became available in watershed W1. This change was likely caused by the drainage pattern of the paved road, which had a much higher centerline than the unpaved road. Runoff water that flowed over the unpaved road to watershed W2 was blocked by the higher paved road.

3.3. Sediment

3.3.1. Sediment Load

The annual soil loss per unit area followed the same overall pattern as the direct runoff. The sediment loss was calculated by summing the product of discharge and sediment concentration and dividing by the area of the watershed. On average, 21 storms were recorded in each rainy year from all weirs. The soil losses were greatest during the first two years (Table 2). For example, the maximum yield for the entire watershed W5 in 2010 was 70 Mg ha^{-1} and then decreased to values ranging from 0.2 to 16 Mg ha^{-1}. The sediment yield per unit area was much greater from the entire watershed than its parts. The difference was larger during wet years than in dry years. In almost all the study years, watershed W2 had greater soil losses than the W1, W3, and W4 due to the road drainage and the formation of a gully.

Table 2. Annual and average soil loss (Mg ha^{-1} a^{-1}) and standard deviation (st dev) for the Debre Mawi watershed. Soil and water conservation practices (SWCPs) were installed from 2012 to 2014.

Watershed	2010	2011	2012	2013	2014	2015	2016	2017	2018	Average	St dev
W1	3.1	3.4	2.6	1.2	1.7	0.4	1.2	1.9	0.2	1.7	1.1
W2	18.5	13.7	4.3	-	8.2	1.1	2.2	3.4	0.3	6.5	6.5
W3	5.2	8	2.4	2.6	2.7	0.2	0.5	2.9	0.2	2.7	2.5
W4	12	19.9	5.1	1.8	4.7	0.4	0.8	4.1	1.4	5.6	6.4
W5	70.3	53.9	9.0	13.3	12.5	0.3	4.1	15.8	2.8	20.2	24.6
Average	21.8	19.8	4.7	4.7	6.0	0.5	1.8	5.6	1.0		
St dev	27.8	20.1	2.7	5.7	4.4	0.4	1.5	5.7	1.1		

3.3.2. Sediment Concentration

The average annual suspended sediment concentration (Table 3) decreased over the nine years to concentrations that were approximately half of the 2010 concentrations. Sediment concentrations were the smallest in 2015, which was a dry year at the end of the period of implementing SWCPs. Concentrations at the outlet of the entire watershed were greater than its upland parts, indicating that sediment was picked up in the valley bottom. Except for watershed W2 in 2011, watershed W4 (which had an active gully in the first years) had a greater concentration than the nested watersheds W1, W2, and W3 during these years. The W4 gully became stable over the course of the experimental period. In the next sections, we will further analyze the reasons for the change in sediment concentrations over the nine years.

Table 3. Suspended sediment concentration (g L^{-1}) for the main watershed W5 and the nested watersheds (W1–W4).

Watershed	2010	2011	2012	2013	2014	2015	2016	2017	2018
W1	3.3	3.7	2.4	2.3	2.8	1.5	2.9	1.6	0.4
W2	5.7	6.3	3.2	-	3.6	1.3	3.0	3.0	1.4
W3	3.5	4.3	3.1	3.3	3.8	1.0	2.8	2.2	1.2
W4	6.2	5.5	3.7	3.3	3.6	2.0	2.8	2.0	0.6
W5	12.7	13.0	13.1	11.5	8.2	4.2	5.8	7.7	5.7

3.4. Discharge–Sediment Concentration Hysteresis Patterns

For the five watersheds, a total of 590 runoff events with a minimum of five observations at ten-minute intervals of discharge and suspended sediment concentration pairs were analyzed for determining the relationship between the suspended sediment concentration and discharge. By plotting the sediment discharge pairs, three types of loops were identified: clockwise, counter-clockwise, and mixed. Examples are given in Figure 9. In general, clockwise and counter-clockwise loops are characterized by systematic offsets in the time of peak discharge and the time of peak sediment concentrations. In mixed loops, the concentrations during the storm did not vary greatly.

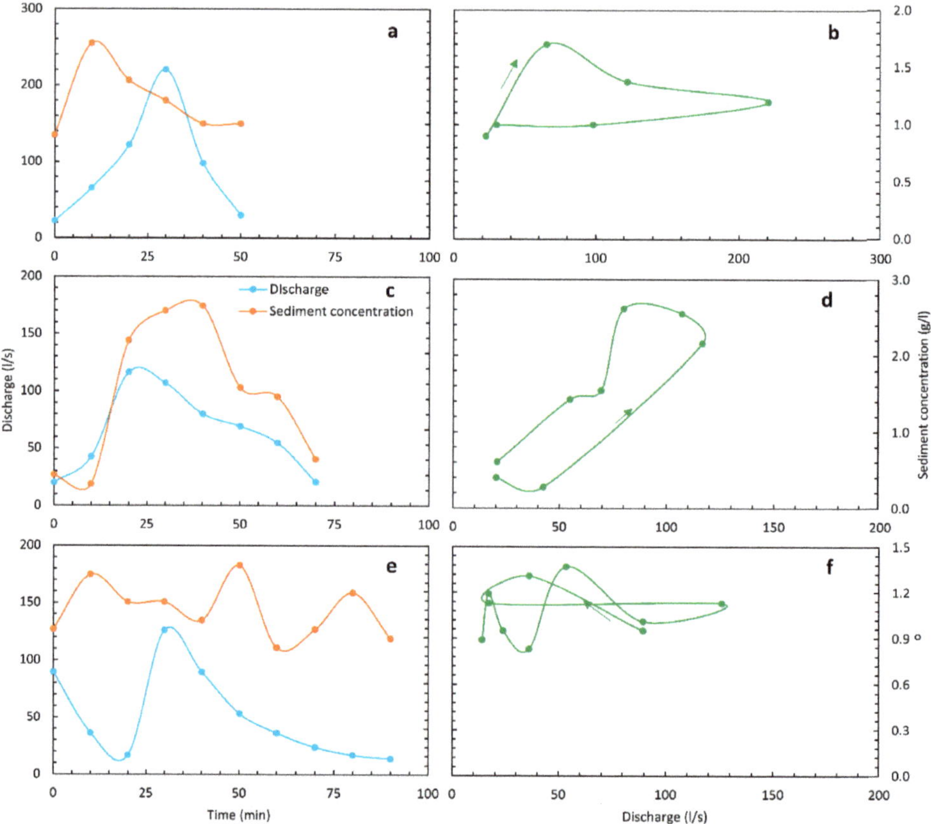

Figure 9. Representative examples of hydrographs and sedigraphs (**left**) and hysteresis loops (**right**) for the Debre Mawi watershed during the rainy season: (**a,b**), clockwise pattern at the outlet of the main watershed W5 on 27 August 2017; (**c,d**), counter-clockwise pattern at the outlet of the nested watershed W1 on 1 August 2012; and (**e,f**), mixed loop pattern, watershed W2 on 6 July 2016.

The percentage occurrence of each typical loop for all the years in each watershed is shown in Figure 10, and the total number of loops for each year for all five watersheds is in Figure S1 in the Supplementary Material. The number of loops decreased over the years, because the discharge and the duration of the storms decreased so that there were fewer storms with five or more paired measurements of discharge and sediment concentrations. Of the storms analyzed, the mixed loops with no clear pattern between runoff and sediment concentration (Figure 10c) were most common for the early years and became approximately equal to the number of clockwise loops in 2015 and later (compare Figure 10a,c). The number of counter-clockwise loops decreased over the study period (Figure 10b).

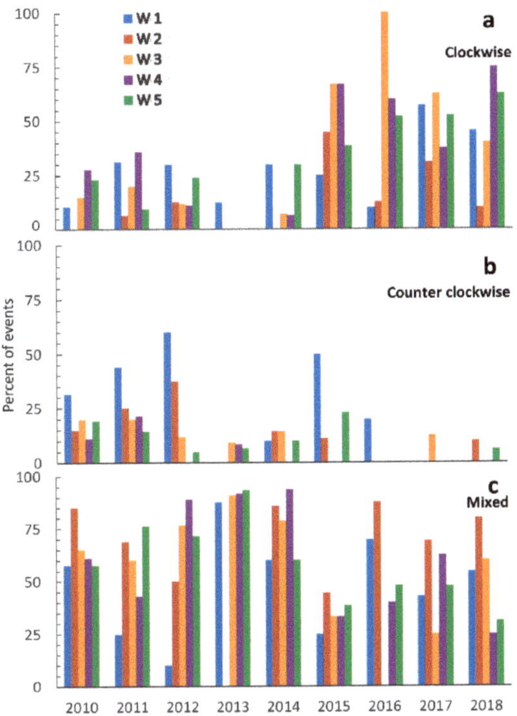

Figure 10. Hysteresis in the sediment concentration and discharge at the outlet of the four nested watersheds W1–W4) and the main watershed W5 of Debre Mawi watershed. (**a**) Clockwise loops; (**b**) Counter-clockwise loops, and (**c**) Mixed loops. The location of the watersheds is given in Figure 1.

3.5. Effect of Changes in the Watershed on Sediment Transport

3.5.1. Effect of SWCPs and Gullies on Sediment Transport

SWCPs: After the implementation of SWCPs, sediment concentrations and load significantly decreased in all the five watersheds. The decrease in sediment load, especially during the SWCPs implementation years (2012–2014), was caused by the decrease in runoff that has been trapped by the infiltration furrows. This agrees with [10] and [37] that indicated SWCPs were effective in the short-term in humid watersheds. However, in the long term, infiltration furrows were not effective in reducing soil loss because most infiltration furrows were filled with sediment, as shown in Figure 4a. The bunds concentrated the flow and caused rill formation downstream of the bunds (Figure 4c).

Gullies can affect sediment load and concentration. The sediment concentrations in the main watershed W5 were generally greater than the other nested watersheds because of the pickup of sediment in the gullies [16]. The 1500 m^2 gully, especially in watershed W5, was contributing to the large portion of catchment soil loss (Table 1). A recent study in the Debre Mawi watershed found that that 92% of the sediment at the outlet of a gully originated from the gully itself [23].

3.5.2. Effect SWCPs and Gullies on Discharge–Sediment Concentration Relationship

To understand the effect of SWCPs on discharge–sediment concentration relationship, we divided the storm event sediment loops during the nine-year period (shown in Figure 10) into three blocks: Period I (2010 and 2011) before SWCPs implementation; Period II, consisting of the years 2012, 2013, and 2014 during implementation of SWCPs; and Period III, covering the last four years 2015, 2016, 2017, and 2018 after SWCPs implementation and during road construction (Table 4). In Period I, before 2012, out of the 185 recorded storm events with five or more sediment concentration observations for all five watersheds, 17% were clockwise, 22% counter-clockwise, and 61% showed mixed patterns (Table 4). In Period II, of the 212 storm events, the mixed loops increased to 72% with approximately the same number of clockwise and counter-clockwise loops. After the implementation of SWCPs, of 193 events, the mixed loops reduced to 48%, clockwise 45%, and 7% counter-clockwise patterns.

Table 4. Percent of loop types before, during, and after installation of SWCPs.

Type of Loops	2010–2011	2012–2014	2015–2018
Clockwise	17	14	45
Counter-clockwise	22	15	7
Mixed	61	72	48

The large number of mixed hysteresis loops (Table 4) during the implementation of the soil and water conservation practices (Period II) indicates that unconsolidated soil from the newly constructed bunds was available throughout the rainstorm that could be transported by runoff. Thus, the sediment concentration was elevated during both the rising and falling limbs of the hydrograph, as shown in Figure 9e. In addition, the mixed loops were associated with the largest runoff events. It is not incidental that the longest-lasting storm in Figure 9 had a mixed loop.

The increase in the clockwise loops and the decrease in counter-clockwise loops from Period I to Period III in Table 4 indicates that a change in timing of when transport occurs from later in storm to earlier in storm that is partly due to less sediment available for transport as indicated by the temporally decreasing sediment concentrations (Table 3). The increase in clockwise can also be explained by the stored sediment in the channel from the previous rainstorm that becomes the source of sediment on the rising limb of the new event [24,43,44].

The counter-clockwise hysteresis pattern occurs when the sediment source is at a distance from the measurement location. Watershed W1 had the greatest number of counter-clockwise loops in Period I and the beginning of Period II (Figure 10b; Table 4). This watershed has a large grassy area in front of the weir, and agricultural land begins at 175 m above the weir. We hypothesize that the first runoff to reach the weir during the rainstorm is generated from the grassy area, which is perpetually saturated during the rainy season. This runoff water has a low sediment concentration because of the grass. The saturated area expands during the rainstorm and will expand in the cropland during a large enough storm, increasing the sediment concentration in the water. If the intensity of the rain decreases towards the end of a storm, the runoff will decrease, but the sediment concentration still will be high, generating a peak in the sediment concentrations after the peak of the runoff. The other watersheds do not show this pattern because their cropped area is much larger than any grassy area. Thus, the sediment signal of the grassy area is hidden by the rill erosion of cropped land.

Bottomland gullies could also affect discharge–sediment patterns. For instance, the sediment available through gully bank failure by the previous storm can result in a clockwise loop in that the sediment peak reaches the outlet before the discharge peak [23].

In addition to the SWCPs and gullies, another likely reason for our finding that soil loss was reduced over the course of the study period might be the hysteresis analysis method itself. For example, if we take a closer look at Figure 9e, it is obvious that the sediment supply becomes limiting at the end of the rainstorm because the flow increased while the sediment concentration remained low. This is unlike at the beginning of the storm. According to our classification scheme, this should be a clockwise loop, while the result of the analysis is a mixed loop (Figure 9f).

4. Conclusions

This study investigated, over nine years, the impact of soil and water conservation practices and the expansion of acreage of eucalyptus trees on the hydrology and soil loss in an agricultural watershed in the 95 ha Debre Mawi watershed in the Ethiopian highlands. The watershed is of volcanic origin, and lava dikes block the flow of subsurface water at several locations forcing the subsurface flow to the surface and causing periodically saturated areas. The saturated areas are the source of the surface runoff. Soil and water conservation practices consisted of 50 cm deep infiltration furrows with bunds downhill. Maintenance was poor, and infiltration furrows were filled up at the end of the study. The results show that the direct runoff and sediment decreased by a factor of two or three over the nine years. Although further studies are needed, it seemed that the evapotranspiration of eucalyptus during the dry season was mainly responsible for the reduction in the direct runoff by creating storage for the rainwater to infiltrate. This is in accordance with local knowledge that wetlands dry up after eucalyptus were planted. Soil loss reductions were mainly related to the smaller amounts of runoff.

The implication of the research is that soil and water conservation practices are effective in the short term but likely not as effective in the long-term after they are filled up with sediment. Expansion of eucalyptus trees in a watershed reduces direct runoff and erosion from saturated areas in the watersheds of the sub-humid Ethiopian highlands.

Supplementary Materials: The following are available online at http://www.mdpi.com/2073-4441/11/11/2299/s1. Figure S1. Number of loop types in time in the Debre Mawi Watershed over a nine-year period from 2010–2019.

Author Contributions: D.A.M., D.C.D., T.C.A., C.D.G., and S.A.T.; collected; analyzed; wrote the manuscript. D.C.D., S.A.T., B.F.Z., and T.S.S.; analyzed; supervised and advised all the research work that led to this paper. T.S.S. and B.F.Z. revised and conducted language editing.

Funding: Funding for this research was provided in part by the Belmont Forum NILE-NEXUS project through the United States NSF award ICER-1624335. Additional funding was also obtained from the Blue Nile Water Institute, Bahir Dar University.

Acknowledgments: We are grateful for the data made available to this research from the PEER Science project of Bahir Dar University.

Conflicts of Interest: The authors declare that they do not have a conflict of interest.

References

1. Paudyal, K.; Baral, H.; Bhandari, S.P.; Bhandari, A.; Keenan, R.J. Spatial assessment of the impact of land use and land cover change on supply of ecosystem services in Phewa watershed, Nepal. *Ecosyst. Serv.* **2019**, *36*, 100895. [CrossRef]
2. Romijn, E.; Coppus, R.; De Sy, V.; Herold, M.; Roman-Cuesta, R.M.; Verchot, L. Land Restoration in Latin America and the Caribbean: An Overview of Recent, Ongoing and Planned Restoration Initiatives and Their Potential for Climate Change Mitigation. *Forests* **2019**, *10*, 510. [CrossRef]
3. Taye, G.; Poesen, J.; Vanmaercke, M.; van Wesemael, B.; Martens, L.; Teka, D.; Nyssen, J.; Deckers, J.; Vanacker, V.; Haregeweyn, N.; et al. Evolution of the effectiveness of stone bunds and trenches in reducing runoff and soil loss in the semi-arid Ethiopian highlands. *Z. Geomorphol.* **2015**, *59*, 477–493. [CrossRef]
4. Haregeweyn, N.; Poesen, J.; Nyssen, J.; De Wit, J.; Haile, M.; Govers, G.; Deckers, S. Reservoirs in Tigray (Northern Ethiopia): Characteristics and sediment deposition problems. *Land Degrad. Dev.* **2006**, *17*, 211–230. [CrossRef]
5. Tamene, L.; Vlek, P.L. Soil erosion studies in northern Ethiopia. In *Land Use and Soil Resources*; Springer: Dordrecht, The Netherlands, 2008; pp. 73–100.
6. El-Swaify, S.A.; Hurni, H. Transboundary effects of soil erosion and conservation in the Nile basin. *Land Husb.* **1996**, *1*, 7–21.
7. Hurni, H.; Tato, K.; Zeleke, G. The implications of changes in population, land use, and land management for surface runoff in the upper Nile basin area of Ethiopia. *Mt. Res. Dev.* **2005**, *25*, 147–154. [CrossRef]

8. Nyssen, J.; Veyret-Picot, M.; Poesen, J.; Moeyersons, J.; Haile, M.; Deckers, J.; Govers, G. The effectiveness of loose rock check dams for gully control in Tigray, northern Ethiopia. *Soil Use Manag.* **2004**, *20*, 55–64. [CrossRef]

9. Hurni, H. Degradation and conservation of the resources in the Ethiopian highlands. *Mt. Res. Dev.* **1988**, *8*, 123–130. [CrossRef]

10. Dagnew, D.C.; Guzman, C.D.; Zegeye, A.D.; Tibebu, T.Y.; Getaneh, M.; Abate, S.; Zimale, F.A.; Ayana, E.K.; Tilahun, S.A.; Steenhuis, T.S. Impact of conservation practices on runoff and soil loss in the sub-humid Ethiopian Highlands: The Debre Mawi watershed. *J. Hydrol. Hydromech.* **2015**, *63*, 210–219. [CrossRef]

11. Zimale, F.A.; Moges, M.A.; Alemu, M.L.; Ayana, E.K.; Demissie, S.S.; Tilahun, S.A.; Steenhuis, T.S. Budgeting suspended sediment fluxes in tropical monsoonal watersheds with limited data: The Lake Tana basin. *J. Hydrol. Hydromech.* **2018**, *66*, 65–78. [CrossRef]

12. Tebebu, T.Y.; Steenhuis, T.S.; Dagnew, D.C.; Guzman, C.D.; Bayabil, H.K.; Zegeye, A.D.; Collick, A.S.; Langan, S.; MacAlister, C.; Langendoen, E.J.; et al. Corrigendum: Improving efficacy of landscape interventions in the (sub) humid Ethiopian highlands by improved understanding of runoff processes. *Front. Earth Sci.* **2015**, *3*, 57. [CrossRef]

13. Herweg, K.; Ludi, E. The performance of selected soil and water conservation measures—Case studies from Ethiopia and Eritrea. *Catena* **1999**, *36*, 99–114. [CrossRef]

14. Zegeye, A.D.; Langendoen, E.J.; Stoof, C.R.; Tilahun, S.A.; Dagnew, D.C.; Zimale, F.A.; Guzman, C.D.; Steenhuis, T.S. Morphological dynamics of gully systems in the subhumid Ethiopian Highlands: The Debre Mawi watershed. *Soil* **2016**, *2*, 443–458. [CrossRef]

15. Guzman, C.D.; Tilahun, S.A.; Dagnew, D.C.; Zegeye, A.D.; Tebebu, T.Y.; Yitaferu, B.; Steenhuis, T.S. Modeling sediment concentration and discharge variations in a small Ethiopian watershed with contributions from an unpaved road. *J. Hydrol. Hydromech.* **2017**, *65*, 1–17. [CrossRef]

16. Tebebu, T.Y.; Abiy, A.Z.; Zegeye, A.D.; Dahlke, H.E.; Easton, Z.M.; Tilahun, S.A.; Collick, A.S.; Kidnau, S.; Moges, S.; Dadgari, F.; et al. Surface and subsurface flow effect on permanent gully formation and upland erosion near Lake Tana in the northern highlands of Ethiopia. *Hydrol. Earth Syst. Sci.* **2010**, *14*, 2207–2217. [CrossRef]

17. Abiy, A.Z. Geological Controls in the Formations and Expansions of Gullies over hillslope Hydrological Processes in the Highlands of ETHIOPIA, Northern Blue Nile Region. MPS Thesis, Cornell University, Ithaca, NY, USA, 2009.

18. Tebebu, T.Y.; Bayabil, H.K.; Stoof, C.R.; Giri, S.K.; Gessess, A.A.; Tilahun, S.A.; Steenhuis, T.S. Characterization of degraded soils in the humid Ethiopian highlands. *Land Degrad. Dev.* **2017**, *28*, 1891–1901. [CrossRef]

19. Seeger, M.; Errea, M.P.; Beguería, S.; Arnáez, J.; Martı, C.; Garcıa-Ruiz, J.M. Catchment soil moisture and rainfall characteristics as determinant factors for discharge/suspended sediment hysteretic loops in a small headwater catchment in the Spanish Pyrenees. *J. Hydrol.* **2004**, *288*, 299–311. [CrossRef]

20. Williams, G.P. Sediment concentration versus water discharge during single hydrologic events in rivers. *J. Hydrol.* **1989**, *111*, 89–106. [CrossRef]

21. Dickinson, A.; Bolton, P. A Programme of monitoring sediment transport in north central Luzon, the Philippines. In *Erosion and Sediment Transport Monitoring Programmes in River Basins*; IAHS Publ.: Wallingford, UK, 1992; pp. 483–492.

22. Wood, P.A. Controls of variation in suspended sediment concentration in the River Rother, West Sussex, England. *Sedimentology* **1977**, *24*, 437–445. [CrossRef]

23. Zegeye, A.D.; Langendoen, E.J.; Guzman, C.D.; Dagnew, D.C.; Amare, S.D.; Tilahun, S.A.; Steenhuis, T.S. Gullies, a critical link in landscape soil loss: A case study in the subhumid highlands of Ethiopia. *Land Degrad. Dev.* **2017**, *29*, 1222–1232. [CrossRef]

24. Bacča, P. Hysteresis effect in suspended sediment concentration in the Rybárik basin, Slovakia/Effet d'hystérèse dans la concentration des sédiments en suspension dans le bassin versant de Rybárik (Slovaquie). *Hydrol. Sci. J.* **2008**, *53*, 224–235. [CrossRef]

25. Walling, D.E. Assessing the accuracy of suspended sediment rating curves for a small basin. *Water Resour. Res.* **1977**, *13*, 531–538. [CrossRef]

26. Goodwin, T.H.; Young, A.R.; Holmes, M.G.; Old, G.H.; Hewitt, N.; Leeks, G.J.; Packman, J.C.; Smith, B.P. The temporal and spatial variability of sediment transport and yields within the Bradford Beck catchment, West Yorkshire. *Sci. Total Environ.* **2003**, *314*, 475–494. [CrossRef]

27. Steegen, A.; Govers, G.; Nachtergaele, J.; Takken, I.; Beuselinck, L.; Poesen, J. Sediment export by water from an agricultural catchment in the Loam Belt of central Belgium. *Geomorphology* **2000**, *33*, 25–36. [CrossRef]

28. Ayele, G.K.; Gessess, A.A.; Addisie, M.B.; Tilahun, S.A.; Tebebu, T.Y.; Tenessa, D.B.; Langendoen, E.J.; Nicholson, C.F.; Steenhuis, T.S. A biophysical and economic assessment of a community-based rehabilitated gully in the Ethiopian highlands. *Land Degrad. Dev.* **2016**, *27*, 270–280. [CrossRef]

29. FAO. *Highlands Reclamation Study: Ethiopia*; Final Report, Executive Summary and 2 vols; Food and Agriculture Organization of the United Nations: Rome, Italy, 1986.

30. Tilahun, S.A.; Guzman, C.D.; Zegeye, A.D.; Engda, T.A.; Collick, A.S.; Rimmer, A.; Steenhuis, T.S. An efficient semi-distributed hillslope erosion model for the sub-humid Ethiopian Highlands. *Hydrol. Earth Syst. Sci.* **2013**, *17*, 1051–1063. [CrossRef]

31. Alemie, T.C.; Tilahun, S.A.; Ochoa-Tocachi, B.F.; Schmitter, P.; Buytaert, W.; Parlange, J.-Y.; Steenhuis, T.S. Predicting shallow groundwater tables for sloping highland aquifers. *Water Resour. Res.* **2019**, in press.

32. IUSS-WRB; 2015 IUSS (International Union of Soil Sciences) Working Group; WRB (World Reference Base). *World Reference Base for Soil Resources 2014, Update 2015: International Soil Classification System for Naming Soils and Creating Legends for Soil Maps*; World Soil Resources Reports No. 106; FAO: Rome, Italy, 2015.

33. Tilahun, S.A.; Guzman, C.D.; Zegeye, A.D.; Dagnew, D.C.; Collick, A.S.; Yitaferu, B.; Steenhuis, T.S. Distributed discharge and sediment concentration predictions in the sub-humid Ethiopian highlands: The Debre Mawi watershed. *Hydrol. Process.* **2015**, *29*, 1817–1828. [CrossRef]

34. Chanson, H. *Hydraulics of Open Channel Flow*; Elsevier: Amsterdam, The Netherlands, 2004.

35. Mann-Kendall, M.G. *Rank Correlation Measures*; Charles Griffin: London, UK, 1975; Volume 202, p. 15.

36. Tilahun, S.A.; Ayana, E.K.; Guzman, C.D.; Dagnew, D.C.; Zegeye, A.D.; Tebebu, T.Y.; Yitaferu, B.; Steenhuis, T.S. Revisiting storm runoff processes in the upper Blue Nile basin: The Debre Mawi watershed. *Catena* **2016**, *143*, 47–56. [CrossRef]

37. Guzman, C.D.; Zimale, F.A.; Tebebu, T.Y.; Bayabil, H.K.; Tilahun, S.A.; Yitaferu, B.; Rientjes, T.H.; Steenhuis, T.S. Modeling discharge and sediment concentrations after landscape interventions in a humid monsoon climate: The Anjeni watershed in the highlands of Ethiopia. *Hydrol. Process.* **2017**, *31*, 1239–1257. [CrossRef]

38. Enku, T.; Melesse, A.; Ayana, E.; Tilahun, S.; Abate, M.; Steenhuis, T. April. Response of Groundwater table to Eucalyptus Plantations in a Tropical Monsoon Climate, Lake Tana Basin, Ethiopia. In Proceedings of the EGU General Assembly Conference Abstracts 2017, Vienna, Austria, 23–28 April 2017; Volume 19, p. 4652.

39. Chanie, T.; Collick, A.S.; Adgo, E.; Lehmann, C.J.; Steenhuis, T.S. Eco-hydrological impacts of Eucalyptus in the semi humid Ethiopian Highlands: The Lake Tana Plain. *J. Hydrol. Hydromech.* **2013**, *61*, 21–29b. [CrossRef]

40. Jaleta, D.; Mbilinyi, B.P.; Mahoo, H.F.; Lemenih, M. Effect of Eucalyptus expansion on surface runoff in the central highlands of Ethiopia. *Ecol. Process.* **2017**, *6*, 1. [CrossRef]

41. Tilahun, S.A.; Mukundan, R.; Demisse, B.A.; Engda, T.A.; Guzman, C.D.; Tarakegn, B.C.; Easton, Z.M.; Collick, A.S.; Zegeye, A.D.; Schneiderman, E.M.; et al. A saturation excess erosion model. *Trans. ASABE* **2013**, *56*, 681–695. [CrossRef]

42. Liu, B.M.; Collick, A.S.; Zeleke, G.; Adgo, E.; Easton, Z.M.; Steenhuis, T.S. Rainfall-discharge relationships for a monsoonal climate in the Ethiopian highlands. *Hydrol. Process.* **2008**, *22*, 1059–1067. [CrossRef]

43. House, W.A.; Warwick, M.S. Hysteresis of the solute concentration/discharge relationship in rivers during storms. *Water Res.* **1998**, *32*, 2279–2290. [CrossRef]

44. Asselman, N.E. Suspended sediment dynamics in a large drainage basin: The River Rhine. *Hydrol. Process.* **1999**, *13*, 1437–1450. [CrossRef]

Article

Geomorphodynamics in Argan Woodlands, South Morocco

Mario Kirchhoff [1,*]**, Lars Engelmann** [1]**, Lutz Leroy Zimmermann** [1]**, Manuel Seeger** [1]**, Irene Marzolff** [2]**, Ali Aït Hssaine** [3] **and Johannes B. Ries** [1]

[1]   Department of Physical Geography, Trier University, DE-54286 Trier, Germany; s6laenge@uni-trier.de (L.E.); s6luzimm@uni-trier.de (L.L.Z.); seeger@uni-trier.de (M.S.); riesj@uni-trier.de (J.B.R.)

[2]   Department of Physical Geography, Goethe University Frankfurt am Main, DE-60438 Frankfurt am Main, Germany; marzolff@em.uni-frankfurt.de

[3]   Department of Geography, Université Ibn Zohr, MA-80060 Agadir, Morocco; aithssaine55@gmail.com

*   Correspondence: kirchhoff@uni-trier.de

Received: 30 August 2019; Accepted: 16 October 2019; Published: 22 October 2019

Abstract: The endemic argan tree (*Argania spinosa*) populations in South Morocco are highly degraded due to their use as a biomass resource in dry years and illegal firewood extraction. The intensification and expansion of agricultural land lead to a retreat of the wooded area, while the remaining argan open woodlands are often overgrazed. Thus, canopy-covered areas decrease while areas without vegetation cover between the argan trees increase. In total, 36 rainfall simulation experiments as well as 60 infiltration measurements were conducted to investigate the potential difference between tree-covered areas and bare intertree areas. In addition, 60 soil samples were taken under the trees and in the intertree areas parallel to the contour lines. Significant differences using a *t*-test were found between tree and intertree areas for the studied parameters K_{sat}, K_h, pH, electric conductivity, percolation stability, total C-content, total N-content, K-content, Na-content, and Mg-content. Surface runoff and soil losses were not as conclusive but showed similar trends. The results showed that argan trees influence the soil underneath significantly, while the soil in intertree areas is less protected and more degraded. It is therefore reasonable to assume further degradation of the soil when intertree areas extend further due to lack of rejuvenation of argan trees.

Keywords: soil erosion; argan; South Morocco; soil degradation; tree; intertree

---

1. Introduction

Soil erosion is a serious issue which endangers sustainable land use strategies, especially in arid and semiarid regions [1–3]. Water and wind as erosive factors degrade the land, their importance varies depending on the region [4,5]. In Morocco, water is often the crucial erosive force [5]. Most of the time only a few strong events are responsible for the majority of soil losses [6,7]. Due to the small amount of precipitation coupled with its great variability, rangeland is more common in these arid regions than farmland. Therefore, agrosilvopastoral systems are the most typical land uses. These forms of land use often become established in the dryland forests which occur naturally in these areas [3,7]. Forest degradation as a consequence of erosion and mismanagement is a common phenomenon in these dryland forests. It has been recorded all over the world (e.g., in Spanish dehesas [8], Algerian oak forests [9], Moroccan argan woodlands [3], and Mongolian forest steppes [10]). While the degradation dynamics are well understood in some regions, there is a lack of knowledge elsewhere. For example, in the Spanish dehesas, there are studies regarding the interrelations between rainfall intensity, vegetation cover, grazing and soil loss [6], the influence of vegetation on moisture conditions in the soil [7] and the connection between soil conditions, vegetation cover, and pasture production [2]. Furthermore, the influence of patchy vegetation on soil conditions beneath is highlighted around the world [11–16]. Typically, there is an enrichment of organic matter and nutrients under plants. Therefore the term "fertile islands" is often used for patchy vegetation in dry regions [14,15]. It is likely that these "fertile islands" are better protected against soil loss, because the input of organic matter in the soil is crucial for its resistance against erosion [17–20].

Corresponding studies for the Moroccan argan woodlands investigating the interaction between soil, vegetation, and erosion processes are missing, although protection of argan trees and land-use strategies around them have been stipulated by UNESCO since 2014. Reasons for this are the complex socioeconomic structures behind the argan economy and argan oil which is in growing demand around the world [21]. A massive decline of argan tree density from 27 to 15 trees ha^{-1} between 1970 and 2007 shows that protection of these dryland forests is necessary to preserve them [3]. *Argania spinosa* is endemic in southern Morocco and has a small dispersal area of ~950,000 ha [3] which is mostly limited to the Souss basin [22]. Although *Argania spinosa* is well adapted to the dry conditions of Morocco's southwest, the population is endangered because it is the only source of forage in dry periods [3,22]. In consequence, there is a high pressure of use by the local population and their goat herds, as well as by nomadic tribes and their goat and dromedary herds [23]. In addition, the fruits of *Argania spinosa* are collected to produce argan oil, which is sold for cosmetic and culinary usage [3]. In other areas, access for animals is not permitted and speculative rainfed agriculture is practiced [23]. Especially overgrazing and deforestation lead to varying tree-crown sizes and growth forms, smaller numbers of trees per hectare, and a lower proportion of covered soil [23,24] (Figure 1). As a result, there is less protection against surface runoff, desiccation is smaller, soils become more degraded, young sprouts cannot grow, and thus, the vegetation cover further declines.

Figure 1. (**a**,**b**) Varying argan tree densities as seen in vertical aerial photographs (each image in (**a**) and (**b**) shows 1 ha) as well as different growth forms of *Argania spinosa*: (**c**) round crown; (**d**) umbrella-shaped crown with basal cushion; (**e**) cone-shaped cushion.

While various studies have investigated the degradation of argan trees [3,22,25,26], there is a lack of knowledge regarding the connection between vegetation degradation and soil degradation as well as soil loss. Other studies have already shown that there are strong erosive processes in the Souss valley [27]. Usage on the sites is not only limited to the area between the trees. The soil is often cultivated between the trees and under the tree crowns. Herds of goats, sheep, and dromedaries graze and browse the tree area most of all, thus possibly compacting the soil beneath the crown as was demonstrated by Mulholland and Fullen on loamy sands and Stavi et al. [28,29]. It is unknown if the soils beneath argan trees are already so degraded that, should the tree be removed, young sprouts would not be able to grow. If this is already the case in the intertree areas, a regeneration even by afforestation is unlikely to be successful. The main aim of this study is to analyze the interaction between argan trees and multiple soil parameters, such as C-content and N-content, as indicators for erosion stability. Our main hypothesis is that the soils in the tree area are more protected from rain, splash erosion, and surface wash and thus show less degraded soils than the unprotected intertree areas. Consequently, the securing of adequate tree density would be an important step towards sustainable land management in southern Morocco. Yet, the soils beneath the trees have been the subject of potentially degrading factors. If a difference between the soil parameters exists, the extent to which they differ is an important indicator for the degradation of soils in the intertree and tree areas. To test this hypothesis, measurements of erodibility and soil characteristics were taken under trees and in corresponding intertree areas at 30 test sites of 1 ha each and statistically analyzed.

2. Materials and Methods

2.1. Study Area

The study areas are located in the western part of the Souss basin in southwest Morocco (Figure 2).

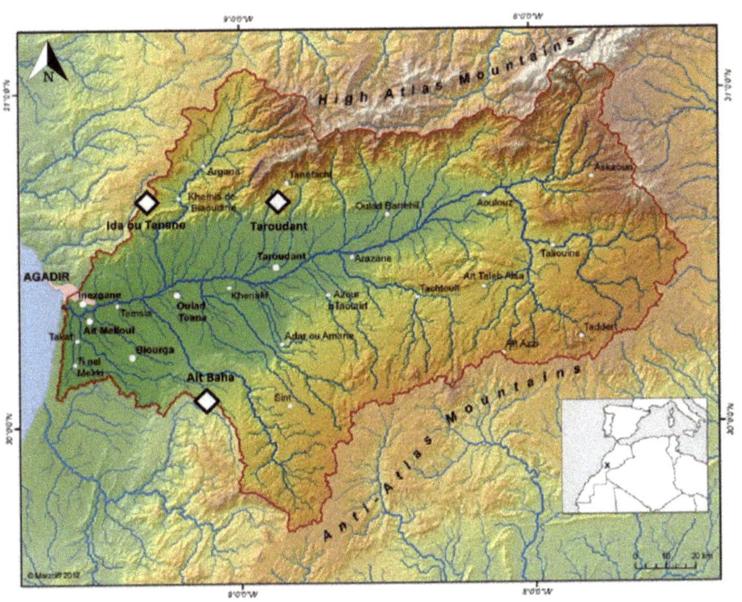

Figure 2. Study areas in the Souss basin (modified from d'Oleire-Oltmanns [30]).

The Souss basin, between 30° and 31° northern latitude and 9° and 7° western longitude, is contained by the High Atlas Mountains in the North, the Anti-Atlas Mountains in the South, and the connection of both mountain regions in the Siroua formation in the East and the Atlantic Ocean in the

West. Its east–west extension stretches for about 200 km with a total catchment area of approximately 16,000 km^2 [31], making it a transition zone between the Atlantic coastal landscape and the Sahara Desert [32]. Palaeozoic and Mesozoic structures of the High Atlas and Precambrian structures of the Anti-Atlas are separated by the Atlasic fault in the Siroua region. Igneous and metamorphic rocks crop out in the western and central Anti-Atlas (Siroua region). On the piedmont of the High Atlas, as well as in the Souss basin, a Cretaceous–Eocene succession is overlain by thick Pliocene and Quaternary deposits, including significant fluvial–lacustrine sequences [33], which again are overlain by coarse-grained, often carbonate-encrusted, alluvial deposits on the fans [34]. A precise geomorphological characterization of this area was published by Aït Hssaine and Bridgland [35] and a further detailed description of the geomorphological evolution of the foothills of the High Atlas near Taroudant was given by Aït Hssaine [36]. Intense morphodynamics are permanently reshaping the substrate. Mostly immature profound raw soils with mixed substrate and slow pedogenesis are found, with high proportions of fine sand, silt, and clay [37]. Ghanem [38] differentiated three different soil types using the French soil classification. According to the World Reference Base (IUSS-WRB 2015) they are classified as Fluvisols, Regosols as well as a mixture of the first soil type and Regosols—soils from different origins, phases, and soil associations—which contain a mixture of sand, silt, and clay [37,38]. Loams and sandy loams are found most often on the test sites.

Vegetation cover consists of subtropic, desert, and Mediterranean species. In addition to *Argania spinosa*, also *Acacia*, *Euphorbia*, *Artemisia herba-alba*, and *Ziziphus* are common in the south Moroccan bush and shrub landscape. Gramineous species include *Dactylus glomerata*, *Cynodo dactylon*, and *Andropogon hirtus*. In years with high precipitation *Tamarix*, *Salicornia*, and *Salsola* can also grow. For the last several decades, a highly dynamic land-use change in the Souss basin has been taking place, accompanied by labor migration [39]. Land use is dominated by citrus fruit and banana plantations which are irrigated by deep wells [37]. To use the percolating water efficiently, irrigated fields are often additionally surrounded by trees such as apple, walnut, almond, olives, and date palms [40].

The three study areas are all part of the Souss region. They are called Ida-Outanane, Taroudant, and Aït Baha. In these study areas, 30 test sites were chosen with an extent of 100 × 100 m. There are 6 in Ida-Outanane, 11 in Taroudant, and 13 in Aït Baha. Since argan trees grow in the High Atlas, as well as the Anti-Atlas, and in continental or maritime climate, the study areas were chosen to encompass these different settings. Ida-Outanane and Taroudant are both situated on the southern foothills of the High Atlas yet they differ in climate. In general, climate is characterized by a hot dry summer and a mild winter with short rain periods with high variability, as typical for B climates in the Köppen and Geiger climate classification [41]. Precipitation rates range from less than 300 mm y^{-1} in the center plains of the Souss basin to more than 500 mm y^{-1} toward the High-Atlas and the Siroua summits. Like rainfall intensities, the temperature range highly depends on the altitude with a mean annual temperature of 20 °C [31]. Ida-Outanane is influenced by a maritime climate with a mean annual temperature of 14 °C and nearly 400 mm precipitation annually, while Taroudant is further inland with a more continental climate and 20 °C annual temperature and 220 mm precipitation [42]. Aït Baha lies in the northern foothills of the Anti-Atlas. Although closer to the Atlantic Ocean than Taroudant it is also influenced by a more continental climate with 16 °C annual temperature and nearly 300 mm precipitation. This is accounted for by a higher dependence on altitude for precipitation and temperature rather than distance to the ocean as national meteorological data shows [43]. Precipitation varies unevenly in between years with the most chance of rainfall in the late autumn to early spring. Kirchhoff et al. showed precipitation data from 2011–2014 for the study site of Taroudant with varying rainfall [27]. To ensure comparability across the three study areas in spite of the differences in altitude, climate conditions, and soil types, land-use/environmental characteristics were included as classification parameters for grouping the test sites (see Figure 3; Table 1). These land-use/environmental attributes were used to differentiate two or three neighboring test sites by their principal characteristics. The attributes were cultivated (c; silvo-agricultural land use with recent or anterior signs of ploughing; 5 test sites), afforestation (a; 2 test sites), excluded (x; fenced, no grazing allowed; 3 test sites), browsed (b; excessive browsing leading to cone-shaped growth of trees/bushes; 1 test site), dense (d; dense growth due to large crowns or numerous trees; 2 test sites), gullies (g; site strongly incised by gullies; 1 test site), logged (l; once covered with argan trees, but now bare; 1 test site), rock fragment cover (r; high rock fragment cover >80%; 1 test site), steep slopes (s; steeper than 10°; 1 test site), terraced (t; 2 test sites), and nonspecific (n; silvopastoral land use; 11 test sites). It is important to stress that the test sites do not always conform to only one characteristic (e.g., some excluded test sites can also be situated on steep slopes), but the exclusion is important to differentiate between the two sites.

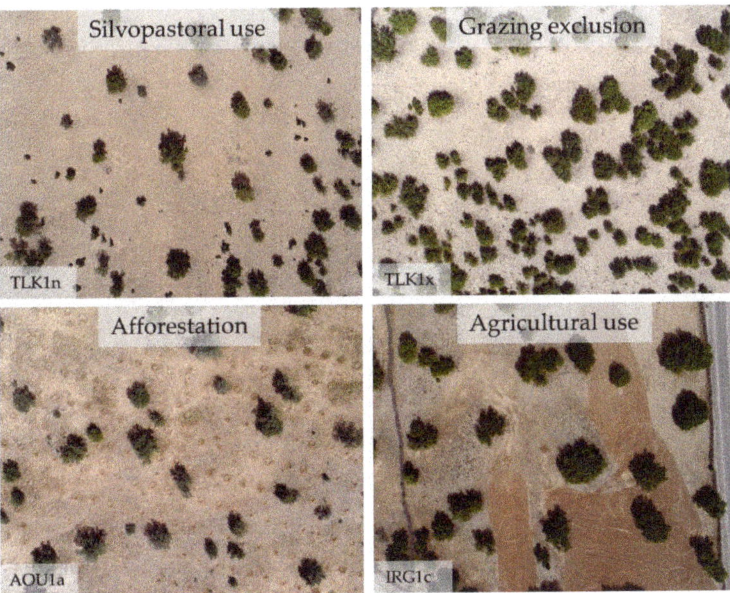

Figure 3. Examples for argan test sites with different land-use/environmental attributes of argan test sites. Vertical small-format aerial photography taken with an unmanned aerial vehicle from a height of 120 m (TLK: Tamaloukt; AOU: Aouarga; IRG: Irguitène).

Table 1. Test sites, principal attributes, and measurements on test sites (ABA: Aït ben Ali; ABH: Aït Baha; AOU: Aouarga; MAO: Maouriga; SHY: Sidi el Haij Yahia; TAS: Tasakat; BOU: Boulaajlate; IRG: Irguitène; TLK: Tamaloukt; AZR: Azrarag; OUF: Tamaït Oufella, X: method was applied to site).

Site	Principal Attribute	Study Area	Rainfall Simulations	Infiltration Measurements	Soil Analysis
ABA1n	silvopastoral	Aït Baha	X	X	X
ABA1s	steep slope >10°	Aït Baha	X	X	X
ABH1n	silvopastoral	Aït Baha	X	X	X
ABH1x	grazing exclusion	Aït Baha	X	X	X
ABH2c	cultivated	Aït Baha	X	X	X
ABH2t	terraced	Aït Baha	X	X	X
AOU1a	afforestation	Aït Baha		X	X
AOU2a	afforestation	Aït Baha		X	X
AOU2x	grazing exclusion	Aït Baha		X	X
MAO1n	silvopastoral	Aït Baha	X	X	X
MAO2t	terraced	Aït Baha	X	X	X
SHY1g	gullies	Aït Baha	X	X	X
TAS1r	rock fragment cover	Aït Baha	X	X	X
BOU1b	browsed	Taroudant	X	X	X
BOU1n	silvopastoral	Taroudant	X	X	X

Table 1. *Cont.*

Site	Principal Attribute	Study Area	Rainfall Simulations	Infiltration Measurements	Soil Analysis
IRG1c	cultivated	Taroudant		X	X
IRG1n	silvopastoral	Taroudant		X	X
IRG2d	dense	Taroudant		X	X
IRG2n	silvopastoral	Taroudant		X	X
IRG3c	cultivated	Taroudant		X	X
IRG3l	logged	Taroudant		X	X
IRG3n	silvopastoral	Taroudant		X	X
TLK1n	silvopastoral	Taroudant	X	X	X
TLK1x	grazing exclusion	Taroudant	X	X	X
AZR1c	cultivated	Ida-Outanane	X	X	X
AZR1n	silvopastoral	Ida-Outanane	X	X	X
AZR2c	cultivated	Ida-Outanane	X	X	X
AZR2n	silvopastoral	Ida-Outanane	X	X	X
OUF1d	dense	Ida-Outanane		X	X
OUF1n	silvopastoral	Ida-Outanane		X	X

To test the hypothesis, rainfall simulation and infiltration experiments were conducted, and soil samples were taken in pairs (in tree and intertree areas). Each sample pair consists of one sample beneath an argan tree and one beyond this tree's crown in the intertree area. The location for a tree sample was between the trunk and the canopy edge, while the intertree sample was taken halfway between the sample tree and its closest neighbor, parallel to the contour lines. The trees chosen for analysis were as representative as possible for the surrounding area regarding their size, degradation status, and the distance between the sampling tree and its neighbor. The intertree areas were mostly bare with a rather high stone cover, while the tree areas showed vegetation cover in various percentages.

2.2. Rainfall Simulations

A small modified nozzle-type rainfall simulator was used [44], for which a detailed description is available in Iserloh et al. [45]. The test plots were circular with a diameter of about 60 cm and the total area encompassed 0.28 m^2. The plots were sprayed from a height of 2 m using a nozzle of the type Lechler 460.608. A rainfall intensity of 40 mm h^{-1} was calibrated and maintained during the experiment by managing the flow control. This rainfall intensity (including kinetic energy and drop size distribution) and calibration were described by Iserloh et al. [45,46] to obtain reproducible and comparable results to regions with differing natural rainfall intensities. Calibrations were carried out in the beginning and end of the experiment. The intensity needed to be constant (the difference should not exceed 5–10%) to qualify as a successful experiment. The duration of each experiment was 30 min. A total of 36 rainfall simulation experiments were carried out between October 2018 and February 2019; corresponding tree and intertree tests for any site were conducted on the same day to minimize changes in external conditions. Of the 30 test sites, 18 sloped test sites were chosen (>1°), since on the remaining 12 flat test sites with slopes from 0–1° surface runoff would not have moved downslope; rather small changes in roughness would have made measurement of surface runoff and soil erosion difficult and not representative of the test site. Simulating rainfall under trees the same way as on bare intertree areas was not possible, yet the results of the rainfall simulations give information about the erodibility of the soils underneath tree cover and in between trees.

Before the experiment, plot characteristics such as slope, vegetation, and stone cover, and antecedent soil moisture were measured. Soil surface roughness was measured with the chain method. To obtain the chain roughness (Cr) index, the following equation (Equation (1)) was used [47,48]:

$$Cr = (1 - L2/L1) \times 100, \tag{1}$$

where L1 (m) is the distance over the surface and L2 (m) is the distance from one end of the plot to the other measured with a ruler.

The 30 min experiment was divided into 6 intervals of 5 min. The total runoff and eroded sediment were sampled in PE (polyethylene) bottles, which were changed at the start of every interval. The bottles were weighed in the laboratory, the runoff was calculated by subtracting the weight of the empty bottle and the weight of the sediment. Each bottle was filtered separately with circular fine-meshed filter papers to obtain the amount of eroded sediment for each interval.

2.3. Infiltration Measurements

In addition to the rainfall simulations, two different methods of infiltration measurements were used to obtain information about both infiltration rates [49] and hydraulic conductivity in saturated and unsaturated states. These measurements were carried out in the nearest possible distance and at the same time or directly before/after the rainfall simulations for a high comparability of soil properties. On 30 test sites 60 infiltration measurements with one constant-head single-ring infiltrometer and 480 hydraulic conductivity measurements with two tension-disc infiltrometers in total were conducted over two measurement periods of several weeks in autumn/winter 2018 and spring 2019.

2.3.1. Constant-Head Single-Ring Infiltrometer

The infiltrometer used (Figure 4a) consists of an iron ring with a diameter of 15 cm that is carefully inserted vertically at least 5 cm into the soil to minimize the disturbance of the plot. The ring is topped with a float assembly linked to a water column to assure a constant ponding height of 5 cm. The ponding height of 5 cm is generally the chosen height for infiltration measurements due to the required correction calculations (single-ring measurements overestimate actual infiltration rates primarily due to unaccounted lateral seepage) as given by Tricker [50]. The constant water output due to the float assembly differs from the commonly used single-ring infiltrometers after Hills [51]; a detailed description of this method is given by Link [52]. The experiment starts when the ponding height is reached and takes 60 min. The plot cover was recorded before (Figure 4b) and profile and soil samples were taken after the experiment (Figure 4c). Regular tap water was used. Measurements of the sinking water column were done in intervals differing between 5 s at the beginning to 5 min in the last 40 min of the experiment, as was done by Peter and Ries [49]. Steady-state flow rates were attained mostly within the first 10 min of the experiment. Knowing the amount of infiltrated water over the time intervals with the dimensions of the ring and water tank, infiltration rates can be easily calculated [53]. With these recorded infiltration rates the coefficient of permeability could be calculated using the method of Elrick and Reynolds [54].

(a) (b) (c)

Figure 4. (a) Single-ring infiltrometer; (b) plot record before the experiment; (c) profile of the infiltration plot after experiment.

2.3.2. Tension-Disc Infiltrometer

To measure the unsaturated hydraulic conductivity of the soils, the "minidisk" tension-disc infiltrometer by Decagon Devices, METER Group Inc. (METER Group Inc., Munich, Germany) was used (Figure 5a). The tube-shaped device is divided into two water chambers, the lower part ending in a sintered steel disc using the principle of a Mariotte's bottle and the upper part or bubble chamber controls the negative pressure. By being able to adjust the pressure or suction it is possible to eliminate the effect of macropores with an air entry volume smaller than the suction. With the water under tension it does not enter the macropores, but rather flows deeper into the soil as determined by the hydraulic forces [55]. The sintered steel disc has a diameter of 4.5 cm, and the 20-cm high water reservoir can hold up to 135 mL water. Regular tap water was used as well. Other than single-ring measurements, an experiment with the minidisk infiltrometer just takes 15 min for one suction rate and uses a smaller amount of water; therefore, it was possible to carry out more runs in the same amount of time. A total of 480 measurements were recorded, with suctions of −4, −2, −1, and −0.5 cm on every plot. Two measurements were conducted for every plot and every suction for tree and intertree areas on all 30 test sites. It is worth mentioning that a thin layer of fine sand (Figure 5b) was added to level the plot underneath the disc and guarantee full contact and an optimal hydraulic connection to the soil surface (see also Perroux and White [56]). The change in water column was read every minute in the beginning to every 3 min at the end of the experiment. To determine the hydraulic conductivity from the collected data the method of Zhang [55] was chosen. It requires measuring cumulative infiltration versus time, which uses the slope of the curve of the cumulative infiltration versus the square root of time C_1 divided by the van Genuchten parameter A [57] for the sampled soil:

$$k = \frac{C_1}{A} \tag{2}$$

The van Genuchten parameters for 12 different texture classes were obtained from Carsel and Parrish [58].

(a)   (b)

Figure 5. (**a**) Minidisk infiltrometer; (**b**) plot record after the experiment, with a thin layer of fine sand added for improved hydraulic connection.

The single-ring infiltrometer measures maximum infiltration capacity and saturated hydraulic conductivity of the soil using a constant ponding head, pressing water into the soil in a three-dimensional way [49], while the tension-disk infiltrometer measures soil matric potential with a negative suction excluding the influence of macropores in an unsaturated state. With these methods it is possible to determine the unsaturated hydraulic conductivity from tension-disc infiltrometer data and thus to compare saturated and unsaturated hydraulic conductivities of the soils (see also Reynolds et al. [59]).

2.4. Soil Analyses

Disturbed paired soil samples were taken at the surface up to a depth of 5 cm. The soil samples were air-dried. Afterwards the fine soil was separated from the coarse soil by dry sieving to 2 mm. The soil material referred to in the following section is this separated fine fraction. The pH value was measured in a solution with 0.01 M $CaCl_2$. For the determination of the electrical conductivity of each soil sample the method described in DIN ISO 11265:1997-06 [60] has been followed. The WTW Multi 3410 Set Tetra Con was used as a measuring device for the pH value and electrical conductivity. Furthermore, the particle size distribution of the fine fraction was determined, using the pipette analysis as described by Köhn [61].

The percolation stability was measured as proposed by Becher and Kainz [62] and described further by Auerswald [20] and Becher [63]: 10 g of air-dried, 1–2 mm aggregates were put in plexiglass tubes. Both ends of the tube were filled with a thin layer of sand. To ensure tight and homogenous packing, the tube was tapped 20 times from a height of 2 cm onto a hard surface. Demineralized water was used at a constant head of 20 hPa, which was maintained by a Mariotte bottle. This bottle sat on a balance, continuously measuring and sending data to a connected computer. With the inflow of water into the tube, inflow rates could be determined every 10 s for a total runtime of 10 min. When measuring a high throughflow the aggregates remained stable and maintained bigger pores; with low throughflow aggregates broke apart and started slaking, causing the water to only flow through the smallest pores [20,63]. Afterwards, the percolation stability was corrected for total sand as proposed by Mbagwu and Auerswald [64].

The total soil–carbon concentration and the nitrogen concentration were analyzed with the Euro CHNS Elemental Analyzer 3000 built by HEKAtech. The dry combustion took place in a concentrated oxygen atmosphere with the addition of helium as a carrier gas at 1010 °C. To ensure accurate results, each sample was measured twice and after 10 measurements, two standards with known carbonate content were measured to verify the calibration. Other studies have already shown a link between the presence of vegetation and the concentration of total carbonate and nitrogen (e.g., in Spanish and Chinese soils [15,65]). Considering random preliminary investigations, the determination of the organic carbon content was dispensed with in this paper. Most of the measured total carbonate has an organic origin; inorganic carbon ranged from 0%–13% of the total carbon.

The content of exchangeable cations was determined using the flame-emission spectrophotometer AA240 built by Varian. To measure the Ca^{2+}, Mg^{2+}, K^+, and Na^+ content, these exchangeable cations were replaced with NH_4Cl. A determination of the exchangeable H^+ ions did not take place. To ensure accurate results once again, each sample was measured twice and for each cation three blank values were determined. The cation-exchange capacity was calculated by summing up all cation contents.

2.5. Statistical Analyses

To find potential differences between the means for tree and intertree areas for each studied parameter, a *t*-test was used. Significant differences between the two areas are present when the *p*-value is <0.05. The *t*-test analysis was carried out using the software Microsoft Excel 2016 (Microsoft, Redmond, WA, USA).

Furthermore, a two-step cluster analysis was used to recognize potential patterns in the data. The software IBM SPSS Statistics 25 (IBM, Armonk, USA) was used to carry out this analysis. All data was scaled metrically. Since clustering of all the variables only resulted in one cluster, the results were split between rainfall simulation results and soil analysis/infiltration results. The variables were chosen in a way so that at least two clusters could be differentiated. To cluster the results of the rainfall simulations, the variables suspended sediment load, surface runoff, suspended sediment concentration and slope were used. For a cluster analysis of the soil and infiltration data, the variables total carbon content, total nitrogen content, corrected percolation stability, K_{sat} value, vegetation cover, stone cover, pH value, electrical conductivity, and cation concentrations of K, Mg, and Na were used. Each cluster was assigned a number (e.g., cluster 1 = 1, cluster 2 = 2) to facilitate further analysis and find the "explaining" variables for the differences between the clusters.

The results are represented by box plots which show the medians as a solid line, means as a dashed line, as well as outliers.

3. Results

3.1. Rainfall Simulations

3.1.1. Environmental Plot Characteristics

Several differences can be noted regarding the plot characteristics. In Table 2 tree and intertree areas are compared through slope, vegetation cover, stone cover, and soil moisture. The slope is quite similar in both tree-covered and bare intertree areas, as is the surface roughness. Vegetation cover on the ground (undergrowth for tree area) shows a significant difference in the means between tree and intertree areas with 36.8 ± 28.4% and 10.2 ± 14.1%, respectively. Stone cover is significantly different as well, with higher percentages in the intertree plots than tree plots. The soil moisture before the rainfall simulations varied between 14.6% and 0.1% for the tree areas and between 12.9% and 0.1% for the intertree areas, leading to a slightly higher soil moisture mean for the tree areas.

Table 2. Environmental plot characteristics of tree and intertree areas (* significant difference between tree and intertree area with *p*-value < 0.05).

Type	Slope (°)	Roughness (Cr)	Vegetation Cover (%)	Stone Cover (%)	Soil Moisture (%)
Tree	4.3 ± 4.4	6.8 ± 4.4	36.8 ± 28.4 *	35.0 ± 21.4 *	4.4 ± 4.5
Intertree	4.2 ± 4.2	7.5 ± 4.0	10.2 ± 14.1 *	51.9 ± 20.5 *	3.4 ± 3.8

3.1.2. Soil Loss, Surface Runoff, and Sediment Concentration

The results of the 36 rainfall simulations are presented in Figure 6. They are shown as boxplots with the total averages as well as median and 5th/95th percentile for the tree (T) and intertree (IT) areas.

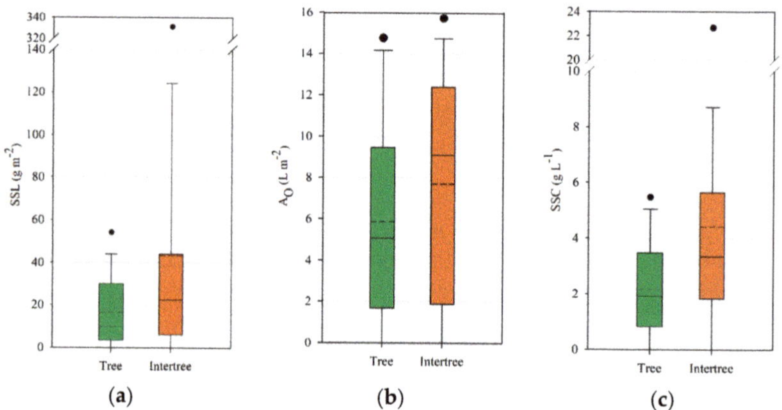

Figure 6. Total average of (**a**) suspended sediment load (SSL), (**b**) surface runoff (A_O), and (**c**) suspended sediment concentration (SSC) (short dash: mean line, solid line: median, dots: outliers). Results for the tree-covered areas (left box) and intertree areas (right box).

The rainfall simulations in the tree-covered areas show a total average suspended sediment load (SSL) of 16.35 g m^{-2} with a maximum value of 54.16 g m^{-2}. The surface runoff (A_O) shows a mean value of 5.86 L m^{-2} and the maximum runoff in the tree area is 14.75 L m^{-2}. Therefore, the suspended sediment concentration (SSC) averages 2.18 g L^{-1}, showing a maximum value of 5.48 g L^{-1}. The mean runoff coefficient (RC) is 28.9% and the maximum value amounts to 76.1% of simulated rain on the plot as runoff. Thereby, the infiltration coefficient (IC) averages 71.1% and the minimum coefficient of infiltration shows only 23.9% of rainfall infiltrating into the soil under tree-covered areas.

The mostly bare intertree areas average suspended sediment loads of 43.25 g m^{-2} with a maximum value of 331.19 g m^{-2}. The mean runoff is 7.69 L m^{-2} and the maximum runoff amounts to 15.72 L m^{-2}. The mean SSC reaches 4.42 g L^{-1} and its maximum shows 22.65 g L^{-1}. A mean of 39.08% of the simulated rainfall was collected as runoff while an average of 60.92% infiltrated into the soil. The maximum RC shows 78.44%, with the resulting IC at a minimum of 21.56%.

Over the 30 min experiment, six out of 36 simulations produced no runoff and suspended sediment, four of these were situated in tree-covered areas, while the remaining two were conducted in the intertree areas. A *t*-test was used to identify potential significant differences of SSL, A_O, SSC, RC, and IC between tree and intertree areas. For none of the parameters was a *p*-value < 0.05 reached. A *p*-value < 0.1 was observed for SSC. Differences in the data can be observed visually, showing a wider range of data for the intertree areas compared to the tree areas especially for SSL and SSC (amplitudes for T SSL: 54.16 g m^{-2}; IT SSL: 331.19 g m^{-2}; T SSC: 5.48 g L^{-1}; IT SSC: 22.65 g L^{-1}). If the rainfall simulations of ABH2c (IT) and BOU1b (T) are removed as outliers, this difference in amplitudes is not as clearly distinguishable (amplitudes for T SSL: 42.90 g m^{-2}; IT SSL: 100.99 g m^{-2}; T SSC: 5.00 g L^{-1}; IT SSC: 7.16 g L^{-1}). Although not statistically significant, possibly due to inherent disparity of the study areas, a difference between each test site's tree and intertree areas may be observed. Figure 7 shows each test site's tree and intertree SSC. Out of 18 test sites, 12 showed higher SSC values in the intertree area, while five showed higher SSC values in the tree area. Only at test site AZR1c was no runoff and suspended sediment load, and thus no SSC, measured.

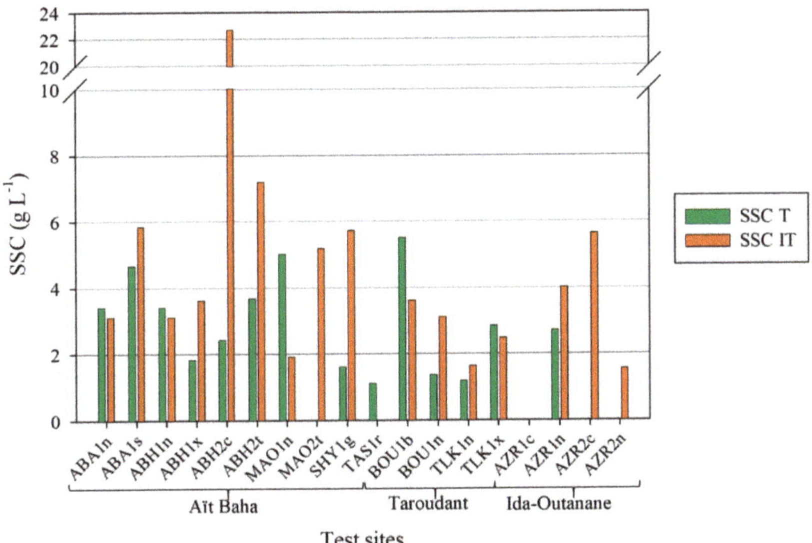

Figure 7. Suspended sediment concentration (SSC) for tree (T, black bars) and intertree areas (IT, grey bars) for each test site. Break between 10 and 20 g L^{-1} because of test site ABH2c.

Figure 7 also arranges the test sites to each study area to show potential differences between test sites. Although the study area of Ida-Outanane does not show any SSC in half of the experiments, the other half show similar values as the other study areas. Aït Baha shows the highest SSC values, but otherwise does not differ much from the rest of the study areas. The mean and median values for

Aït Baha are 4.22 g L^{-1} and 3.41 g L^{-1}, for Taroudant 2.70 g L^{-1} and 2.66 g L^{-1}, and for Ida-Outanane 1.73 g L^{-1} and 0.77 g L^{-1}, respectively.

3.1.3. Cluster Analysis for Rainfall Simulations

The two-step cluster analysis for the rainfall simulations shows a good cluster quality with two clusters found out of the four variables used. The predictor influence ranges from 1.0 to 0.06. SSC is the most important predictor (1.0), followed by SSL (0.64), and A$_O$ (0.20). The predictor with the least influence is slope (0.06). Out of 36 test sites, 19 belong to the first cluster. As presented in Table 3, cluster 1 shows the lower values for all the variables. The last column (tree/intertree) indicates the relationship of tree or intertree area test sites in each cluster. Cluster 1 contains 15 tree areas and four intertree areas, while cluster 2, with higher values for SSC, SSL, A$_O$ and in parts slope, includes 14 intertree and three tree areas.

Table 3. Means of the inputs for the two-step cluster analysis in order of predictor influence. A tree/intertree column was separately added to indicate potential cluster predominance by tree or intertree area. Parameters are suspended sediment concentration (SSC), suspended sediment load (SSL), surface runoff (A$_O$), and slope.

Cluster	SSC (g L^{-1})	SSL (g m^{-2})	A$_O$ (L m^{-2})	Slope ($°$)	Tree/Intertree
Cluster 1	1.35	13.73	4.55	6.26	15/4
Cluster 2	8.95	53.26	46.91	7.38	3/14

3.2. Infiltrations

The maximum or potential infiltration rates could be measured by means of the constant-head single-ring infiltrometer. In total, 60 infiltration measurements were carried out on all 30 test sites, one for the tree area and the other one for the intertree area. Mean infiltration rates were measured as 452.57 mm h^{-1} for tree areas and 229.56 mm h^{-1} for intertree areas. Minimum values did not differ greatly with 52.60 mm h^{-1} in the tree area (test site ABA1n) and 54.13 mm h^{-1} in the intertree area (test site ABH2c). Maximum values of 1556.14 mm h^{-1} were almost certainly outliers due to interflow activation on a steep gravelly slope. The maximum value for the intertree areas with 566.26 mm h^{-1} was measured on a slope of 0°, eliminating this possibility.

The saturated hydraulic conductivity (K$_{sat}$) measured with the single-ring infiltrometer is depicted in Figure 8a. Although the minimum and maximum values of the tree area are respectively lower and higher than any values in the intertree area, the tree areas (203.19 mm h^{-1}) average a higher K$_{sat}$ value than the intertree areas (144.39 mm h^{-1}). In comparison to the amplitude of 410 mm h^{-1} in the tree areas, the intertree areas have a rather small amplitude of resulting K$_{sat}$ values with 133.18 mm h^{-1}. This high variability of K$_{sat}$ has been noted in previous studies [66,67], yet it is only visible in tree areas. However, there is a significant difference of the means between tree and intertree area ($p < 0.01$).

In Figure 8b the differences between the study areas are shown. The study area Aït Baha displays the lowest K$_{sat}$ values (mean: 166.43 mm h^{-1}), followed by Taroudant (mean: 171.40 mm h^{-1}), while Ida-Outanane displays the highest values (mean: 193.74 mm h^{-1}) and the highest range (excluding outliers). All three have similar minimum values (Aït Baha 93.54 mm h^{-1}, Taroudant 93.91 mm h^{-1}, Ida-Outanane 106.60 mm h^{-1}), yet Aït Baha shows the lowest mean and the lowest minimum, but the highest maximum with 503.64 mm h^{-1}. Although there are differences between the means and the amplitudes of the K$_{sat}$ values in the three study areas, these are not statistically significant for the means.

Figure 9 shows the unsaturated hydraulic conductivity values for the different suctions used. All four suctions were used on each test site as well as tree and intertree areas. As can be seen, the range of values (excluding outliers) in both the tree and intertree areas increases from higher to lower suctions. This increase in range is not too surprising, because of the differing soil material. While suctions on sandy soils should normally be lower and higher on clayey soils, high suctions on sandy soils lead to a higher throughflow during the infiltration experiments and thus to higher K$_h$ values. The mean K$_h$

values for suction -4 cm are 19.13 mm h^{-1} for tree areas and 15.81 mm h^{-1} for intertree areas. They are raised with increasing suction up to 28.89 mm h^{-1} for tree areas at suction -0.5 cm, at the same suction for intertree areas to 19.65 mm h^{-1}. Significant differences between tree and intertree areas cannot be found for all applied suctions. At the suction -4 cm mean K_h values between tree and intertree areas did not differ significantly, yet at suctions -2 cm, -1 cm, and -0.5 cm these differences were statistically significant ($p < 0.05$).

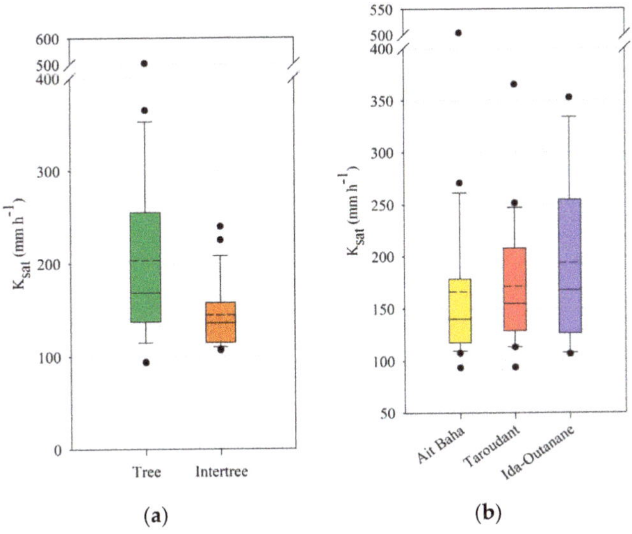

(a) (b)

Figure 8. (**a**) Saturated hydraulic conductivity (K_{sat}, $p < 0.05$) for tree and intertree areas as well as (**b**) for the three study areas (short dash: mean line, solid line: median, dots: outliers).

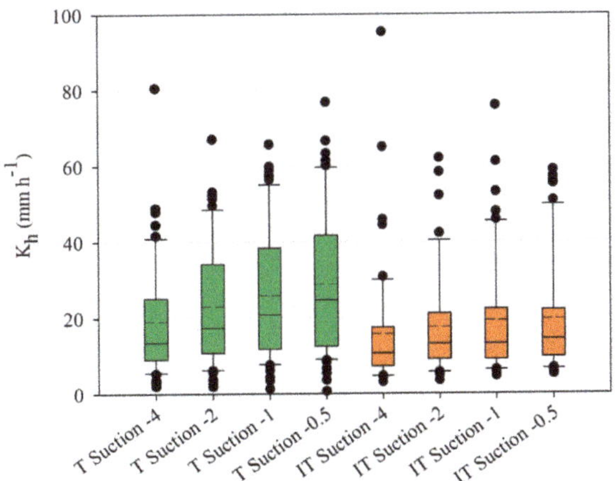

Figure 9. Unsaturated hydraulic conductivity (K_h) for different suctions (-4 to -0.5) and tree (T) and intertree (IT) areas (short dash: mean line, solid line: median, dots: outliers).

Figure 10 displays the differences of the K_h values between the three study areas. As was the case for the K_{sat} values, the study area of Ida-Outanane shows the highest range and has the highest means. Unlike the K_{sat} values, the K_h values for Aït Baha are not the lowest but actually display higher means, minimums, and maximums than the study area of Taroudant. Although there are no significant differences in the means between Aït Baha and Taroudant at suctions −4 cm and −2 cm, there are at suctions −1 cm and −0.5 cm ($p < 0.05$). Significant differences occur between Ida-Outanane and the other two study areas ($p < 0.01$).

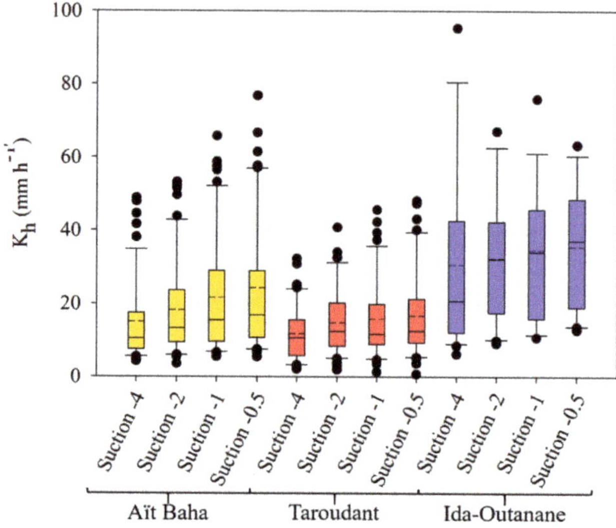

Figure 10. Unsaturated hydraulic conductivity (K_h) for different suctions (−4 to −0.5) for the three study areas (short dash: mean line, solid line: median, dots: outliers).

3.3. Soil Analyses

The results of the analyses for the different soil parameters (pH, electrical conductivity, percolation stability, N-content, total carbon (TC)-content, and mean grain size) are shown in Figure 11. The mean pH value for the tree and intertree areas is 7.41 and 7.32, respectively. The values range between 7.67 and 6.93 for tree and 7.58 and 6.85 for the intertree areas. Although the values do not seem to differ much, there is a significant difference in the means (p-value < 0.05). Significant differences between tree and intertree areas are also present for the parameters electrical conductivity (EC, $p < 0.001$), percolation stability (PS, $p < 0.001$), N-content ($p < 0.001$), and TC-content ($p < 0.001$). The mean grain size does not differ significantly between tree and intertree areas. The mean grain size of all test sites in the tree area is 0.22 mm; in the intertree area it is slightly lower with 0.18 mm. It is obvious that most of the analyzed soil parameters show a higher range of results in the tree area than in the intertree area. While this amplitude is not as pronounced for the mean grain size (0.34 mm and 0.27 mm for tree and intertree areas, respectively), the parameters EC, PS, N-content, and TC-content show a much higher amplitude of values. The EC results in the tree area range between 539 and 190 µS cm⁻¹, while the intertree areas show maximum and minimum values of 283 and 169 µS cm⁻¹, respectively. The values of the percolation stability show means of 183.69 mL 10 min⁻¹ for the tree and 43.23 mL 10 min⁻¹ for the intertree areas, while the amplitudes for tree and intertree areas are 392.27 and 147.43 mL 10 min⁻¹, respectively. The N-content is on average 0.34% in the tree and 0.11% in the intertree areas, while maximum and minimum values are 0.79% and 0.14% for tree-covered areas and 0.42 and 0.05% for intertree areas. TC-content shows similar behavior one order of magnitude higher with a mean of 4.79% (T) and 1.77% (IT). The maximum TC-content in the tree areas is 10.57% and 6.43% in the intertree areas; minimum values are 1.58% and 0.53%, respectively.

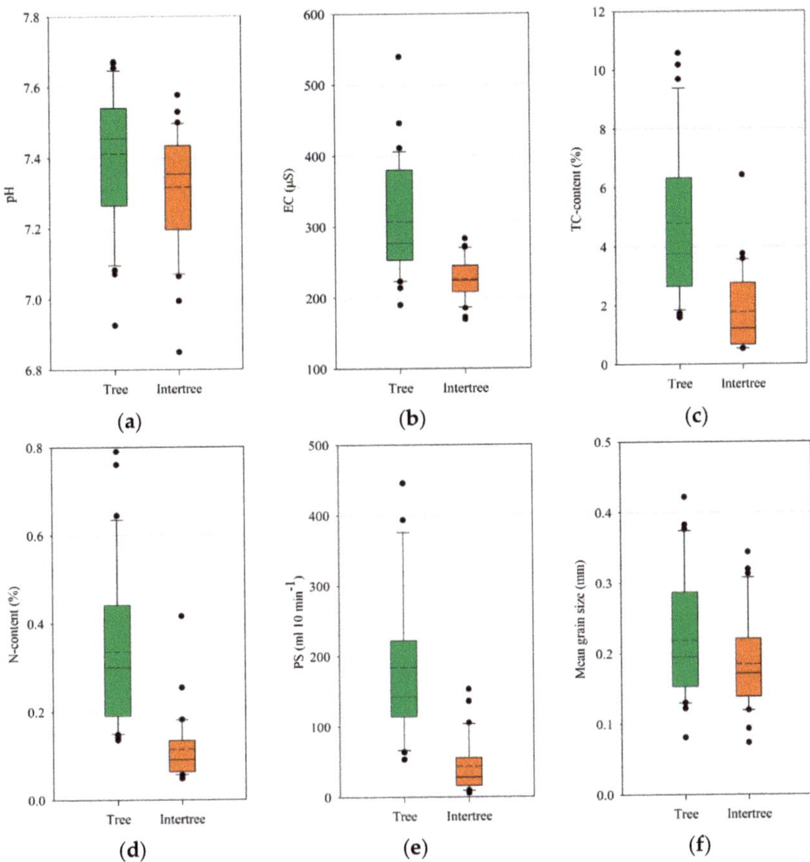

Figure 11. (**a**) pH ($p < 0.05$), (**b**) electrical conductivity (EC) ($p < 0.05$), (**c**) total carbon content (TC-content, $p < 0.05$), (**d**) nitrogen content (N-content, $p < 0.05$), (**e**) percolation stability (PS, $p < 0.05$), and (**f**) mean grain size (short dash: mean line, solid line: median, dots: outliers). Results for the tree-covered areas (left box) and intertree areas (right box).

The differences in the study areas are given in Table 4. The highest means of pH, EC, TC-content, and N-content are all found in the study area of Ida-Outanane, while Taroudant shows the highest PS and mean grain size. The study areas Aït Baha and Taroudant are similar regarding pH, EC, and mean grain size, yet Aït Baha shows much lower values in TC-content and N-content. Ida-Outanane shows significant differences in the means ($p < 0.05$) for pH values compared to the other two study areas and for EC, TC-content, and N-content compared to Aït Baha.

Table 4. Statistical summary of soil parameters pH, electric conductivity (EC), total carbon content (TC-content), nitrogen content (N-content), percolation stability (PS), and mean grain size for each study area (bold values show significant differences ($p < 0.05$) between one study area compared to the other two; italic values show significant differences between two study areas).

Study Area	pH	EC (μS cm^{-1})	TC-Content (%)	N-Content (%)	PS (mL 10 min^{-1})	Mean Grain Size (mm)
Aït Baha	7.34 ± 0.21	*250.42 ± 73.01*	2.33 ± 1.67	*0.17 ± 0.13*	108.77 ± 104.47	0.20 ± 0.09
Taroudant	7.33 ± 0.17	264.00 ± 64.87	3.73 ± 3.17	0.25 ± 0.22	121.39 ± 119.43	0.21 ± 0.06
Ida-Outanane	**7.47 ± 0.11**	*299.14 ± 70.53*	4.32 ± 2.48	*0.28 ± 0.16*	102.32 ± 71.86	0.17 ± 0.09

The concentrations of basic cations potassium (K), sodium (Na), magnesium (Mg), and calcium (Ca) are presented in Figure 12. The highest contents of the measured cations are found for Ca with mean values of 38.75 cmol$_c$ kg^{-1} for tree sites and 36.83 cmol$_c$ kg^{-1} for intertree sites. Significant differences between the means cannot be observed. However, in spite of smaller values, all other cations measured show significant differences between tree and intertree areas. The means for K are 5.40 and 4.36 cmol$_c$ kg^{-1} for tree and intertree areas, respectively. Na means are even lower with 1.12 cmol$_c$ kg^{-1} (T) and 0.76 cmol$_c$ kg^{-1} (IT). Mg shows similar values as K, with means for the tree area of 6.03 and 5.07 cmol$_c$ kg^{-1} for the intertree area. It is evident from Figure 12 that the amplitude excluding outliers of the intertree area is higher for K and Ca, while this is not as obvious the case for Na and Mg.

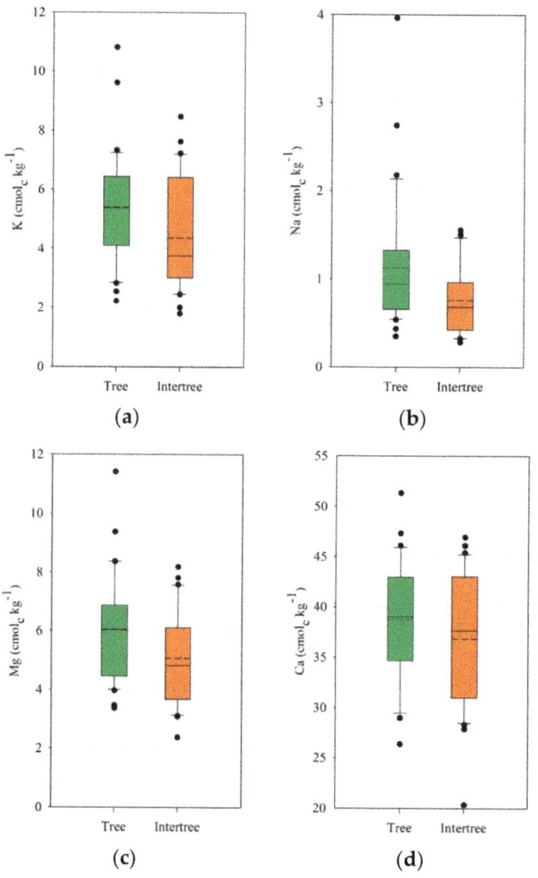

Figure 12. (**a**) Potassium (K) content ($p < 0.05$), (**b**) sodium (Na) content ($p < 0.05$), (**c**) magnesium (Mg) content ($p < 0.05$), and (**d**) calcium (Ca) content (short dash: mean line, solid line: median, dots: outliers). Results for the tree-covered areas (left box) and intertree areas (right box).

Table 5 structures the measured cation concentration contents according to their study area. The study area of Ida-Outanane shows the highest concentrations for K, Ca, and Na and the lowest for Mg. The concentrations of K, Ca, and Na are the lowest in the study area of Aït Baha, while the Mg concentration is highest in this study area. The cation concentrations in the study area of Taroudant are mostly similar to Aït Baha, but a little higher except for Mg. Significant differences ($p < 0.05$) between the study areas could only be found between Ida-Outanane and the two other areas for Ca, and also between Aït Baha and Ida-Outanane for K.

Table 5. Statistical summary of cation concentrations for potassium (K), sodium (Na), magnesium (Mg); and calcium (Ca) for each study area (bold values show significant differences ($p < 0.05$) between one study area compared to the other two; italic values show significant differences between two study areas).

Study Area	K (cmol$_c$ kg^{-1})	Na (cmol$_c$ kg^{-1})	Mg (cmol$_c$ kg^{-1})	Ca (cmol$_c$ kg^{-1})
Aït Baha	*4.27 ± 1.63*	0.80 ± 0.40	5.66 ± 1.57	35.76 ± 5.58
Taroudant	4.95 ± 2.04	0.98 ± 0.48	5.57 ± 1.67	36.45 ± 6.52
Ida-Outanane	*5.87 ± 1.96*	1.15 ± 1.01	5.32 ± 2.23	**43.66 ± 3.66**

3.4. Cluster Analysis for Infiltration/Soil Analysis Results

The two-step cluster analysis for the infiltration and soil analysis results shows a medium cluster quality with two classified clusters out of 11 variables used. The predictor influence ranges from 1.0 to 0.13. The principal predictors for the clusters are TC-content (1.0), EC (0.98), and N-content (0.93). Vegetation cover and PS (both 0.75) follow, then K$_{sat}$ (0.48), pH (0.39), and stone cover (0.25) are less important, while the cation concentrations are the least influential (Na (0.24), Mg (0.16), and K (0.13)). Table 6 shows the means of the inputs of the two clusters with an extra column indicating the relationship of tree or intertree area test sites in each cluster. The seven most influential variables are presented in Table 6. Two intertree areas and three tree areas cannot be attributed to either cluster. Cluster 1 shows the lower mean values for the variables, with six tree areas and 27 intertree areas being attributed to this cluster. Cluster 2 includes 22 tree areas and two intertree areas with the higher mean values for all predictors.

Table 6. Means of the inputs for the two-step cluster analysis in order of predictor influence. A tree/intertree column was separately added to indicate potential cluster predominance by tree or intertree area.

Cluster	TC (%)	EC (µS cm^{-1})	N (%)	Vegetation Cover (%)	PS (mL 10 min^{-1})	K$_{sat}$ (mm h^{-1})	pH	Tree/Intertree
Cluster 1	1.74	223.42	0.12	5.88	60.10	141.19	7.30	6/27
Cluster 2	5.20	315.46	0.36	34.29	186.03	211.17	7.46	22/2

4. Discussion

A comparison between tree and intertree areas in argan woodlands was carried out using rainfall simulations, infiltration measurements, as well as soil analyses. Since this study did not focus on specific tree degradation stages but tried to encompass a multitude of different land uses as well as tree and soil degradation stages, the range of values is not surprising. Disturbance of vegetation, such as loss of herbaceous cover, can lead to an increase in sediment transport and erosion. This is especially true when there is a decrease in woody plant canopy cover from woodlands to shrublands [68], which is partly the case in the argan woodlands with different stages of degradation and a change in canopy cover due to overgrazing [69,70]. The argan woodlands have been degraded because of grazing pressure, rainfed agriculture in silvo-agricultural land uses, as well as fuelwood extraction [3,71], leading to a wider and more disturbed intertree area with less and wider spaced trees [3].

Although significant differences in the *t*-tests could not be found, most of the test sites showed lower SSC in the tree areas compared to the neighboring intertree areas. Other studies showed higher runoff and erosion rates on bare intercanopy patches compared to canopy-covered areas [72,73], whereas vegetated intercanopy patches showed medium rates of runoff and erosion [72]. Ceballos et al. obtained differing results, where vegetated grassy areas showed little runoff and erosion rates, yet tree-covered areas showed much higher rates due to soil hydrophobicity hindering infiltration [74]. Since rainfall simulations in our study were carried out the same way under trees as in the intertree areas, the rainfall intensity was much higher than under natural conditions, when rainfall may be reduced by 6.6% to 82.7% due to interception [75]. Nevertheless, splash erosion by throughfall drops may be 2.59 times higher than on open fields [76], so the rainfall simulations under the trees should not be equated with natural conditions. It is possible, however, to compare the erodibility [77,78] between tree and intertree areas. The cluster analyses in Tables 3 and 6 show the tendency of clusters being attributed to tree or intertree areas. Intertree areas with lower SSC, SSL, and A_O could all be assigned to active agricultural use with better infiltration potential due to ploughing. The three tree areas belonging to cluster 2 showed degraded tree areas with little to no vegetation cover or organic litter.

Other studies in Morocco, also in the study area of Taroudant, found mean suspended sediment concentrations on fallow land and orange plantations of 2.7 g L^{-1} [37] compared to lower tree area and higher intertree area means of 2.18 g L^{-1} and 4.42 g L^{-1}, respectively. Ceballos et al. also showed suspended sediment concentrations for tree-covered areas in the Spanish dehesas which were lower both under dry (mean 0.37 g L^{-1}) and wet conditions (mean 0.65 g L^{-1}) than either tree or intertree areas. Even sheep trails showed lower average SSC (dry: 0.45 g L^{-1}, wet: 0.76 g L^{-1}) [74]. This indicates the higher erodibility of the soils in the Moroccan study areas, especially in the intertree areas that are not protected by vegetation or leaf litter. This extra layer has been found to strongly reduce runoff and sediment loss [79]; it is missing in some test sites such as MAO1n, where higher SSC was measured in the tree area (see Figure 7). Since some test sites are more disturbed and show little to no vegetation in the intertree and tree areas (undergrowth), there are cases where SSC can also be higher in tree areas. Ludwig et al. [80] showed higher runoffs and sediment yields on disturbed areas as compared to undisturbed or only slightly grazed areas. Wind erosion has not yet been studied in these regions, but due to a high crust cover and little loose sediment on the surface, it is assumed to be secondary.

Soil erodibility is, of course, also dependent on aggregate stability. The results differ significantly between the tree and intertree areas. This could also be due to the higher carbon contents in the tree areas, which can lead to the formation of larger aggregates [81,82]. Under agricultural use organic matter content and aggregate stability can decline [83]. This was the case on all test sites that were under agricultural use in the intertree areas, which partly explains the higher SSC on the intertree areas. In some cases, there was nearly no surface runoff until the end of the experiment, which was partly due to infiltration into the loosened soil or enhanced surface roughness by ploughing. As Mbagwu and Auerswald noted, high interrill erosion rates should be expected on soils with a PS <250 mL 10 min^{-1} [64]. The measured PS was mostly below this value in tree or intertree areas. This could explain the much higher SSC compared to the values of Ceballos et al. [74]. In comparison to the values published by Mbagwu and Auerswald the measured PS ranked very low in many of the studied test sites, with a mean of 43.23 mL 10 min^{-1} in the intertree area, which is just slightly higher than the minimum value published by the aforementioned authors. Their values for fallows, pastures or (secondary) forests (>250 mL 10 min^{-1}) cannot be repeated in our measurements for most intertree and tree areas. Intertree areas are partly cultivated or unmulched plots which should show medium (250–150 mL 10 min^{-1}) to low PS (<150 mL 10 min^{-1}). They could also be compared to secondary forest subsoil with low PS (<150 mL 10 min^{-1}), since topsoil could have been eroded, thus explaining the low percolation stabilities [64]. Goebel et al. [84] obtained similar results regarding higher aggregate stability in topsoil, and also found higher aggregate stabilities in forest soils than arable soils, which matches the results with higher aggregate stabilities under trees compared to the crusted and mostly eroded intertree areas. Since aggregate breakdown leads to crusting of soil, the infiltration rate decreases with lower aggregate stability and runoff, and sediment detachment occurs [85,86].

The infiltration measurements have shown a significant difference between tree and intertree areas for both unsaturated and saturated hydraulic conductivity, as well as infiltration rates, indicating a better infiltrability into the soil in vegetation patches [80,87]. Lichner et al. [88] noted high K_{sat} values in forest soils due to the greater presence of macropores, yet found the lowest K_h values for forest sites possibly due to a higher water repellency caused by the coating of soil particles by pine-needle waxes.

A study conducted in the southern part of the study area of Taroudant found much lower infiltration rates than the values presented here with a mean of 84 mm h^{-1} and a maximum of 265 mm h^{-1} [49]. This study focused on cultivated and leveled landscapes with high soil-crust cover in contrast to the less crusted tree areas or the intertree areas with a higher stone-fragment cover on the surface. The influence of stone cover on infiltration rates may be either positive or negative, depending on the measurement context [89,90]. If the measurements occur between large shrubs, or in this case trees, the influence is positive, yet when shrub and intershrub are compared, the correlation with stone cover becomes negative since the finer sediments under shrubs have a positive effect on infiltration [89]. Thus, compared to the results of Peter and Ries [49], stone cover influences the infiltration in the intertree areas positively, yet tree area infiltration is still mostly higher. Although there are high percentages of sand, mostly in the coarse fraction, in both tree and intertree areas, differences in infiltration rates and hydraulic conductivities could also be due to the plant cover which could maintain macropores on the surface and conduct water into the soil [91], while in the intertree areas vegetation cover is rather low (10.17 ± 14.10%). A higher density of soil underneath the argan trees due to compaction by livestock was not measured in this study but would affect infiltration into the soil negatively [92]. Since the measured infiltration values are higher in the tree area, it suggests that the soil is rather stable against compaction by livestock. The grain size distribution for corresponding tree and intertree areas does not vary much, they are mostly classified as loam, sandy loam, or sandy clay loam.

With a higher vegetation cover and more leaf litter in the tree areas in most cases, the difference in the content of total carbon between tree and intertree areas can be explained. The same is true for the content of total nitrogen, which could also be influenced by the droppings of goats and sheep. During dry years the goats and sheep feed from the trees, which are the only biomass storage in these areas, and are shepherded from one tree to another. Thus, they spend most time in the tree area, possibly influencing the N-content in the soil. Soil microbial communities profit from the higher N-content under the plants [15]. Micro-, meso-, and macro-fauna are drawn to canopy-covered areas, since they are the places with the lowest temperatures in the daytime and the highest soil moisture, even in these very dry climates (see Table 2), and therefore the places where food is most abundant [13]. The higher carbon contents under the trees originate from the litter as well as the undergrowth. Tree or shrub litter has a positive influence on soil carbon as well as on aggregate stability, shear strength, as well as infiltration rates, which in turn leads to greater resistance against runoff and erosion [14,93]. The removal or absence of litter can therefore have a negative influence on the studied soil parameters.

The cluster analysis classifying soil parameters shows that an attribution of tree and intertree areas to specific clusters is in many cases possible. Since total carbon content is the most influential predictor, areas with low total carbon content are often classified into the cluster associated with lower values. On the other hand, agriculturally used intertree areas where organic litter was ploughed into the soil can be classed into the cluster with higher values. Yet, the majority of tree areas is classed into the cluster with generally higher values, while the majority of intertree sites can be associated with the cluster with lower values.

Many studies focus on shrubs and intershrub differences [14,15,94] where the shrub covers the soil well and there is no great distance between the soil surface and the leaf cover of the shrub. It can thus act as an obstacle for runoff and transported sediment. The distance between the canopy of trees and the soil surface is much higher though, resulting in splash erosion due to throughfall and interception, as noted. The influence of leaf litter can prevent soil erosion, yet without litter the canopy-covered area is unprotected against splash erosion as well as from runoff from further upslope, leading to a degradation of the soil in the tree area. Here, different degradation stages of the trees could show a high influence on the results. This depends on the tree architecture. Round, healthy-looking crowns (see Figure 1c) produce a lot of leaf litter, yet the thickness of the litter layer also depends on the height of the tree. The litter from higher trees might be scattered more widely by the wind and therefore not as concentrated as litter from lower trees. Rather low trees can lead to a greater number of sheep (and goats) that could erode the soil further with their hooves [95]. Since sheep do not climb the trees as goats do, they browse only the trees they can reach from the ground. With more open crowns throughfall is higher, while litter concentration is lower. In the case of the argan as a cushion-type shrub with a very dense network of branches and twigs (Figure 1e), the most degraded 'tree', the distance between canopy and soil surface is lowest, thus the litter and soil underneath are most protected, although only in a very small area.

Significant differences between tree and intertree areas were also observed for the cation concentrations of K, Mg, and Na. Berthrong et al. [96] noted decreases of Ca, Mg, and K as well as pH and an increase of Na in response to afforestation. For argan afforestation projects in Morocco this potential decrease should be kept in mind. Organic matter also increases the cation exchange capacity and thus the available cations [14].

However, another possible explanation for the differences could be due to coastal fog which is caught by the argan trees [97] and precipitates in the tree area. It is a major source of precipitation for the argan trees, since rainfall is very variable in the study areas [22]. Although fog has lower concentration of nutrients than rain, as shown by Azevedo and Morgan [98], it could still lead to a higher accumulation of cations in the tree area, since precipitation by rainfall influences both tree and intertree areas, but fog predominantly precipitates in the tree area. The occult deposition rates of nutrients by fog decrease with the distance from the ocean [99]. This is shown in the results. Ida-Outanane is the closest study area to the sea and has the highest cation concentrations for K, Ca, and Na. Although the study areas of Taroudant and Aït Baha are situated further from the ocean, fog can occur on the foothills of the High Atlas and Anti-Atlas and be collected by argan plants, as shown in Figure 13; this can therefore lead to a higher concentration of the cations K, Ca, and Na in the tree areas.

Figure 13. Collection of fog precipitation by the argan tree.

The differences in the pH between tree and intertree areas are potentially caused by the source of nitrogen in the soil. If a lot of nitrate (NO_3^-) is available, the plants will absorb more anions and thus the pH rises. If cations are absorbed the pH decreases in the vicinity of the roots [100,101].

Inter-plant soil properties such as pH, EC, and C:N ratio are all influenced by the adjacent vegetation [15]. This shows the importance of a closer distance between trees rather than an expanding intertree area. Since the distances between the studied trees and their intertree areas are all different depending on the tree density of each test site, variabilities in the values could stem from these differences. The reintroduction of specific tree species into the degraded intertree areas could lead to an increase in carbon and nutrients [102]. Afforestation projects are in process, yet only with limited success, since young argan sprouts need to be protected from grazing animals. This is mostly done by fenced exclusion areas, although the exclusion is in many cases not successful. Break-ins are punishable, but in dry years there is no other food source for the many herds. In addition, young sprouts need a lot of water for several years to develop their roots and be able to reach water deeper in the soil. Dry years often lead to withering of the sprouts, since planting companies are often only engaged for two to three years, which does not give the young sprouts enough time to grow.

5. Conclusions

Tree and intertree areas on 30 test sites covered by argan trees in three study areas in South Morocco were analyzed using rainfall simulations, infiltration measurements, and soil analyses. Significant differences could be found between tree and intertree areas for many soil parameters although the three study areas show (partly significantly) differing results. Two cluster analyses were conducted, one focusing on erosion parameters SSC, SSL, A_O, and slope, the other focusing on the soil parameters total carbon content, total nitrogen content, corrected percolation stability, K_{sat} value, vegetation cover, stone cover, pH value, electrical conductivity, and the cation concentrations of K, Mg, and Na. Each resulted in two clusters which mostly include tree areas in one and intertree areas in the other. However, agriculturally used intertree areas show less erosion in one analysis or high carbon contents in the other and are thus classed into the 'tree' cluster, while more degraded tree areas are included in the 'intertree' cluster. There are several parameters, like K_{sat} value, pH, EC, PS, TC-content, N-content, K-content, Na-content, and Mg-content that show that tree and intertree areas in argan woodlands differ significantly, even under various usages and in different study areas. With an expanding intertree area due to overgrazing and expansion of agriculture among others, the soils in these argan woodlands face further degradation. Further research is needed on how different growth forms of the argan trees influence the soil as well as how far the influence of the tree area extends into the intertree area.

Author Contributions: Conceptualization, M.K., M.S., I.M., and J.B.R.; methodology, M.S., I.M., and J.B.R.; investigation, M.K., L.E., and L.L.Z.; resources, M.S., I.M., A.A.H., and J.B.R.; data curation, M.K., L.E., and L.L.Z.; writing—original draft preparation, M.K., L.E., and L.L.Z.; writing—review and editing, M.K. and I.M.; visualization, M.K. and L.L.Z.; supervision, M.S., I.M., A.A.H., and J.B.R.; project administration, I.M. and J.B.R.; funding acquisition, I.M. and J.B.R.

Funding: This research was funded by Deutsche Forschungsgemeinschaft (DFG), grant numbers RI 835/24-1 & MA 2549/6.

Acknowledgments: We hereby acknowledge the support of the laboratories of Trier University as well as of the Université Ibn Zohr Agadir. We also would like to thank Tobias Buchwald, Laura Kögler, Sabrina Reichelt, Robin Stephan, Stefan Weiss, and Nicole Weymann for their considerable help during data acquisition in the field.

Conflicts of Interest: The authors declare no conflicts of interest.

References

1. Cerdan, O.; Govers, G.; Le Bissonnais, Y.; Van Oost, K.; Poesen, J.; Saby, N.; Gobin, A.; Vacca, A.; Quinton, J.; Auerswald, K.; et al. Rates and spatial variations of soil erosion in Europe: A study based on erosion plot data. *Geomorphology* **2010**, *122*, 167–177. [CrossRef]
2. Pulido, M.; Schnabel, S.; Lavado Contador, J.F.; Lozano-Parra, J.; Gómez-Gutiérrez, Á. Selecting indicators for assessing soil quality and degradation in rangelands of Extremadura (SW Spain). *Ecol. Indic.* **2017**, *74*, 49–61. [CrossRef]
3. Le Polain de Waroux, Y.; Lambin, E.F. Monitoring degradation in arid and semi-arid forests and woodlands: The case of the argan woodlands (Morocco). *Appl. Geogr.* **2012**, *32*, 777–786. [CrossRef]
4. Breshears, D.D.; Whicker, J.J.; Johansen, M.P.; Pinder, J.E., III. Wind and water erosion and transport in semi-arid shrubland, grassland and forest ecosystems: Quantifying dominance of horizontal wind-driven transport. *Earth Surf. Process. Landf.* **2003**, *28*, 1189–1209. [CrossRef]
5. Dahan, R.; Boughlala, M.; Mrabet, R.; Laamari, A.; Balaghi, R.; Lajouad, L. *A Review of Available Knowledge on Land Degradation in Morocco*; The International Center for Agricultural Research in the Dry Areas (ICARDA): Aleppo, Syria, 2012; pp. 10–11.
6. Schnabel, S.; Gómez Gutiérrez, A.; Lavado Contador, J.F. Grazing and soil erosion in dehesas of SW Spain. In *Advances in Studies on Desertification*; Romero Diaz, A., Belmonte Serrato, F., Alonso Sarria, F., Lopez Bermudez, F., Eds.; Ediciones de la Universidad de Murcia: Murcia, Spain, 2009; pp. 725–728.

7. Schnabel, S.; Lozano-Parra, J.; Gutiérrez, A.; Alfonso Torreño, A. Hydrological dynamics in a small catchment with silvopastoral land use in SW Spain. *Cuadernos de Investigación Geográfica* **2018**, *44*, 557–580. [CrossRef]
8. Pulido, M.; Schnabel, S.; Lavado Contador, J.F.; Lozano-Parra, J.; González, F. The impact of heavy grazing on soil quality and pasture production in rangelands of SW Spain. *Land Degrad. Dev.* **2016**, *29*, 219–230. [CrossRef]
9. Wojterski, T.W. Degradation stages of oak forests in the area of Algiers. *Vegetatio* **1990**, *87*, 135–143. [CrossRef]
10. Chantsallkham, J.; Reid, S.; Fernandez-Gimenez, M.; Tsevlee, A.; Yadamsuren, B.; Heiner, M. Applying a dryland degradation framework for rangelands: The case of Mongolia. *Ecol. Appl.* **2018**, *28*, 1–21. [CrossRef]
11. Bochet, E.; Rubio, J.L.; Poesen, J. Modified topsoil islands within patchy Mediterranean vegetation in SE Spain. *Catena* **1999**, *38*, 23–44. [CrossRef]
12. Escudero, A.; Giménez-Benavides, L.; Iriondo, J.M.; Rubio, A. Patch Dynamics and Islands of Fertility in a High Mountain Mediterranean Community. *Arctic Antarct. Alp. Res.* **2004**, *36*, 518–527. [CrossRef]
13. Garner, W.; Steinberger, Y. A proposed mechanism for the formation of 'Fertile Islands' in the desert ecosystem. *J. Arid Environ.* **1989**, *16*, 257–262. [CrossRef]
14. Pérez, F.L. Plant Organic Matter Really Matters: Pedological Effects of Kūpaoa (Dubautia menziesii) Shrubs in a Volcanic Alpine Area, Maui, Hawai'i. *Soil Syst.* **2019**, *3*, 31. [CrossRef]
15. Qu, L.; Wang, Z.; Huang, Y.; Zhang, Y.; Song, C.; Ma, K. Effects of plant coverage on shrub fertile islands in the Upper Minjiang River Valley. *Sci. China Life Sci.* **2018**, *61*, 340–347. [CrossRef]
16. Schlesinger, W.H.; Pilmanis, A.M. Plant-soil interactions in deserts. *Biogeochemistry* **1998**, *42*, 169–187. [CrossRef]
17. Aguiar, M.R.; Sala, O.E. Patch structure, dynamics and implications for the functioning of arid ecosystems. *Trends Ecol. Evol.* **1999**, *14*, 273–277. [CrossRef]
18. Cerdà, A. Aggregate stability under different Mediterranean vegetation types. *Catena* **1998**, *32*, 73–86. [CrossRef]
19. Cerdà, A. Aggregate stability against water forces under different climates on agriculture land and scrubland in southern Bolivia. *Soil Tillage Res.* **2000**, *57*, 159–166. [CrossRef]
20. Auerswald, K. Percolation Stability of Aggregates from Arable Topsoils. *Soil Sci.* **1995**, *159*, 142–148. [CrossRef]
21. UNESCO. Argan, Practices and Know-How Concerning the Argan Tree. 2015. Available online: www.unesco: Culture/ich/en/RL/00955 (accessed on 22 July 2019).
22. Ehrig, F.R. Die Arganie. Charakter, Ökologie und wirtschaftliche Bedeutung eines Tertiärreliktes in Marokko. *Petermanns Geographische Mitteilungen* **1974**, *118*, 117–125.
23. Müller-Hohenstein, K.; Popp, H. *Marokko. Ein Islamisches Entwicklungsland mit Kolonialer Vergangenheit*, 1st ed.; Klett: Stuttgart, Germany, 1990; pp. 860–862.
24. Benabid, A. Forest Degradation in Morocco. In *The North African Environment at Risk (State, Culture, and Society in Arab North Africa)*; Swearingen, W.D., Bencherifa, A., Eds.; Westview Press: Boulder, CO, USA, 1996; pp. 175–190.
25. Culmsee, H. Vegetation und Weidenutzung im Westlichen Hohen Atlas (Marokko). Eine Nachhaltigkeitsbewertung aus geobotanischer Sicht. *Dissertationes Botanicae* **2004**, *389*, 1–244.
26. McGregor, H.V.; Dupont, L.; Stuut, J.-B.W.; Kuhlmann, H. Vegetation change, goats, and religion: A 2000-year history of land use in southern Morocco. *Quat. Sci. Rev.* **2009**, *28*, 1434–1448. [CrossRef]
27. Kirchhoff, M.; Peter, K.D.; Aït Hssaine, A.; Ries, J.B. Land use in the Souss region, South Morocco and its influence on wadi dynamics. *Zeitschrift für Geomorphologie* **2019**, *62*, 137–160. [CrossRef]
28. Mulholland, B.; Fullen, M.A. Cattle trampling and soil compaction on loamy sands. *Soil Use Manag.* **1991**, *7*, 189–192. [CrossRef]
29. Stavi, I.; Ungar, E.D.; Lavee, H.; Sarah, P. Livestock modify ground surface microtopography and penetration resistance in a semi-arid shrubland. *Arid Land Res. Manag.* **2009**, *23*, 237–247. [CrossRef]
30. D'Oleire-Oltmanns, S.; Marzolff, I.; Peter, K.D.; Ries, J.B. Unmanned aerial vehicle (UAV) for monitoring soil erosion in Morocco. *Remote Sens.* **2012**, *4*, 3390–3416. [CrossRef]
31. Elmouden, A.; Bouchaou, L.; Snoussi, M. Constraints on alluvial clay mineral assemblages in semiarid regions. The Souss Wadi Basin (Morocco, Northwestern Africa). *Geologica Acta* **2005**, *3*, 3–13. [CrossRef]
32. Mensching, H. *Marokko—Die Landschaften im Maghreb*; Keysersche Verlagsbuchhandlung: Heidelberg, Germany, 1957; pp. 210–221.
33. Ambroggi, R. Etude géologique du versant meridional du Haut-Atlas occidental et de la plaine du Souss. *Notes et Mémoires du Service Géologique* **1963**, *157*, 322.

34. Bhiry, N.; Rognon, P.; Occhietti, S. Origine et diagénèsedes sédiments quaternaires de la vallée moyenne du Souss. *Sciences Géologiques Mémoire* **1989**, *84*, 135–144.

35. Aït Hssaine, A.; Bridgland, D. Pliocene–Quaternary fluvial and aeolian records in the Souss Basin, southwest Morocco: A geomorphological model. *Glob. Planet. Chang.* **2009**, *68*, 288–296. [CrossRef]

36. Aït Hssaine, A. Evolution géomorphologique du piémont sud-atlasique dans la région de Taroudant (SW-Maroc) au cours du Tertiaire et du Pléistocène inférieur. *Bulletin de l'Institut Scientifique* **2000**, *22*, 17–28.

37. Peter, K.D.; d'Oleire-Oltmanns, S.; Ries, J.B.; Marzolff, I.; Aït Hssaine, A. Soil erosion in gully catchments affected by land-levelling measures in the Souss Basin, Morocco, analysed by rainfall simulation and UAV remote sensing data. *Catena* **2014**, *113*, 24–40. [CrossRef]

38. Ghanem, H. Monographie pédologique de la plaine du Souss. *Serv. Rech. Ecol. Direction de la Recherche Agronomique* **1974**, *5*, 101.

39. Rössler, M.; Kirscht, H.; Rademacher, C.; Platt, S. Demographic development in Southern Morocco: Migration, urbanization and the role of institutions in resource management. In *Impacts of Global Change on the Hydrological Cycle in West and Northwest Africa*; Speth, P., Christoph, M., Diekkrüger, B., Eds.; Springer: Berlin, Germany, 2010; pp. 305–314.

40. Barrow, C.J.; Hicham, H. Two complimentary and integrated land uses of the western High Atlas Mountains, Morocco: The potential for sustainable rural livelihoods. *Appl. Geogr.* **2000**, *20*, 369–394. [CrossRef]

41. Köppen, W.; Geiger, G. *Handbuch der Klimatologie*; Gebrüder Borntraeger: Berlin, Germany, 1930; pp. 176–184.

42. Endlicher, W.; Weischet, W. *Regionale Klimatologie: Die Alte Welt: Europa, Afrika, Asien*; Teubner: Leipzig, Germany, 2000; pp. 200–202.

43. Saidi, M.E. Contribution à L'hydrologie Profonde et Superficielle du Bassin du Souss (Maroc). Climatologie, Hydrologie, Crues et Bilans Hydrologiques en Milieu Sub-Aride. Ph.D. Thesis, Cadi Ayyad University, Sorbonne, Paris, France, 1995.

44. Cerdà, A. Parent material and vegetation affect soil erosion in eastern Spain. *Soil Sci. Soc. Am. J.* **1999**, *63*, 362–368. [CrossRef]

45. Iserloh, T.; Fister, W.; Seeger, M.; Willger, H.; Ries, J.B. A small portable rainfall simulator for reproducible experiments on soil erosion. *Soil Tillage Res.* **2012**, *124*, 131–137. [CrossRef]

46. Iserloh, T.; Ries, J.B.; Cerdà, A.; Echeverría, M.T.; Fister, W.; Geißler, C.; Kuhn, N.J.; León, F.J.; Peters, P.; Schindewolf, M.; et al. Comparative measurements with seven rainfall simulators on uniform bare fallow land. *Zeitschrift für Geomorphologie* **2013**, *57*, 11–26. [CrossRef]

47. Saleh, A. Soil roughness measurement: Chain method. *J. Soil Water Conserv.* **1993**, *48*, 527–529.

48. Thomsen, L.M.; Baartman, J.E.M.; Barneveld, R.J.; Starkloff, T.; Stolte, J. Soil surface roughness: Comparing old and new measuring methods and application in a soil erosion model. *SOIL* **2015**, *1*, 399–410. [CrossRef]

49. Peter, K.D.; Ries, J.B. Infiltration rates affected by land levelling measures in the Souss valley, South Morocco. *Zeitschrift für Geomorphologie* **2013**, *57*, 59–72. [CrossRef]

50. Tricker, A.S. The infiltration cylinder: Some comments on its use. *J. Hydrol.* **1978**, *36*, 383–391. [CrossRef]

51. Hills, R.C. The determination of the infiltration capacity of field soils using the cylinder infiltrometer. *Br. Geomorphol. Res. Group Tech. Bull.* **1970**, *3*, 1–24.

52. Link, M. Das Einring-Infiltrometer mit schwimmergeregelter Überstauhöhe—Ein neues Gerät zur Messung von Infiltrationsraten in Böden. *Geoöko* **2000**, *21*, 121–132.

53. Reynolds, W.D.; Elrick, D.E. Ponded Infiltration from a Single Ring: I. Analysis of Steady Flow. *Soil Sci. Soc. Am. J.* **1990**, *54*, 1233–1241. [CrossRef]

54. Elrick, D.E.; Reynolds, W.D. Infiltration from Constant-Head Well Permeameters and Infiltrometers. In *Advances in Measurement of Soil Physical Properties: Bringing Theory into Practice*; Topp, C.G., Reynolds, W.D., Green, R.E., Eds.; SSSA Special Publication 30: Madison, WI, USA, 1992; pp. 1–24.

55. Zhang, R. Determination of soil sorptivity and hydraulic conductivity from the disk infiltrometer. *Soil Sci. Soc. Am. J.* **1997**, *61*, 1024–1030. [CrossRef]

56. Perroux, K.M.; White, I. Designs for disc permeameters. *Soil Sci. Soc. Am. J.* **1988**, *52*, 1205–1215. [CrossRef]

57. Van Genuchten, M.T. A Closed-Form Equation for Predicting the Hydraulic Conductivity of Unsaturated Soils. *Soil Sci. Soc. Am. J.* **1980**, *44*, 892–898. [CrossRef]

58. Carsel, R.F.; Parrish, R.S. Developing joint probability distributions of soil water retention characteristics. *Water Resour. Res.* **1988**, *24*, 755–769. [CrossRef]

59. Reynolds, W.D.; Bowman, B.T.; Brunke, R.R.; Drury, C.F.; Tan, C.S. Comparison of Tension Infiltrometer, Pressure Infiltrometer, and Soil Core Estimates of Saturated Hydraulic Conductivity. *Soil Sci. Soc. Am. J.* **2000**, *64*, 478–484. [CrossRef]

60. DIN ISO 11265:1997-06. *Bodenbeschaffenheit—Bestimmung der Spezifischen Elektrischen Leitfähigkeit*; Beuth Verlag: Berlin, Germany, 1997.

61. Köhn, M. Korngrößenanalyse vermittels Pipettenanalyse. *Tonindustrie-Zeitung* **1929**, *53*, 729–731.

62. Becher, H.H.; Kainz, M. Auswirkungen einer langjährigen Stallmistdüngung auf das Bodengefüge im Lößgebiet bei Straubing. *Zeitschrift für Acker und Pflanzenbau* **1983**, *152*, 152–158.

63. Becher, H.H. Influence of long-term liming on aggregate stability of a loess-derived soil. *Int. Agrophys.* **2001**, *15*, 67–72.

64. Mbagwu, J.S.C.; Auerswald, K. Relationship of percolation stability of soil aggregates to land use, selected properties, structural indices and simulated rainfall erosion. *Soil Tillage Res.* **1999**, *50*, 197–206. [CrossRef]

65. Howlett, D.S.; Moreno, G.; Mosquera Losada, M.R.; Ramachandran Nair, P.K.; Nair, V.D. Soil carbon storage as influenced by tree cover in the Dehesa cork oak silvopasture of central-western Spain. *J. Environ. Monit.* **2011**, *13*, 1897–1904. [CrossRef] [PubMed]

66. Aiello, R.; Bagarello, V.; Barbagallo, S.; Consoli, S.; Di Prima, S.; Giordano, G.; Iovino, M. An assessment of the Beerkan method to determine the hydraulic properties of a sandy loam soil. *Geoderma* **2014**, *235–236*, 300–307. [CrossRef]

67. Alagna, V.; Di Prima, S.; Rodrigo-Comino, J.; Iovino, M.; Pirastru, M.; Keesstra, S.D.; Novara, A.; Cerdà, A. The Impact of the Age of Vines on Soil Hydraulic Conductivity in Vineyards in Eastern Spain. *Water* **2017**, *10*, 14. [CrossRef]

68. Ravi, S.; Breshears, D.D.; Huxman, T.E.; D'Odorico, P. Land degradation in drylands: Interactions among hydrologic-aeolian erosion and vegetation dynamics. *Geomorphology* **2010**, *116*, 236–245. [CrossRef]

69. Culmsee, H. Vegetation and pastoral use in the Western High Atlas Mountains (Morocco). An assessment of sustainability from the geobotanical perspective. In *Pour une Nouvelle Perception des Fonctions des Montagnes du Maroc, Actes du 7eme Colloque Maroco-Allemand, Rabat, Morocco, 2004*; Aït Hamza, M., Popp, H., Eds.; Publications de la Faculté des Lettres et des Sciences Humaines de Rabat. Série: Colloques et Séminaires, Rabat, Morocco, 2005; Volume 119, pp. 67–80.

70. Kenny, L. Biologie de l'Arganier. In *Atlas de l'Arganier*, 1st ed.; Kenny, L., Ed.; Institut agronomique et vétérinaire Hassan II: Rabat, Morocco, 2007; pp. 43–44.

71. Charrouf, Z.; Guillaume, D. Sustainable Development in Northern Africa: The Argan Forest Case. *Sustainability* **2009**, *1*, 1012–1022. [CrossRef]

72. Reid, K.D.; Wilcox, B.P.; Breshears, D.D.; MacDonald, L. Runoff and Erosion in a Piñon-Juniper Woodland: Influence of Vegetation Patches. *Soil Sci. Soc. Am. J.* **1999**, *63*, 1869–1879. [CrossRef]

73. Pierson, F.B.; Williams, C.J.; Hardegree, S.P.; Clark, P.E.; Kormos, P.R.; Al-Hamdan, O.Z. Hydrologic and Erosion Responses of Sagebrush Steppe Following Juniper Encroachment, Wildfire, and Tree Cutting. *Rangel. Ecol. Manag.* **2013**, *66*, 274–289. [CrossRef]

74. Ceballos, A.; Cerdà, A.; Schnabel, S. Runoff production and erosion processes on a dehesa in western Spain. *Geogr. Rev.* **2002**, *92*, 333–353. [CrossRef]

75. Sheng, H.; Cai, T. Influence of Rainfall on Canopy Interception in Mixed Broad-Leaved-Korean Pine Forest in Xiaoxing'an Moundains, Northeastern China. *Forests* **2019**, *10*, 248. [CrossRef]

76. Geißler, C.; Kühn, P.; Böhnke, M.; Bruelheide, H.; Shi, X.; Scholten, T. Splash erosion potential under tree canopies in subtropical SE China. *Catena* **2012**, *91*, 85–93. [CrossRef]

77. Ries, J.B.; Iserloh, T.; Seeger, M.; Gabriels, D. Rainfall simulations—Constraints, needs and challenges for a future use in soil erosion research. *Zeitschrift für Geomorphologie* **2013**, *57*, 1–10. [CrossRef]

78. Kamphorst, A. A small rainfall simulator for the determination of soil erodibility. *Neth. J. Agric. Sci.* **1987**, *35*, 407–415.

79. Bochet, E.; Poesen, J.; Rubio, J.L. Runoff and soil loss under individual plants of semi-arid Mediterranean shrubland: Influence of plant morphology and rainfall intensity. *Earth Surf. Process. Landf.* **2006**, *31*, 536–549. [CrossRef]

80. Ludwig, J.A.; Wilcox, B.P.; Breshears, D.D.; Tongway, D.J.; Imeson, A.C. Vegetation Patches and Runoff-Erosion as Interacting Ecohydrological Processes in Semiarid Landscapes. *Ecology* **2005**, *86*, 288–297. [CrossRef]

81. Gispert, M.; Emran, M.; Pardini, G.; Doni, S.; Ceccanti, B. The impact of land management and abandonment on soil enzymatic activity, glomalin content and aggregate stability. *Geoderma* **2013**, *202–203*, 51–61. [CrossRef]

82. Plaza-Bonilla, D.; Cantero-Martínez, C.; Viñas, P.; Álvaro-Fuentes, J. Soil aggregation and organic carbon protection in a no-tillage chronosequence under Mediterranean conditions. *Geoderma* **2013**, *193–194*, 76–82. [CrossRef]

83. Douglas, J.T.; Goss, M.J. Stability and organic matter content of surface soil aggregates under different methods of cultivation and in grassland. *Soil Tillage Res.* **1982**, *2*, 155–175. [CrossRef]

84. Goebel, M.; Woche, S.K.; Bachmann, J. Quantitative analysis of a liquid penetration kinetics and slaking of aggregates as related to solid-liquid interfacial properties. *J. Hydrol.* **2012**, *442–443*, 63–74. [CrossRef]

85. Le Bissonais, Y. Aggregate stability and assessment of soil crustability and erodibility: I. Theory and methodology. *Eur. J. Soil Sci.* **1996**, *47*, 425–437. [CrossRef]

86. Nciizah, A.D.; Wakindiki, I.I.C. Soil sealing and crusting effects on infiltration rate: A critical review of shortfalls in prediction models and solutions. *Arch. Agron. Soil Sci.* **2015**, *61*, 1211–1230. [CrossRef]

87. Guzmán, G.; Perea-Moreno, A.-J.; Gómez, J.A.; Cabrerizo-Morales, M.Á.; Martínez, G.; Giráldez, J.V. Water Related Properties to Assess Soil Quality in Two Olive Orchards of South Spain under Different Management Strategies. *Water* **2019**, *11*, 367. [CrossRef]

88. Lichner, L.; Orfánus, T.; Nováková, K.; Šír, M.; Tesař, M. The Impact of Vegetation on Hydraulic Conductivity of Sandy Soil. *Soil Water Res.* **2007**, *2*, 59–66. [CrossRef]

89. Abrahams, A.D.; Parsons, A.J. Relation between infiltration and stone cover on a semiarid hillslope, Southern Arizona. *J. Hydrol.* **1991**, *122*, 49–59. [CrossRef]

90. Brakensiek, D.L.; Rawls, W.J. Soil containing rock fragments: Effects on infiltration. *Catena* **1994**, *23*, 99–110. [CrossRef]

91. Eldridge, D.J.; Freudenberger, D. Ecosystem wicks: Woodland trees enhance water infiltration in a fragmented agricultural landscape in eastern Australia. *Austral Ecol.* **2005**, *30*, 336–347. [CrossRef]

92. Castellano, M.J.; Valone, T.J. Livestock, soil compaction and water infiltration rate: Evaluating a potential desertification recovery mechanism. *J. Arid Environ.* **2007**, *71*, 97–108. [CrossRef]

93. Jeddi, K.; Cortina, J.; Chaieb, M. Acacia Salicina, Pinus halepensis and Eucalyptus occidentalis improve soil surface conditions in arid southern Tunisia. *J. Arid Environ.* **2009**, *73*, 1005–1013. [CrossRef]

94. Fu, H.; Pei, S.; Chen, Y.; Wan, C. Influence of Shrubs on Soil Chemical Properties in Alxa Desert Steppe, China. In *Proceedings: Shrubland Dynamics—Fire and Water*; Lubbock, USA, 10–12 August 2004; Sosebee, R.E., Wester, D.B., Britton, C.M., McArthur, E.D., Kitchen, S.G., Eds.; U.S. Department of Agriculture, Forest Service, Rocky Mountain Research Station: Fort Collins, CO, USA, 2007; pp. 117–122.

95. Ries, J.B.; Andres, K.; Wirtz, S.; Tumbrink, J.; Wilms, T.; Peter, K.D.; Burczyk, M.; Butzen, V.; Seeger, M. Sheep and goat erosion—Experimental geomorphology as an approach for the quantification of underestimated processes. *Zeitschrift für Geomorphologie* **2014**, *58*, 23–45. [CrossRef]

96. Berthrong, S.T.; Jobbágy, E.G.; Jackson, R.B. A global meta-analysis of soil exchangeable cations, pH, carbon, and nitrogen with afforestation. *Ecol. Appl.* **2009**, *19*, 2228–2241. [CrossRef] [PubMed]

97. Alba-Sánchez, F.; López-Sáez, J.A.; Nieto-Lugilde, D.; Svenning, J.-C. Long-term climate forcings to assess vulnerability in North Africa dry argan woodlands. *Appl. Veg. Sci.* **2015**, *18*, 283–296. [CrossRef]

98. Azevedo, J.; Morgan, D.L. Fog precipitation in coastal California forests. *Ecology* **1974**, *55*, 1135–1141. [CrossRef]

99. Nyaga, J.M.; Neff, J.C.; Cramer, M.D. The Contribution of Occult Precipitation to Nutrient Deposition on the West Coast of South Africa. *PLoS ONE* **2015**, *10*, 1–21. [CrossRef]

100. Cunningham, R.K. Cation-anion relationships in crop nutrition: III. Relationships between the ratios of sum of the cations: Sum of the anions and nitrogen concentrations in several plant species. *J. Agric. Sci.* **1964**, *63*, 109–111. [CrossRef]

101. Riley, D.; Barber, S.A. Bicarbonate Accumulation and pH Changes at the Soybean (*Glycine max* (L.) Merr.) Root-Soil Interface. *Soil Sci. Soc. Am. J.* **1969**, *33*, 905–908. [CrossRef]

102. Boulmane, M.; Oubrahim, H.; Halim, M.; Bakker, M.R.; Augusto, L. The potential of Eucalyptus plantations to restore degraded soils in semi-arid Morocco (NW Africa). *Ann. For. Sci.* **2017**, *74*, 57. [CrossRef]

Article

Triggering Influence of Seasonal Agricultural Irrigation on Shallow Loess Landslides on the South Jingyang Plateau, China

Rui-Xin Yan [1,2,3], Jian-Bing Peng [1,2,*], Qiang-Bing Huang [1,2], Li-Jie Chen [2], Chen-Yun Kang [2] and Yan-Jun Shen [4]

[1] Key Laboratory of Western Mineral Resources and Geological Engineering of Ministry of Education, Chang'an University, Xi'an 710054, China
[2] College of Geology Engineering & Geomatics, Chang'an University, Xi'an 710054, China
[3] College of Architecture and Civil Engineering, Xi'an University of Science and Technology, Xi'an 710054, China
[4] College of Geology and Environment, Xi'an University of Science and Technology, Xi'an 710054, China
* Correspondence: dicexy_1@126.com; Tel.: +86-139-9288-1198

Received: 8 June 2019; Accepted: 14 July 2019; Published: 16 July 2019

Abstract: Since large-scale agricultural irrigation began in the 1980s, 92 landslides have occurred around the South Jingyang Plateau during the past 40 years. The geological disaster and soil erosion have caused numerous casualties and substantial property loss. In this work, several field investigations are carried out to explore the soil erosion and mechanical mechanism of these irrigated shallow loess landslides on the South Jingyang Plateau. (1) We investigated the spatial distributions, types and developmental characteristics of loess landslides. (2) We surveyed and monitored seasonal agricultural irrigation features and groundwater changes in the area since the 1980s and found that irrigation is a significant factor influencing groundwater changes, soil erosion and even causing landslides to occur. (3) Based on the field investigation, the occurrence of these irrigated shallow loess landslides was generalized, and it was found that the core process was due to the liquefaction of softening zone. We carried out a static liquefaction test and verified that the natural loess was prone to liquefaction. (4) The three main reasons for shallow loess landslides in the South Jingyang Plateau were discussed. This study provides a valuable reference for achieving an understanding of the relationship between seasonal agricultural irrigation and the occurrence of loess landslides in the area as well as similar irrigated agricultural areas.

Keywords: loess landslide; agricultural irrigation; field investigation; static liquefaction; soil erosion

---

1. Introduction

The loess plateau, a special geomorphologic landscape located in the northwestern area of China, covers approximately 4.4% (620,000 km^2) of China's total land area [1,2]. Due to the continental monsoon climate limit, seasonal irrigation has become a necessary approach for maintaining agricultural production in this area [3]. Loess has typical features including macro-pores, vertical joints, weak cementation and high sensitivity to water [4–6]. These characteristics are prone to lead to soil erosion, structural collapse under long-term immersion water and may even lead to instability [7–9]. Thus, seasonal agricultural irrigation has become an important triggering factor inducing serious landslides on the loess plateau of northwest China [10–12]. As a matter of fact, agricultural activity has become an important threat to the natural process around the world. For example, man-made terraces in many areas of the world lead to an increase in instability hazards according to the previous reports [13–15]. Irrigation concrete-ditch prevented the evaporation of groundwater, resulting in the "canopy-cover" effect [16,17]. Focusing on the loess plateau, many loess landslides caused by seasonal irrigation have been also occurred [2,18–20]. For instance, there have been 120 loess landslide events in the Heifangtai area, which have killed approximately 70 people, due to irrigation, since 1984 [21]. In addition, the phenomenon of agricultural irrigation influencing the local climate change should be also investigated [22–24]. Therefore, it is necessary to carry out the systematic study on the influence of seasonal agricultural irrigation on natural changes in the loess plateau.

In the study area of the south Jingyang platform of Shaanxi Province, China, the large-scale water irrigation performed in farmland since the 1980s has caused the loess to suffer the soil erosion under its own weight, which has damaged buildings, and slope instability has occurred due to infiltration of irrigation water in some areas. According to statistics, there have been as many as 92 multi-cyclical landslides in this region, resulting in the fact that at least 17 places have experienced at least two slides, and more than 30 people have been reported dead. One of these landslides occurred on 2 December 1984, killed 20 people, and led to 20 people being seriously injured. In addition, a high, steep slope has formed at the edge of the south Jingyang platform in Shaanxi Province, due to the considerable lateral erosion near the Jinghe River. A pumping station built along the river increased the irrigation volume and seriously affected the stability of the slope, resulting in frequent landslides along the edges of the plateau [25]. With tableland farmland collapsing and declining, the cultivated farmland area has decreased continuously, in turn, the development of local agriculture has been seriously affected and great property losses have been caused to farmers on the plateau. In addition, Xianyang International Airport in the largest city of Xi'an in the Northwest in China is located on the tableland, and the change in the drainage system on the raft due to soil erosion will also affect normal operations of the airport [26].

The mechanism of the loess landslides on the south Jingyang platform has been studied by many authors due to their disastrous effects [25–28]. Some authors hypothesize that irrigation water infiltrates into the slope along vertical joints, which leads to an increase in the degree of saturation of the soil. The local loess then becomes saturated, and instability occurs [11,20]. At present, studies on the mechanism of irrigation-induced loess landslides are mostly carried out through approaches such as laboratory tests, field tests, and numerical simulations. For example, Puri [29], Zhang et al. [10], Li et al. [27], Xu et al. [19] and Li et al. [18] discussed how irrigation induces loess landslides step by step through a series of indoor triaxial tests and stated the generation of pore water pressure and shear deformation during loess saturation. These authors suggested that with the infiltration of irrigation water into the slope, the perched water level rises, leading to failure to the upper slope and then fluidization, which induces high-speed and long-distance landslides. Here, the key point is that the liquefaction of loess is an important cause of rapid, long-distance loess flow slides induced by irrigation. Cui et al. [30] adopted a centrifugal model test to simulate the mechanical behavior of the loess slope on the Heifangtai platform under irrigation conditions. The characteristics of loess slope deformation, pore water pressure and soil pressure were systematically tested, and the characteristics of the evolution of the instability of the loess slope caused by irrigation were discussed. It has been inferred from laboratory shear tests that static liquefaction of the loess slope foundation is one of the main reasons for the instability of the loess slope [29–31].

In addition, the ring shear test has been applied to test the undrained shear characteristics of loess to some extent. Peng et al. [26] carried out a consolidated undrained shear test and ring shear test, revealing the characteristics and mechanism of loess sliding-flow landslides on the south Jingyang platform. It was concluded that the surface friction of landslides caused by irrigation was deeper than that of landslides caused by the Wenchuan earthquake and that saturated loess samples exhibited a high liquefaction ability. Zhang et al. [32] carried out undrained ring shear tests on loess samples with different initial porosities, hypothesizing that soil densification caused by irrigation may lead to the occurrence and mobility of landslides along the plateau. Some authors have also carried out irrigation simulation tests using the farmland field conditions or similar indoor models to analyze the effect of irrigation on the stability of tableland slopes. Xu et al. [11] conducted laboratory tests and large-scale field tests on a typical cracked plateau in Heifangtai to simulate the impact of irrigation on slope stability. The test showed that cracks had a significant impact on the irrigation water flowing into the ground. With the infiltration of irrigation water into the main cracks, the pore water pressure on the loess slope increases rapidly, causing local instability of the slope. Numerical simulations of irrigation-induced landslides in loess areas have also been studied to some extent. For example, Lian et al. [33] applied Phase2 software to numerically simulate the typical profile of Huangci landslides, considering the stability evolution law of landslides under conditions of a long-term rise and fall of irrigation water levels, and these authors revealed the general law of irrigation-induced landslides in this area through numerical simulation. Similarly, Pan et al. [34] used a numerical simulation method to analyze the failure mechanism of loess landslides on the Heifangtai platform caused by irrigation. Li et al. [27] applied the improved Sassa K model to simulate the sliding of landslides.

In general, previous studies on agricultural irrigation-type landslides in this region have mostly adopted approaches such as indoor tests, field tests, and numerical simulations. By carrying out tests and simulations on the relationship between the mechanical properties of loess and water, the induced effects of irrigation on loess landslides are indirectly reflected. This study focuses on a systematic investigation of the hydrogeological conditions of loess landslides in this area, through statistical analysis of the relationship between irrigation, groundwater and intuitively establishes the internal relationship between agricultural irrigation and landslides. Then, a generalized model of landslide occurrence is proposed to reveal the internal mechanism of shallow surface landslides induced by seasonal agricultural irrigation in this region.

2. Materials and Methods

2.1. Geographic, Geomorphologic and Climatic Features on the South Jingyang Plateau

The South Jingyang Plateau, located on the south bank of Jinghe River, in Jingyang county, Shaanxi Province, extends from east to west in terms of overall direction, with a total extension of 28 km, mainly passing through the three administrative townships of Taiping, Jiangliu and Gaozhuang (shown in Figure 1). The Quaternary loess was deposited on the southern uplift of the Jinghe River, influenced by the hidden faults of Jinghe River, and eventually formed the loess tableland whose area is approximately 70 km^2. The difference in elevation between its top and bottom is approximately 30–90 m, and the elevation is approximately 420–490 m above sea level. The tableland surface is open and flat, suitable for cultivation. In addition, due to the effect of long-term lateral erosion of the Jinghe River, the edge of the tableland is 30–90 m high, and the slope is 45°–80° [25]. The south Jingyang platform has been affected by large-scale agricultural irrigation since 1980 and had experienced 92 loess landslide accidents as of April 2016. Among these events, many slips have occurred in 17 places, causing numerous casualties and substantial property losses [35].

Furthermore, Jingyang belongs to a temperate continental monsoon climate zone with four distinct seasons and average annual precipitation of 548.7 mm, but rainfall is extremely uneven. The annual precipitation is mainly concentrated in July–September, accounting for 71% of the total annual precipitation. Figure 2 shows the monthly average precipitation distribution in the region from 2000 to 2017. Rainfall in July–September is relatively heavy, exceeding 80 mm, and especially that in July and September is over 90 mm. In other months, rainfall is significantly lower. From December to February of the second year, the monthly average precipitation did not exceed 15 mm. To ensure the normal operation of agricultural production, irrigation has become an inevitable choice.

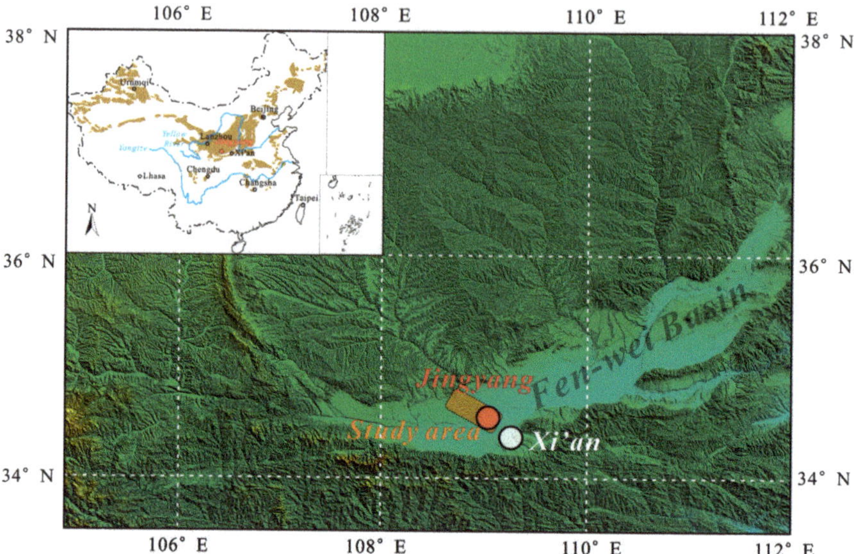

Figure 1. Study location of the South Jingyang Plateau in the northwestern of China.

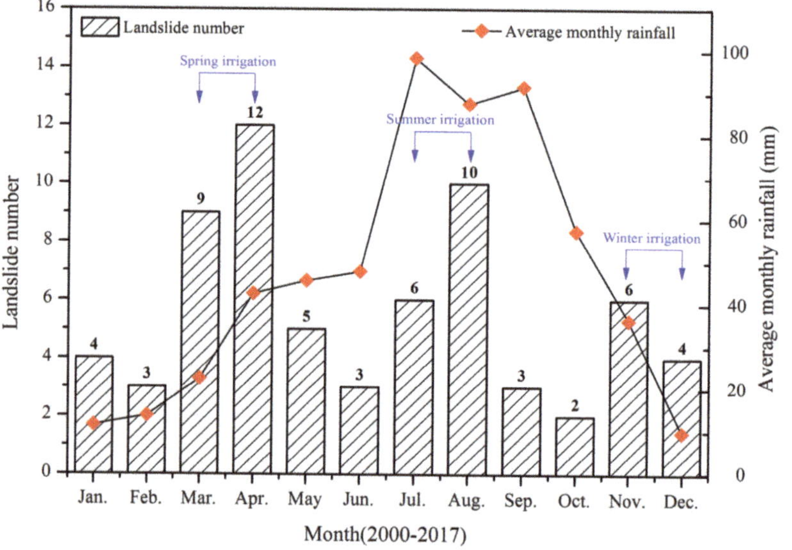

Figure 2. Monthly average precipitation distribution on the South Jingyang Plateau from 2000 to 2017.

2.2. Landslide Development Survey on the South Jingyang Plateau

To further understand the developmental characteristics of landslides in the region, a detailed survey of the development characteristics of the different landslide groups in the region was carried out, and the following main features were found. Due to the strong lateral erosion of the Jinghe River, the slope of the trailing edge of the landslide group is generally steep. The average slope of the trailing edge of the landslide group can reach 50°, and the earlier it slides, the steeper the slope of the corresponding trailing edge. The steep slope along the edge of the platform creates appropriate gravity conditions for the formation of a landslide. There are many cracks after a landslide occurs on tableland. The cracks are mostly arc-shaped, and the overall trend is parallel to the side of the plateau. Such cracks are concentrated in the areas where landslides occur frequently (such as Miaodian or Jiangliu), and the penetration is good. According to statistics, there are 36 cracks in the study area with a length of more than 5 m. The distance between a crack and the edge is generally less than 30 m, accounting for more than 90% of the total number of cracks. The farther away from the edge, the lower the number of cracks. Some cracks show vertical dislocation. Parts of cracks are accompanied by sinkholes. The cracks are formed by the enrichment of irrigation water to form water holes of different scales. A landslide is mostly cut out from the slope foot, and it then slides along the terrace over a long distance, potentially extending for 200–300 m. The resulting landslide body is mostly tongue-shaped, within which depression of the trailing edge is "crescent-shaped". That is, the ends are narrow and shallow, and the center is wide and deep. In the center, the drums and the depressions are arranged to form a wavy terrain, and the drums and depressions extend in a direction perpendicular to the sliding direction. Most of the landslide front is accompanied by mud extrusion, which is like the mudflow accumulation, and the leading edge tends to extend over a long distance.

2.3. Agricultural Irrigation Survey on the South Jingyang Plateau

According to the agricultural production rules of this region, it can generally be divided into three stages of irrigation. Spring irrigation: It is concentrated from early March to mid-April every year. The spring crops recover and grow, and rainfall is obviously insufficient. To ensure the growth of spring seedlings, spring irrigation is mostly provided by diverting water from the Jinghe River. Summer irrigation: It is concentrated from early July to late August each year. Although rainfall is high at this stage, it is not uniform, and most rainfall events are concentrated and drastically reduced. At this stage, the growth of crops is nearing the harvest stage, and the water requirement is high. In addition, evaporation in summer is high, and summer irrigation is needed in Jingyang. According to historical data, summer evaporation on the South Jingyang Plateau is greater than 150 mm. The amount of evaporation in summer accounts for more than half of the total annual evaporation. The maximum evaporation per month in summer can reach 275 mm [36]. The amount of rainfall in summer is much less than the amount of evaporation. Due to a large amount of evaporation in summer, irrigation contributes less to the rise of the groundwater level. Winter irrigation: Due to the freezing away effect after winter irrigation, the surface soil forms 1–2 cm-thick agglomerates, which can reduce evaporation from the ground and loosen the soil. This situation is conducive to the storage of water. Therefore, winter irrigation is often carried out on the South Jingyang Plateau. Winter irrigation is mainly concentrated in early November to late December. However, considering that the region is cold in winter, irrigation water tends to freeze in the ground, and the maximum frozen soil depth can reach 44 cm [37]. Therefore, the actual infiltration water volume and depth will be affected by freezing.

At present, current agricultural irrigation in the region is mostly based on traditional flood irrigation. Because the topography of the tableland side is slightly tilted toward the inner area of the tableland, to facilitate the flow of water in a ditch, the main channel on the South Jingyang Plateau is arranged along the rim. Due to leakage of the main canal and broad irrigation, groundwater recharging on the plateau is artificially increased, which provides important conditions for the occurrence of landslides.

2.4. Experimental Method of Loess Liquefaction Tendency Due to Irrigation

To explore the effect of irrigation on the slope, we also systematically carried out static liquefaction tests on typical soil samples (from middle Pleistocene Q2 undisturbed loess) from the South Jingyang Plateau. In this work, all of loess samples were collected from the investigation adit located in the rear of Zhaitou landslide. The buried depth of the adit is about 22 m from the tableland roof. In order to avoid the disturbance of natural loess, each sampled natural loess was trimmed to a cuboid-shaped block with 30 cm × 20 cm at the sample site. The block was cut into a cylindrical sample with a diameter of 12 cm and a height of 15 cm and sealed in a steel cylindrical bucket by wrapping tape and liquid paraffin to avoid water desorption. Each one was cut a standard cylinder as natural undisturbed loess sample with a diameter of 5 cm and a height of 10 cm in the laboratory. The basic physical parameter tests including moisture content, dry density, porosity ratio, plastic and liquid limits of the prepared loess samples were performed firstly respectively. The measured basic physical parameters are listed in Table 1.

Table 1. Basic physical parameters of Q2 loess samples from a landslide located in the South Jingyang Plateau.

Sample Site	Natural Moisture Content (%)	Dry Density (g/cm^3)	Porosity Ratio	Specific Gravity	Liquid Limit (%)	Plastic Limit (%)
Q2 loess	17	1.44	0.88	2.7	25.34	19.61

The particle size distribution of the loess sample in the area was also tested by a laser particle size analyzer (Type: Bettersize2000) to reflect the relationship between the static liquefaction and the particle size distribution. The particle distribution curve is illustrated in Figure 3. It can be observed that the silt (size: 0.005–0.05 mm) is the primary grain size which occupies about 67%. While the sand (size: 0.05–2 mm) content and clay (size: <0.005 mm) content are 23% and 10%, respectively. It can be regarded as low-plasticity clay(CL) according to the Casagrande classification.

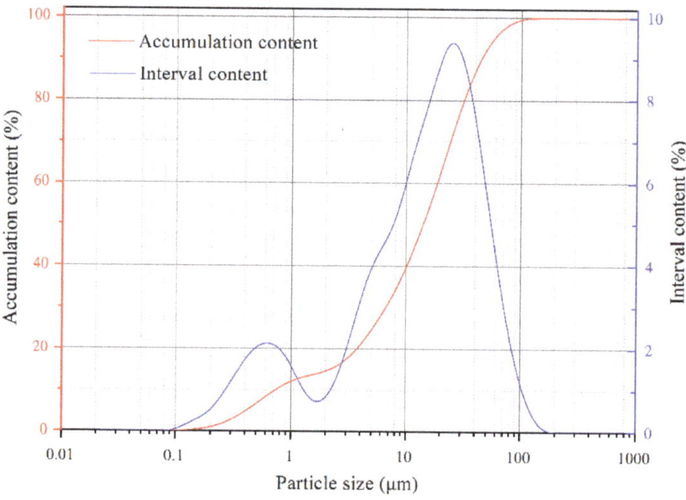

Figure 3. The granulometry chart of intact loess sample.

In this experiment, the undrained triaxial tests were carried out on the saturation intact loess samples by GDS triaxial instruments (Wykeham Farrance company, Hertfordshire, UK), the shear evolution characteristics of saturated loess due to irrigation water infiltration were observed.

3. Results and Analysis

3.1. Distribution Condition of Landslides on the South Jingyang Plateau

Figure 3 shows the distribution of 92 landslides on the South Jingyang Plateau from 1980 to 2016. It can be seen from the figure that the distribution of the groups of landslides in this region has the following typical characteristics:

(1) Group occurrence

The loess landslides in the area are distributed in a "bead necklace"-like pattern, which is coherently developed along the edges of the plateau but locally presents concentrated cluster features, as observed in Jiangliu, Shutangwang, Zhaitou-Miaodian, Taiping and Niujiazui. The concentration of the landslide group is much greater than that of the surrounding areas. For example, in the Jiangliu–Miaodian area, affected by long-term irrigation and lateral erosion of the Jinghe River, there have been 64 loess landslides along the edge over a distance of nearly 10 km. Additionally, there have been 34 loess landslides along the edge of the Zhaitou–Miaodian area, which is only approximately 3.6 km long. These landslides are very dense and concentrated in such a limited platform-edge.

(2) Multiple sliding events

Loess landslides in the study area exhibit multiple failure characteristics. That is to say, due to the influence of irrigation, the same landslide undergoes many slips. Therefore, the soil around the plateau is continuously slipping, and the tableland surface continues to shrink. The multi-sequence developmental characteristics reflect the relationship between landslides in time and space to some extent, which has the following characteristics: The more times that a landslide undergoes sliding, the smaller the landslide volume that is produced, especially when the first landslide volume is large. The horizontal slip of the landslide decreases significantly with an increase in the sliding sequence. When the slope slides for the first time, the shear opening is mostly located at the slope foot and below the terraces. When the number of sliding events increases, the cutting position will gradually increase [36].

(3) Heterogeneity

The loess landslides in the study area also exhibit the characteristics of differentiation, which is mainly affected by the differences in micro-geomorphology. There are significant differences in the development types of landslides in the eastern and western sections. It can be seen from the boundary line in Figure 3 bordered by Dongzhufan Village, Taiping Town, that the eastern section is mainly dominated by landslides with long-distance flows, while the western section is dominated by collapse. By comparison, it was found that the irrigation conditions in the eastern and western sections of the plateau were the same. The regional geological background was basically the same. The slope morphology and material composition were similar, but the topographical features were obviously different. The morphology of the eastern section was relatively complete, while the western section was broken, where the gully was extremely developed. As a result, the relative drainage boundary of the surface water and groundwater in the western section was longer than in the eastern section, while the sunshine range and evaporation area were relatively greater in the western section. Therefore, long-term irrigation has less influence on the surface water and underground water level in the western tableland, which results in the differentiation of landslide development types.

3.2. Landslide Types on the South Jingyang Plateau

According to the statistics for loess tableland landslides in this region, the loess landslides can be divided into five types referring to the results of Hungr et al. [38]: Mudflows, clay/silt rotational slides, erosion slides, external disturbed slides, and silt topple. The corresponding distribution of these types in this region is indicated in Figure 4. Considering that this study mainly focuses on landslides, only four types of landslide development characteristics are introduced, and their representative types are shown in Figure 5 (Note: The distribution locations are marked in Figure 4). There were 92 landslides in this survey, including 35 mudflows, 37 clay/silt rotational slides, 15 erosion slides, and five externally disturbed slides (artificial loading, cutting slope toe, engineering disturbances, etc.). The proportion of landslide types is shown in Figure 6. The number of mudflows and clay/silt rotational slides in the study area is equivalent, accounting for 38.04% and 40.22% of all landslides, respectively. These numbers are far greater than the numbers of erosion slides and external disturbed slides (representing only 21.73% in total). The development characteristics of each landslide type are described below.

(a) **Mudflows:** This type of landslide mostly occurs in slope areas with abundant underground water at the foot of a slope and is the main landslide type in the area. There have been a total of 35 landslides of this type, which are significantly affected by tableland irrigation. This kind of landslide is mostly cut out from the slope foot (Q2 loess layer). Due to the high-water content of the slope foot and terrace silt layer (or sand and gravel), high pore-water pressure can be generated in the soil near the sliding surface during movement, which causes the soil in the sliding belt to fully liquefy.

Even if the terrace sliding bed is gentle, the landslides show significant characteristics of high-speed and long-distance motion. The plane shape of the sliding body is mostly semi-circular to round. The sliding distance is generally 200–300 m, with a maximum of 419 m (e.g., the Xiushidu landslide). The volume ranges from several hundred thousand square meters to one million square meters, representing medium-sized landslides. This type of landslide shows the greatest risk, the widest threat range and the most serious damage. Typical examples include the Jiangliu landslide, Dongfeng landslide, Xihetan landslide, Zhaitou arsenal landslide, and Miaodian landslide. From the perspective of the landslide period, the characteristics of flow slip mainly occur in the first sliding of the slope. However, in the local section, such as the Jiangliu landslide section, due to the accumulation of the landslide at the slope foot, the groundwater level will rise. The formation of a water seepage zone in the new slope foot section will also cause the flow slip characteristic of landslides to occur in the third and fourth phases. For example, in November 2014, January 2014, and March 2016, three consecutive third- or fourth-stage slide landslides occurred. The scale of this type of landslide is not large, but the sliding distance is far, the sliding body exhibits a high-water content, and the leading edge shows the characteristics of a mudflow.

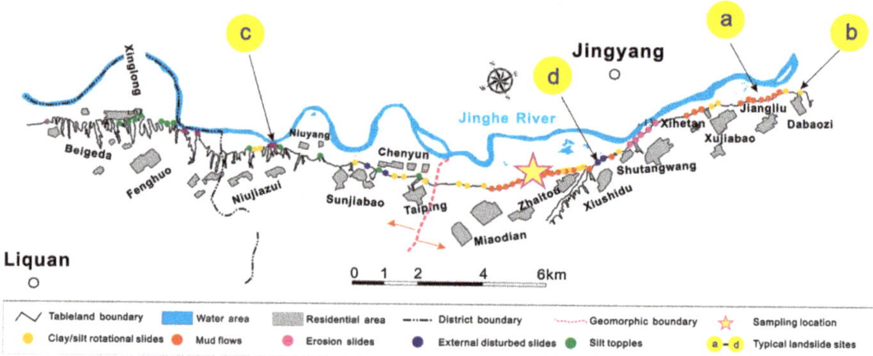

Figure 4. Distribution of loess landslides from 1980 to 2016.

Figure 5. Typical types of loess landslides on the South Jingyang Plateau: (**a**) Mudflow, (**b**) clay/silt rotational slides, (**c**) erosion slide, and (**d**) external disturbed slide.

(b) **Clay/silt rotational slides:** This type of landslide is characterized by shear stress reaching the maximum shear strength of the soil and causing damage. The shear opening associated with such landslides is relatively high and occurs in the unsaturated loess layer, which is caused by a decrease in suction in the unsaturated loess matrix due to surface water infiltration, such as that resulting from farmland irrigation. From the perspective of landslide evolution, sliding mainly occurs in the second and third landslides, and the local section corresponds to the first landslide, due to the deep groundwater. This type of landslide is generally small in scale, the slope is not saturated, and the degree of liquefaction is low during movement. The speed and distance of landslide movement are much smaller than those of the landslide, and there is little damage.

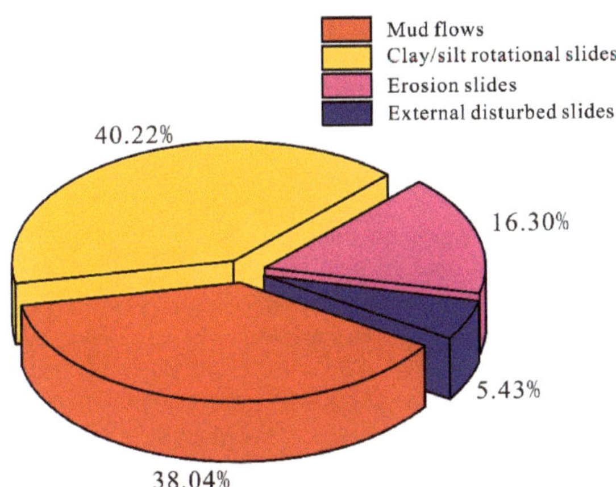

Figure 6. Statistics of the proportions of landslide types on the South Jingyang Plateau.

(c) **Erosion slides:** There are 15 erosion slides. This type of landslide including 15 subclasses mostly develops near the edge of the Jinghe River. Due to the lateral erosion of the Jinghe River, the slope along the edges of the plateau is steep, and the stress at the slope foot is concentrated. Due to the influence of river water infiltration, the saturation intensity of the soil at the slope foot is reduced. The slope is cut from the deep part of the slope to form a slip. Such landslides are generally thick, making them sub-stable within a certain period of time. Such loess landslides generally do not cause casualties, but they exhibit the characteristics of short incubation periods and frequent occurrences, which result in loss of land resources. These events may cause partial blockage of the Jinghe River and flooding of fields as well as affecting residents' travel.

(d) **External disturbed slides:** This type of landslide is closely related to human engineering activities. There have been five external disturbed slides in the area, affected by a cutting foot or large-scale engineering construction disturbance (such as road construction under tableland, or farming on the slope), combined with the impact of river erosion and agricultural irrigation. The distribution of this type of landslide is not regular and shows a clear relationship with the intensity and locations of engineering activities. Engineering disturbance may still play an important role in triggering landslides in this region.

3.3. Changes in Groundwater Levels on the South Jingyang Plateau

The estimated equilibrium phreatic water values in the irrigation area on the South Jingyang Plateau are listed in Table 2 according to a survey report from the Xianyang Water Conservancy Survey and Design Group from 1986. The annual average recharge of atmospheric rainfall precipitation that infiltrates into the soil on the South Jingyang Plateau is 188.7×10^4 m^3, while annual average excretion in the area is 220.8×10^4 m^3. The atmospheric recharge is much less than the excretion. Thus, precipitation alone is obviously not enough to raise the phreatic water level. Since the implementation of irrigation, the annual recharge has increased by 268.4×10^4 m^3, which means that approximately 236.3×10^4 m^3 of water is stored in the loess, not including the amount of excretion. According to the effective irrigation area, the groundwater level is increased by approximately 1 m.

Table 2. Estimated equilibrium phreatic water values in an effective irrigation area of 62.9 km^2 on the South Jingyang Plateau.

Content	Type	Quantity * (10^4 m^3)	Proportion (%)	Total Quantity (10^4 m^3)	Storage (10^4 m^3)
Recharge	Rainfall infiltration	188.7	41.0	457.1	236.3
	Field infiltration	119.3	26.0		
	Channel infiltration	149.1	33.0		
Excretion	Discharge to Jinghe River	111.8	50.0	220.8	
	Pumping to irrigate	50.0	23.0		
	Domestic water	59.0	27.0		

* According to the hydrogeological survey report for the South Jingyang Plateau from Xianyang Water Conservancy Survey and Design Group in 2016.

According to the variation of the groundwater level at a representative point on the South Jingyang Plateau (Table 3) [25,39,40], we can observe that: (1) In 1976, the groundwater depth was relatively great, and the elevation of the groundwater level was close to that of the riverbed, which was generally lower than that of the Jinghe River bed by approximately 1–8 m. (2) Comparing 1992 with 1976, the groundwater level was obviously elevated at the same point, the uplift was approximately 13–37 m, and the groundwater level was approximately 4–30 m higher than the riverbed. Taking the five representative points in Table 3 as a basic reference, the phreatic water level on the South Jingyang Plateau in 1992 was increased by an average of 23.2 m compared with 1976. The average annual increase ratio was 1.45 m. (3) In 2016 compared with 1992, with the exception of a small decrease in the water depth at Dabuzi Village, the other four points exhibited obvious increases, the uplift was approximately 4–24 m, and the groundwater level was approximately 19–37 m higher than the riverbed. In 2016, the phreatic water level on the South Jingyang Plateau was increased by an average of 33.6 m compared with 1976, with an average annual increase of 0.84 m. This finding reflects the fact that after 1976, water diversion was carried out from Baoji Gorge. With the continuous expansion of the irrigation area, the amount of irrigation water has increased, resulting in an increase in the supply of groundwater. The recharge is greater than the discharge, and the groundwater level has increased each year.

Table 3. Changes in groundwater levels at a representative point on the South Jingyang Plateau.

Time	Location	Groundwater Depth (m)	Groundwater Level Elevation (m)	Difference Between the Ground Water Level and the Riverbed (m)
1976 [25]	Yujiabao	81.0	376.0	−8.0
	Jiangliu	74.5	373.5	−7.7
	Dabuzi	61.0	379.5	−0.5
	Miaodian	66.7	387.9	−8.6
	Zhaitou	52.0	389.0	−2.0
1992 [25]	Yujiabu	61.0	396.0	+12.0
	Jiangliu	37.0	411.0	+30.0
1992 [25]	Dabuzi	33.0	407.5	+27.5
	Miaodian	53.5	401.1	+4.6
	Zhaitou	34.5	406.5	+15.5
2016	Yujiabu [40]	42.0	415.0	+19.0
	Jiangliu	28.83	419.17	+37.97
	Dabuzi	37.37	403.13	+23.13
	Miaodian	28.66	425.94	+29.44
	Zhaitou	30.27	410.73	+19.73

3.4. Experimental Results of Loess Static Liquefaction of Irrigation-induced Landslides

Figure 7 shows the stress-strain curve, pore pressure growth curve, effective confining pressure curve and stress path curve of undisturbed loess. It can be seen based on the indoor liquefaction tendency experiments as: (1) At the initial stage of loading, the deviatoric stress rises rapidly to a peak at a small strain. Then, as the strain increases, the deviatoric stress decreases sharply to a relatively stable value. Under different confining pressures, a strong strain softening type is observed. (2) When the strain is very low, pore pressure increases rapidly. After increasing to a larger value, the pore pressure increases very slowly as the strain increases and gradually becomes stable. (3) The effective confining pressure drops sharply at the beginning and then tends to be stable. The steady-state effective confining pressure increases with the increase of the initial confining pressure. At a steady state, the effective confining pressure is low, and the saturated loess is in an unstable state with low confinement, which is prone to plastic flows.

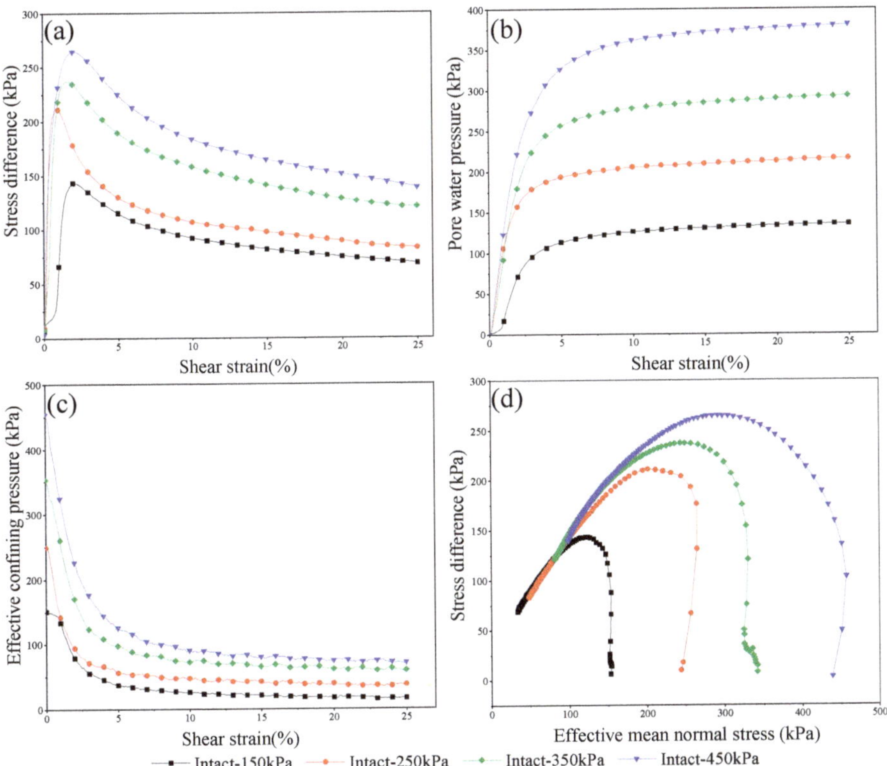

Figure 7. Consolidated undrained triaxial test results for undisturbed saturated loess. (**a**) Stress-strain curve of intact loess, (**b**) pore water pressure curve of intact loess, (**c**) effective confining pressure curve, and (**d**) effective stress paths.

It can be seen from Figure 7 that after the deviatoric stress reaches a peak, the soil structure is rapidly destroyed as the pore water pressure inside the soil sample continues to increase, resulting in a rapid decline in the strength of the structure. After the axial strain reaches approximately 20%, the pore water pressure and the deviatoric stress gradually reach a steady state. Based on the theory of soil plasticity, the yield surface of the specimen shrinks with strain in the stress space. Currently, the original saturated loess presents a possibility of liquefaction slippage. This possibility was verified by experiments showing that the infiltration of irrigation water causes static liquefaction of soil saturation, which leads to landslides.

4. Discussion

4.1. Relationship Between Landslides and Irrigation on the South Jingyang Plateau

According to survey data, since diversion irrigation began in 1976 in this area, tableland margin landslides have been significantly active. From 1982 to 1985, there were four large landslides with a volume of 1.88 million m^3, resulting in 24 deaths, 20 injuries and the destruction of 580 mu of fruit trees. Landslide disasters were particularly prevalent from 2002 to 2006, with landslides occurring continuously in the Zhaitou–Jiangliu section, showing beading characteristics. For example, the Shutangwang landslide occurred on 2 October 2002, the Dongfeng landslide occurred on 23 July 2003, the Xiushidu landslide occurred on 12 October 2002, and the Bridgehead landslide occurred on 15 April 2006. Not only is the scale of the landslides large, their frequency is also high.

According to the statistical distribution of the landslides occurring in 2000–2017, it was found that the number of landslides and the monthly average precipitation do not show the same increasing and decreasing trends, indicating that rainfall is not the main cause of landslides in this area. There is a strong correlation between irrigation and the number of landslides. For example, the number of landslides occurring in April was highest, at 12, followed by August and March, with 10 and 9, respectively, while there were six landslides in both July and November. On the South Jingyang Plateau, March–April are the spring irrigation months. July–August are the summer irrigation months, and November is the winter irrigation month. Thus, landslides occurred most frequently in the spring irrigation period, followed by the summer and winter irrigation periods. The number of landslides outside of the main irrigation months (February, June, September, October) is smaller. These observations indicate that the occurrence of landslides in the study area is closely related to irrigation. According to local survey data, the amount of irrigation is generally highest in spring, so the frequency of landslides is also highest in this period. It can be seen that irrigation is an important trigger for landslides on the South Jingyang Plateau.

4.2. Analysis of Loess Landslides Development Induced by Irrigation

Based on the above-mentioned in situ survey, water infiltration due to irrigation is a significant triggering factor of inducing group-occurring of loess landslide on the loess plateau [41]. Additionally, the fissures or cracks located at the margin of the plateau provide a seepage channel for irrigation water infiltration, which in turn decreases the stability of landslides. At the same time, newly formed landslides will progressively cause new fissures, alter groundwater discharge conditions and increase groundwater levels. All these effects promote a new round of landslides. Therefore, the irrigation-associated landslides in this area present a cyclic character.

According to field survey statistics, the formation and evolution of irrigation-associated landslides can be divided into three stages.

The early stage, in which water retained in farmland after large-scale irrigation slowly migrates to the interior of the loess slope through the thick vapor zone. As the frequency of irrigation increases, the groundwater level in the area increases continuously (Table 3). The irrigation water infiltrates and erodes the loess structure in this stage, leading to continuous expansion and degeneration of vertical fissures in the loess. Continuous soil erosion results in numerous macros-cracks distributed around the top and even through the slope.

The infiltration stage of the preferential seepage channel, in which the groundwater level gradually rises with the development of the above phenomenon, the pore hydrostatic pressure increases and cracks located on the side of the plateau begin to gradually expand. Then, repeated suffusion erosion causes cracks located on the side of the plateau to develop into the preferential seepage channels for surface water infiltration, causing surface water to rapidly infiltrate the interior of the loess slopes. This infiltration further aggravates the increase in the groundwater level, and the uplift of the groundwater level, in turn, promotes the rapid development of cracks, which causes the slope to be unstable. Additionally, according to the on-site drilling profile, there is an obvious hydraulic gradient zone close to the side of the plateau, leading to an increase in hydrodynamic pressure on the slope and a rapid increase in the development of fractures induced by hydrodynamic pressure. Furthermore, dairy irrigation is prone to remove chemical fertilizers and soil solutes, which brings about chemical corrosion and accelerates the development of sinkholes and cracks located on the side of the plateau.

The landslide sliding stage, in which surface water from irrigation continuously sinks down into the interior of the loess slopes through the preferential seepage channel, which accelerates the transfixion of the sliding surface. The water then gradually collects downwards, crossing the paleosol layer (aquifuge) at the bottom of the slope and forming perched water at the top of this layer. The bottom of the upper Lishi loess is soaked with water and gradually becomes saturated. Hence, liquefaction phenomena may occur, and the saturated softening layer is softened to be an easy-sliding layer and form as the bottom sliding surface, which brings about landslides [42].

The general process of irrigation-induced loess landslides occurring on the side of the plateau is summarized in Figure 8.

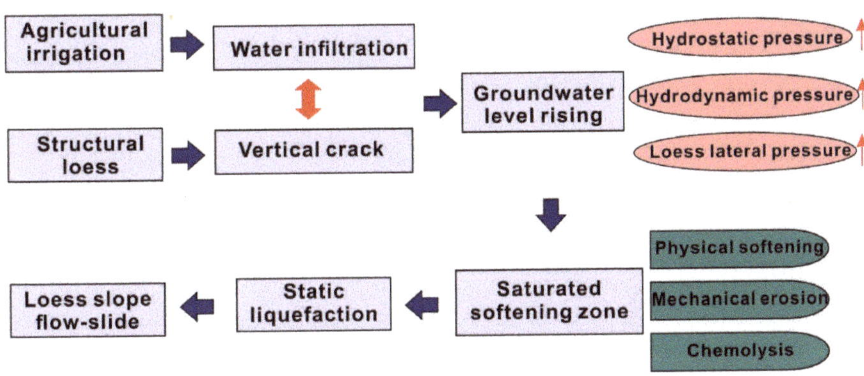

Figure 8. Occurrence of loess landslides due to irrigation on the side of the plateau.

4.3. Discussion on Reasons of Landslides Induced by Irrigation on the Side of the Plateau

Based on the site investigation, indoor tests and previous research findings, the genetic model of landslides induced by irrigation on the South Jingyang Plateau was analyzed. Three main reasons were identified:

(1) **Long-term irrigation leads to significant tension cracks on the side of the plateau.** After farmland irrigation water on the South Jingyang Plateau infiltrates the slope along the vertical joints of the macro-porous loess layer, this layer will undergo wet collapse compression under the action of the seepage force of water [43,44]. Consequently, groundwater on the plateau is slowly uplifted, crosses the paleosol layer (aquifuge) at the bottom of the slope, forms perched water at the top of this layer, and the bottom of the upper Lishi loess is soaked in water and gradually saturated. The water level continues to rise, which causes the saturated layer to thicken and the shear strength to decrease, until water basically submerges part of the sliding surface. Then, deformation failure begins to take place in the soil at the foot of the slope under the action of overburdening by pressure, which in turn causes deformation of the whole slope. Furthermore, soil stress redistribution in the slope increases the shear stress of the upper soil, and splitting failure occurs in the slope soil because shear strength is lower than shear stress. Thus, tension crack occurs at the trailing edge of the slope.

In addition, aquifuge action of the paleosol layer causes the upper loess layer to collapse and uneven subsidence to occur. This phenomenon also causes settlement to be rather great in the loess layer inclined to the free face of the slope, while that of the loess layer on the plateau is small. Thus, the barycenter of the slope moves outwards, which will aggravate the cracks on the side of the plateau.

(2) **Continuous irrigation causes the formation of a saturated softening layer.** Continuous irrigation causes water to infiltrate vertically to the top of the paleosol layer at the foot of the slope (relative to the aquiclude) and to form a saturated layer at this location. Irrigation causes the groundwater level to rise gradually, which increases the thickness of the saturated layer accordingly, and the loess layer relative to the water-repellent layer gradually softens under the action of water, which reduces frictional resistance and forms a saturated softening zone. Moreover, soil in the saturated softening zone is subjected to the combined action of seepage water pressure and pressure overburdening, further destroying its original structure, while soluble salt partially dissolves under the action of water, which aggravates sludging of the sliding zone soil and reduces the frictional resistance of the sliding zone soil.

(3) **The saturated softening zone is prone to flow slides after static liquefaction occurs.** With increasing irrigation, groundwater in the plateau area gradually rises and reaches a certain height, which results in a large hydraulic gradient. This gradient, in turn, leads to an increase in the thickness of the saturated softening zone at the foot of the slope. As the shear strength of the undisturbed loess is significantly reduced and liquefaction potentially occurs, the softening zone of the sliding zone soil gradually penetrates the slope, which causes a flow slide [32,44]. Then, the relative sliding between the loess particles in this layer causes pore pressure to increase sharply, directly leading to reduction of the effective stress in the slope softening zone and finally bringing about a flow-slide type landslide induced by liquefaction.

5. Conclusions

The types and developmental characteristics of loess landslides around the South Jingyang Plateau were first determined. Then, the seasonal agricultural irrigation features and groundwater changes were introduced. Clear evidence was found that irrigation is a significant factor influencing changes in groundwater and even causing the occurrence of landslides. Based on a field investigation and indoor experiment, the intrinsic mechanism of shallow loess landslides induced by seasonal agricultural irrigation was studied in detail. Several main conclusions can be reached.

The shallow loess landslides around the South Jingyang Plateau exhibit three typical characteristics: Group occurrence, multiple times of occurrence, and temporal heterogeneity. Furthermore, they can be divided into five types: Mudflows, clay/silt rotational slides, erosion slides, external disturbed slides and loess topples, among which, mudflows and clay/silt rotational slides account for approximately 78% of the total number of landslides investigated. Seasonal agricultural irrigation on the South Jingyang Plateau can be divided into three stages: Spring irrigation, summer irrigation, and winter irrigation. Current agricultural irrigation in this region is mostly based on traditional flood irrigation, which brings about obvious uplift of the groundwater level in the study area. According to survey data, the occurrence of plateau margin landslides in the study area is closely related to agricultural irrigation.

The formation and evolution of irrigation-associated landslides on the South Jingyang Plateau can be divided into three stages: The early stage, the infiltration stage of preferential seepage channels, and the landslide sliding stage. During this process, the occurrence of landslides is closely related to loess softening, mechanical subsurface corrosion and chemical corrosion caused by groundwater level uplift, and static liquefaction instability finally occurs in the softening zone of loess. Similarly, an indoor experiment confirmed that undisturbed loess under different confining pressures shows a strong stress-softening characteristic. This characteristic is more significant under the condition of low confining pressure, like the "static liquefaction". The genetic model of landslides induced by irrigation on the South Jingyang Plateau consists of three main factors. Long-term irrigation leads to significant tension cracks on the side of the plateau. Continuous irrigation causes the formation of a saturated softening zone, and the saturated softening zone is prone to flow slides after static liquefaction occurs.

Based on the above research, the key to controlling the landslides around the South Jingyang Plateau lies in limiting the recharge–discharge relationship between irrigation water and groundwater. Suggestions for improvement measures are as follows: Move away from the traditional agricultural irrigation method of flood irrigation, and advocate for drip irrigation or well irrigation by pumping groundwater. It is conceivable to carry out grouting and recirculation treatment of cracks on the side of the plateau, but the selection of slurry must take into account ecological and environmental issues. Adopt the rainfall pattern of drilling horizontal holes to reduce the groundwater level at a portion of the plateau at the foot of the slope. Pay attention to the strict inspection and timely repair of irrigation canals.

Author Contributions: R.-X.Y. and J.-B.P. conceived and designed the survey and experiment; R.-X.Y., L.-J.C. and C.-Y.K. performed the field survey and experiments; J.-B. Peng, Q.-B.H. and Y.-J.S. contributed funding supports; R.-X.Y. and Y.-J.S. wrote the paper. J.-B.P. revised the English of this paper.

Funding: This work was supported by the Major Program of National Natural Science Foundation of China (Grant No. 41790441), the National Natural Science Foundation of Shaanxi Province, China (Grant No. 2018JQ5124) and the Foundation of Key Laboratory of Western Mineral Resources and Geological Engineering of Ministry of Education, Chang' an University (Grant No. 300102268503).

Acknowledgments: We thank the entire team for their efforts to improve the quality of the article. At the same time, we would like to thank editor for his timely handling of the manuscripts.

Conflicts of Interest: The authors declare no conflict of interest.

References

1. Shi, H.; Shao, M. Soil and water loss from the loess plateau in China. *J. Arid Environ.* **2000**, *45*, 9–20. [CrossRef]
2. Derbyshire, E. Geological hazards in loess terrain, with particular reference to the loess regions of China. *Earth Sci. Rev.* **2001**, *54*, 231–260. [CrossRef]
3. Ke, H.; Li, P.; Li, Z.; Shi, P.; Hou, J. Soil water movement changes associated with revegetation on the loess plateau of China. *Water* **2019**, *11*, 731. [CrossRef]
4. Jia, Z.; Wang, Z.; Wang, H. Characteristics of dew formation in the semi-arid loess plateau of central Shaanxi province, China. *Water* **2019**, *11*, 126. [CrossRef]
5. Griffiths, E. Loess of banks peninsula. *N. Z. J. Geol. Geophys.* **1973**, *16*, 657–675. [CrossRef]
6. Evstatiev, D. Design and treatment of loess bases in bulgaria. In *Genesis and Properties of Collapsible Soils*; Derbyshire, E., Dijkstra, T., Smalley, I.J., Eds.; Springer: Dordrecht, The Netherlands, 1995; pp. 375–382. [CrossRef]
7. Derbyshire, E.; Dijkstra, T.A.; Smalley, I.J.; Li, Y. Failure mechanisms in loess and the effects of moisture content changes on remoulded strength. *Quat. Int.* **1994**, *24*, 5–15. [CrossRef]
8. Dijkstra, T.A.; Rogers, C.D.F.; Smalley, I.J.; Derbyshire, E.; Li, Y.J.; Meng, X.M. The loess of north-central China: Geotechnical properties and their relation to slope stability. *Eng. Geol.* **1994**, *36*, 153–171. [CrossRef]
9. Acharya, G.; Cochrane, T.; Davies, T.; Bowman, E. Quantifying and modeling post-failure sediment yields from laboratory-scale soil erosion and shallow landslide experiments with silty loess. *Geomorphology* **2011**, *129*, 49–58. [CrossRef]
10. Zhang, D.; Wang, G.; Luo, C.; Chen, J.; Zhou, Y. A rapid loess flowslide triggered by irrigation in China. *Landslides* **2009**, *6*, 55–60. [CrossRef]
11. Xu, L.; Dai, F.C.; Tham, L.G.; Tu, X.B.; Min, H.; Zhou, Y.F.; Wu, C.X.; Xu, K. Field testing of irrigation effects on the stability of a cliff edge in loess, north-west China. *Eng. Geol.* **2011**, *120*, 10–17. [CrossRef]
12. Peng, J.; Zhuang, J.; Wang, G.; Dai, F.; Zhang, F.; Huang, W.; Xu, Q. Liquefaction of loess landslides as a consequence of irrigation. *Q. J. Eng. Geol. Hydrogeol.* **2018**, *51*, 330–337. [CrossRef]
13. Crosta, G.B.; Dal Negro, P.; Frattini, P. Soil slips and debris flows on terraced slopes. *Nat. Hazards Earth Syst. Sci.* **2003**, *3*, 31–42. [CrossRef]
14. Andrea, C.; Pierluigi, B.; Claudia, S.; Ivano, R. Relationships between geo-hydrological processes induced by heavy rainfall and land-use: The case of 25 october 2011 in the Vernazza catchment (Cinque Terre, NW Italy). *J. Maps* **2013**, *9*, 289–298. [CrossRef]

15. Cevasco, A.; Pepe, G.; Brandolini, P. The influences of geological and land use settings on shallow landslides triggered by an intense rainfall event in a coastal terraced environment. *Bull. Eng. Geol. Environ.* **2014**, *73*, 859–875. [CrossRef]

16. McClymont, L.; Goodwin, I.; Whitfield, D.M.; O'Connell, M.G. Effects of within-block canopy cover variability on water use efficiency of grapevines in the Sunraysia irrigation region, Australia. *Agric. Water Manag.* **2019**, *211*, 10–15. [CrossRef]

17. Shen, Y.J.; Wang, Y.Z.; Yang, Y.; Sun, Q.; Luo, T.; Zhang, H. Influence of surface roughness and hydrophilicity on bonding strength of concrete-rock interface. *Constr. Build. Mater.* **2019**, *213*, 156–166. [CrossRef]

18. Li, H.J.; Jin, Y.L. Initiation analysis of an irrigation-induced loess landslide. *Appl. Mech. Mater.* **2012**, *170–173*, 574–580. [CrossRef]

19. Xu, L.; Dai, F.C.; Gong, Q.M.; Tham, L.G.; Min, H. Irrigation-induced loess flow failure in Heifangtai platform, north-west China. *Environ. Earth Sci.* **2012**, *66*, 1707–1713. [CrossRef]

20. Leng, Y.; Peng, J.; Wang, Q.; Meng, Z.; Huang, W. A fluidized landslide occurred in the loess plateau: A study on loess landslide in south Jingyang tableland. *Eng. Geol.* **2018**, *236*, 129–136. [CrossRef]

21. Zhuang, J.; Peng, J.; Wang, G.; Javed, I.; Wang, Y.; Li, W. Distribution and characteristics of landslide in loess plateau: A case study in Shaanxi province. *Eng. Geol.* **2018**, *236*, 89–96. [CrossRef]

22. Alvioli, M.; Melillo, M.; Guzzetti, F.; Rossi, M.; Palazzi, E.; von Hardenberg, J.; Brunetti, M.T.; Peruccacci, S. Implications of climate change on landslide hazard in central Italy. *Sci. Total Environ.* **2018**, *630*, 1528–1543. [CrossRef] [PubMed]

23. Gariano, S.L.; Guzzetti, F. Landslides in a changing climate. *Earth Sci. Rev.* **2016**, *162*, 227–252. [CrossRef]

24. Peres, D.J.; Cancelliere, A. Modeling impacts of climate change on return period of landslide triggering. *J. Hydrol.* **2018**, *567*, 420–434. [CrossRef]

25. Lei, X. The hazards of loess landslides in the southern tableland of Jingyang county, Shaanxi and their relationship with channel water into fields. *J. Eng. Geol.* **1995**, *3*, 56–64.

26. Peng, J.; Wang, G.; Wang, Q.; Zhang, F. Shear wave velocity imaging of landslide debris deposited on an erodible bed and possible movement mechanism for a loess landslide in Jingyang, Xi'an, China. *Landslides* **2017**, *14*, 1503–1512. [CrossRef]

27. Li, T.; Zhao, J.; Li, P.; Wang, F. Failure and motion mechanisms of a rapid loess flowslide triggered by irrigation in the Guanzhong irrigation area, Shaanxi, China. In *Progress of Geo-Disaster Mitigation Technology in Asia*; Wang, F., Miyajima, M., Li, T., Shan, W., Fathani, T.F., Eds.; Springer: Berlin/Heidelberg, Germany, 2013; pp. 421–433. [CrossRef]

28. Wang, W.; Wang, Y.; Sun, Q.; Zhang, M.; Qiang, Y.; Liu, M. Spatial variation of saturated hydraulic conductivity of a loess slope in the south Jingyang plateau, China. *Eng. Geol.* **2018**, *236*, 70–78. [CrossRef]

29. Puri, V.K. Liquefaction Behavior and Dynamic Properties of Loessial (Silty) Soils. Ph.D. Thesis, Missouri University of Science and Technology, Rolla, MO, USA, 1984.

30. Cui, S.-H.; Pei, X.-J.; Wu, H.-Y.; Huang, R.-Q. Centrifuge model test of an irrigation-induced loess landslide in the Heifangtai loess platform, northwest China. *J. Mt. Sci.* **2018**, *15*, 130–143. [CrossRef]

31. Ishihara, K.; Okusa, S.; Oyagi, N.; Ischuk, A. Liquefaction-induced flow slide in the collapsible loess deposit in Soviet Tajik. *Soils Found.* **1990**, *30*, 73–89. [CrossRef]

32. Zhang, F.; Wang, G. Effect of irrigation-induced densification on the post-failure behavior of loess flowslides occurring on the Heifangtai area, Gansu, China. *Eng. Geol.* **2018**, *236*, 111–118. [CrossRef]

33. Lian, B.; Wang, X.; Zhu, R.; Liu, J.; Wang, Y. *A Numerical Simulation Study of Landslides Induced by Irrigation in Heifangtai Loess Area—A Case Study of Huangci*; IOP Publishing Ltd.: Bristol, UK, 2018; Volume 108. [CrossRef]

34. Pan, P.; Shang, Y.-Q.; Lü, Q.; Yu, Y. Periodic recurrence and scale-expansion mechanism of loess landslides caused by groundwater seepage and erosion. *Bull. Eng. Geol. Environ.* **2019**, *78*, 1143–1155. [CrossRef]

35. He, Y. Identification and Monitoring of the Loess Landslide by Using of High Resolution Remote Sensing and Insar. Master's Thesis, Chang'an University, Xi'an, China, 2016.

36. Duan, Z.; Peng, J.; Wang, Q. Characteristic parameter and formation mechanism of repeatedly failure loess landslides. *Mt. Res.* **2016**, *34*, 71–76. [CrossRef]

37. Kang, S.; Zhang, L.; Liang, Y.; Hu, X.; Cai, H.; Gu, B. Effects of limited irrigation on yield and water use efficiency of winter wheat in the loess plateau of China. *Agric. Water Manag.* **2002**, *55*, 203–216. [CrossRef]

38. Hungr, O.; Leroueil, S.; Picarelli, L. The varnes classification of landslide types, an update. *Landslides* **2014**, *11*, 167–194. [CrossRef]

39. Wang, Y. Study on the Hydrogeological Structure Evolution and Disaster Control Mechanism of Loess Slope in the Southern Jingyang Plateau. Master's Thesis, Chang'an University, Xi'an, China, 2017.

40. Dong, J. Study on the Trigger Mechanism and Motion Characteristics of Loess Landslide of Jingyang Village at South Jingyang Plateau. Master's Thesis, Chang'an University, Xi'an, China, 2017.

41. Peng, J.; Wang, S.; Wang, Q.; Zhuang, J.; Huang, W.; Zhu, X.; Leng, Y.; Ma, P. Distribution and genetic types of loess landslides in China. *J. Asian Earth Sci.* **2019**, *170*, 329–350. [CrossRef]

42. Xu, L.; Dai, F.C.; Tu, X.B.; Javed, I.; Woodard, M.J.; Jin, Y.L.; Tham, L.G. Occurrence of landsliding on slopes where flowsliding had previously occurred: An investigation in a loess platform, north-west China. *CATENA* **2013**, *104*, 195–209. [CrossRef]

43. Xu, L.; Qiao, X.; Wu, C.; Iqbal, J.; Dai, F. Causes of landslide recurrence in a loess platform with respect to hydrological processes. *Nat. Hazards* **2012**, *64*, 1657–1670. [CrossRef]

44. Zhou, Y.F.; Tham, L.G.; Yan, W.M.; Dai, F.C.; Xu, L. Laboratory study on soil behavior in loess slope subjected to infiltration. *Eng. Geol.* **2014**, *183*, 31–38. [CrossRef]

Article

Estimation of Soil Erosion in the Chaohu Lake Basin through Modified Soil Erodibility Combined with Gravel Content in the RUSLE Model

Sai Hu [1,2], Long Li [1,2,3], Longqian Chen [1,2,*], Liang Cheng [1,2,4], Lina Yuan [1,2], Xiaodong Huang [5] and Ting Zhang [1,2]

[1]   School of Environmental Science and Spatial Informatics, China University of Mining and Technology, Daxue Road 1, Xuzhou 221116, China
[2]   Engineering Research Center of Ministry of Education for Mine Ecological Restoration, China University of Mining and Technology, Daxue Road 1, Xuzhou 221116, China
[3]   Department of Geography, Earth System Science, Vrije Universiteit Brussel, Pleinlaan 2, 1050 Brussels, Belgium
[4]   College of Yingdong Agricultural Science and Engineering, Shaoguan University, Daxue Road 26, Shaoguan 512005, China
[5]   School of Surveying and Landing Information Engineering, Henan Polytechnic University, Century Road 2001, Jiaozuo 454000, China
*   Correspondence: chenlq@cumt.edu.cn; Tel.: +86-516-8359-1327

Received: 8 July 2019; Accepted: 26 August 2019; Published: 29 August 2019

Abstract: It is generally acknowledged that soil erosion has become one of the greatest global threats to the human–environment system. Although the Revised Universal Soil Loss Equation (RUSLE) has been widely used for soil erosion estimation, the algorithm for calculating soil erodibility factor (K) in this equation remains limited, particularly in the context of China, which features highly diverse soil types. In order to address the problem, a modified algorithm describing the piecewise function of gravel content and relative soil erosion was used for the first time to modify the soil erodibility factor, because it has been proven that gravel content has an important effect on soil erosion. The Chaohu Lake Basin (CLB) in East China was used as an example to assess whether our proposal can improve the accuracy of soil erodibility calculation and soil erosion estimation compared with measured data. Results show that (1) taking gravel content into account helps to improve the calculation of soil erodibility and soil erosion estimation due to its protection to topsoil; (2) the overall soil erosion in the CLB was low (1.78 Mg·ha^{-1}·year^{-1}) the majority of which was slight erosion (accounting for 85.6%) and no extremely severe erosion; and (3) inappropriate land use such as steep slope reclamation and excessive vegetation destruction are the main reasons for soil erosion of the CLB. Our study will contribute to decision-makers to develop soil and water conservation policies.

Keywords: soil erosion; RUSLE; soil erodibility; gravel content; Chaohu Lake Basin

---

1. Introduction

Soil erosion has become a major global environmental hazard [1], posing a serious threat to the ecological environment, natural resources, and socio–economic development [2,3]. It often has a negative impact on downstream areas, including lowered water levels in reservoirs [4], threats from floods and mudslides [5], damage to habitats of species [6], and reduced agricultural productivity [7,8]. For example, the Mediterranean vineyards are among the most degraded agricultural ecosystems affected by extreme soil erosion rates [9,10], and this situation has also been observed in citrus plantations [11,12]. As one of the most severe areas of soil erosion in the world, the Loess Plateau in China has caused great damage to the natural environment, and the economic and social development of the region [13]. It is acknowledged that soil erosion has become a major threat to sustainability due to the immediate damage it causes to the soil system and the acceleration of the land degradation process [12]. Soil erosion is a natural process that is controlled by a variety of environmental factors, such as topography, soil, climate, and vegetation [14]. Human activities, for example, deforestation, agricultural production, and construction, accelerate the rate of soil erosion [2,15]. The United Nations (UN) has highlighted soil and water protection as a key land-use policy issue, which is an effective manner to address the challenges of the UN Sustainable Development Goals [16,17]. In general, estimation of soil erosion provides a basis for soil and water conservation [18].

Quantitative modeling, either physically based or empirically based, has become a widely accepted approach in soil erosion estimation research [19,20]. At present, many empirical models have been developed for estimating soil erosion [1]. Among them, the Revised Universal Soil Loss Equation (RUSLE) has become the most commonly used in different environmental conditions and on varying scales [21–25]. The parameter factors in the RUSLE model include rainfall erosivity, soil erodibility, slope length and steepness, cover fraction, and support practice [26], as soil erosion is the result of a combination of natural and human factors [27].

Topography is one of the factors determining the amount of soil erosion, and slope plays an important role in increasing soil loss [28]. Smith and Wischmeier first found that soil loss was a polynomial function of slope (θ) [29] and later modified this function by creating the slope factor, which is a polynomial function of the sine of slope ($\sin \theta$; see Equation (9)) as it could improve the accuracy of soil erosion prediction on steep areas [30]. McCool et al. proposed an algorithm for calculating the slope factor using a piecewise function and included it in the RUSLE model [31] while Chinese researchers later modified the algorithm of slope factor for slopes above 10° using measured data, which has been widely used in the context of China [32]. However, most of the algorithms were based on measurements obtained from runoff plots below 15°, which may result in less accurate soil erosion prediction on steep areas. Therefore, it is important to improve the slope factor calculation, in order to improve the accuracy of soil erosion prediction, particularly in the areas with high slopes [33]. In this study, we obtained the fitting formula in the form of piecewise function that describes the functional relationship between the sine of slope θ and the slope factor, using the measured soil erosion data with the slopes ranging from 10 to 25° and above 25°.

Soil erodibility is also a key factor related to soil erosion estimation. Among many soil erodibility factor algorithms is the widely used Erosion Productivity Impact Calculator (EPIC) proposed by Williams et al. [34]. However, application of this algorithm to multiple erosion-prone areas of China resulted in soil erodibilities that were greater than measured values for all soil types [35]. This suggests that the EPIC algorithm might not be well suited for soil erodibility estimation in the context of China. Gravel on the soil surface or the top layer of soil has a direct or indirect effect on soil erosion [36]. A series of laboratory-based and runoff plot-based experiments show that rock fragment and gravel content are negatively related to soil erodibility [37–40]. Therefore, it is necessary to modify the prediction of soil erodibility. For improved soil erodibility estimation accuracy, gravel content, an important parameter used to modify the calculation of soil erodibility, should be considered in the RUSLE model. For the purpose of improving the accuracy of soil erodibility estimation, Shi [41] proposed a new algorithm to modify soil erodibility and constructed a piecewise function between relative soil erosion (ratio of soil erosions with different gravel content under the same condition) and gravel content. However, this algorithm has not been tested and applied in soil erodibility and soil erosion studies. In this study, this algorithm was used to modify the soil erodibility for the first time, and then we evaluated its accuracy and estimated soil erosion. China has been tackling soil erosion for decades in the Loess Plateau [42]. However, such an eco–environmental issue also occurs in many other areas, such as the Chaohu Lake Basin (CLB). In recent years, the contradiction between economic development, population, resources, and the eco–environment has become increasingly prominent in this region [43]. The decline of the eco–environmental quality in the CLB, such as eutrophication and increased soil erosion, has seriously affected the sustainability of regional development. It is interesting that the mountainous areas in the eastern part of the CLB have high gravel content, reaching more than 20% in certain places. In this study, we therefore selected the CLB as the study area and used the RUSLE model for soil erosion estimation to contribute to the general goal of water and soil conservation. Specific objectives are as follows:

(1) to modify the algorithms of calculating slope factor and soil erodibility factor in the RUSLE model for estimating the soil erosion in the CLB in 2017;
(2) to examine the spatial distribution of each RUSLE factor over the study area;
(3) to compare the soil erosion estimation results with and without modifying the soil erodibility algorithms.

2. Materials and Methods

2.1. Study Area

At the center of the east Chinese province of Anhui (116°20′–118°0′ E, 29°01′–33°16′ N), the CLB consists of 11 administrative districts—2 urban districts and 9 counties—covering a geographical area of 2.04×10^4 km^2 (Figure 1). In 2017, the CLB had a population of around 11.52 million, with an urbanization rate of 68.5%, and produced a GDP of approximately 8345 billion CNY (Chinese yuan), according to the Statistical Yearbook of Anhui Province. Among the many water systems in the CLB is the Chaohu Lake, which is profiled as one of the five largest freshwater lakes in China, is one of the main drinking water sources in the CLB, and is replenished by surface runoff and rainfall. With an average elevation of 50.35 m, the CLB features highlands in the southwest and lowlands in the northeast. Influenced by geomorphic types and parent materials, the soil types in the CLB are particularly complex. The rainfall is mainly concentrated in summer and autumn, which accounts for more than 60% of the total annual rainfall, according to the rainfall data provided by the National Meteorological Information Center.

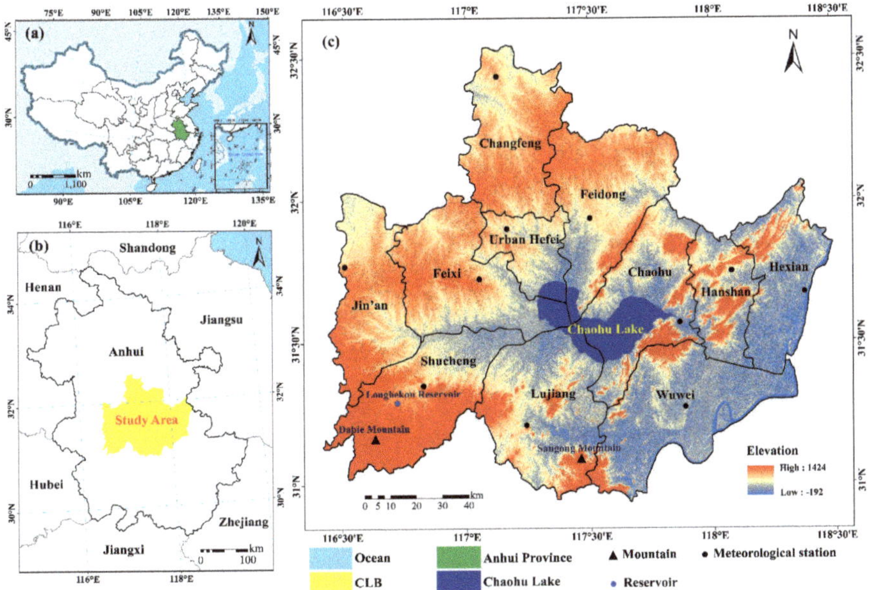

Figure 1. The study area: (**a**) the location of Anhui province in China; (**b**) the location of the Chaohu Lake Basin (CLB) in Anhui province; (**c**) the digital elevation model (DEM) of the CLB.

2.2. Data

Remote sensing images, digital elevation model (DEM) dataset, soil data, rainfall data, and vector data of the study area were used to generate the input variables for the RUSLE model (Table 1).

Table 1. Data used in this study.

Dataset	Description	Resolution	Source
Rainfall	Daily and monthly rainfall data of 13 meteorological stations from 1990 to 2017	0.05 degree	National Meteorological Information Center
Soil	Soil type and soil attribute data (subsoil sand fraction, silt fraction, clay fraction, topsoil organic carbon and gravel content)	1:1,000,000	Cold and Arid Regions Sciences Data Center at Lanzhou
DEM	ASTER GDEM dataset	30 m	Geospatial Data Cloud
Remote sensing imagery	Landsat 8 OLI (Operational Land Imager) data acquired on 21 July 2017 (Path120/Row38) and 28 July 2017 (Path121/Row38)	30 m	Geospatial Data Cloud
Vector	Provincial boundary	1:10,000	National Administration of Surveying, Mapping, and Geo-information

2.3. Methods

In this study, soil erosion was estimated using the RUSLE model, where the soil erodibility factor was modified by incorporating gravel content. The technical flowchart of this study is presented in Figure 2.

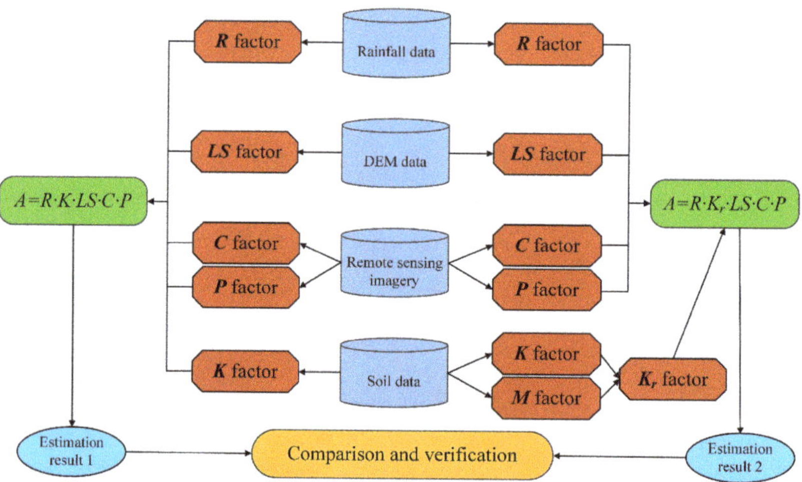

Figure 2. Flowchart depicting methodology of the study.

2.3.1. RUSLE

As a widely used soil erosion prediction model, the Revised Universal Soil Loss Equation (RUSLE) can quantify the soil erosion modulus in different scenarios and reflect the relationship between soil erosion and various impact factors [15]. Use of the RUSLE was made in this study based on the following equation [30,44]:

$$A = R{\cdot}K{\cdot}LS{\cdot}C{\cdot}P, \tag{1}$$

where A is the average annual soil loss (Mg·ha^{-1}·year^{-1}), R is the rainfall–runoff erosivity factor (MJ·mm·ha^{-1}·hr^{-1}·year^{-1}), K is the soil erodibility factor (Mg·ha·h·ha^{-1}·MJ^{-1}·mm^{-1}), LS is the slope length and steepness factor (unitless) accounting the effect of topography on soil erosion, C is the cover fraction factor (unitless), and P is the support practice factor (unitless). These factors are detailed below.

- Rainfall–Runoff Erosivity Factor (R)

Rainfall erosivity is the potential possibility of soil erosion induced by rainfall, which is the most important external driving force and the dynamic indicator of soil separation and transportation [45,46]. Many of the existing classic and straightforward algorithms for estimating this factor are mainly based on annual rainfall, monthly, daily, or hourly rainfall [47–49]. Since using annual and monthly rainfall data results in less accurate estimation and it is often challenging to access hourly rainfall data, daily rainfall data was used in this study to calculate the rainfall erosivity in the CLB through the algorithm presented in the First National Census for Water [50]:

$$\overline{R}_k = \frac{1}{N} \sum_{i=1}^{N} \left(\alpha \sum_{j=1}^{M} P_{d_{ikj}}^{\beta} \right), \tag{2}$$

$$\alpha = 21.293 \beta^{-7.3967}, \tag{3}$$

$$\beta = 0.6243 + \frac{27.346}{\overline{P}_{d_0}}, \tag{4}$$

$$\overline{P}_{d_0} = \frac{1}{N} \sum_{i=1}^{N} \sum_{k=1}^{12} \sum_{j=1}^{M} P_{d_{ikj}}, \tag{5}$$

$$\overline{R} = \sum_{k=1}^{12} \overline{R}_k, \tag{6}$$

where $\overline{R}_k$ is the rainfall erosivity of the kth month (MJ·mm·ha^{-1}·hr^{-1}), N is the sequence length of the calculated data; M is the frequency of erosive rainfall in the kth month of the ith year; $P_{d_{ikj}}$ is the rainfall of the jth erosive rainfall in the kth month of the ith year (mm), and daily rainfall ≥ 12 mm is defined as erosive rainfall; α and β are model parameters calculated using Equations (3) and (4); $\overline{P}_{d_0}$ is the multi-year average of erosive rainfall (mm); and $\overline{R}$ is the average annual rainfall erosivity (MJ·mm·ha^{-1}·hr^{-1}·year^{-1}).

- Soil Erodibility Factor (K)

The soil erodibility factor K reflects the sensitivity of soil to erosion. The K value can be estimated by making observations at a large number of test plots; however, such an approach is difficult to apply to a large watershed [51,52] such as the CLB. To solve this issue, we calculated soil erodibility in this study based on the EPIC model [34] using the following equation:

$$K = \left\{ 0.2 + 0.3 exp\left[-0.0256 W_d \left(1 - \frac{W_i}{100} \right) \right] \right\} \times \left(\frac{W_i}{W_i + W_l} \right)^{0.3}$$
$$\times \left[1 - \frac{0.25 W_c}{W_c + exp(3.72 - 2.95 W_c)} \right] \times \left[1 - \frac{0.7 W_n}{W_n + exp(-5.51 + 22.9 W_n)} \right], \tag{7}$$

where W_d is the fraction of sand (ø 2–0.05 mm) in %, W_i is the fraction of silt (ø 0.05–0.002 mm) in %, W_l is the fraction of clay (ø < 0.002 mm) in %, W_c is the soil total organic carbon content in %, and $W_n = 1 - \frac{W_d}{100}$.

- Slope Length and Steepness Factor (LS)

The slope length and steepness factor (*LS*) can accelerate soil erosion, representing the influence of topographic features on soil erosion [53]. A steeper slope and a longer slope length lead to more serious soil erosion [54]. The *L* value can be calculated by the following equation [55]:

$$L = (cell\ size/22.13)^m,\tag{8}$$

where *cell size* = grid cell size (30 m in this study); the value of m varies between 0.2 and 0.5 (0.2 for slopes less than 1%, 0.3 for 1–3%, 0.4 for 3–4.5%, and 0.5 for slopes exceeding 4.5%).

As mentioned in the Introduction (Section 1), Wischmeier and Smith modified the relationship between soil erosion and slope by creating the slope factor *S*, which is the quadratic function of the sine of slope θ (Equation (9)) and applied it in the USLE (Universal Soil Loss Equation) model [30]:

$$S = 65.4\sin\theta^2 + 4.56\sin\theta + 0.0654.\tag{9}$$

Then, the slope factor formula was modified using observation data in the RUSLE model [32]:

$$S = \begin{cases} 10.8\sin\theta + 0.03 & \theta \leq 5° \\ 16.8\sin\theta - 0.50 & 5° < \theta \leq 10° \\ 21.9\sin\theta - 0.96 & \theta > 10° \end{cases}.\tag{10}$$

This algorithm improves the prediction accuracy of soil loss on slopes above 10°. However, it would not be appropriate for soil erosion estimation in the context of the CLB because soil erosion mainly occurs in mountainous and hilly areas with high slopes in the CLB [56]. As such, it would be necessary to modify this algorithm in the case of slopes greater than 10°. By extraction from scientific research articles, reports, and books, we complied a dataset of measured soil erosion for a number of plots (see Table A1, Appendix A). This dataset covers slopes ranging from 10° to 45°, which are highly representative. We used these sample data to establish the functional relationship between the sine of slope θ and the slope factor.

- Cover Fraction Factor (*C*)

The *C* factor characterizes the restriction of surface vegetation cover on soil erosion as vegetation helps to retain soil and water [1]. Although both spectral vegetation indices and spectral mixture analysis modeled vegetation fractions [57–59] can be used to calculate the *C* factor values [60,61], it is easier to extract spectral vegetation indices than vegetation fractions. In this study, the mostly commonly used vegetation index, Normalized Difference Vegetation Index (*NDVI*), which is the ratio of the difference between spectral reflectance in near infrared and red regions [62,63], was used to calculate *C* factor values according to the following equation [64]:

$$C = exp\left[-\alpha \cdot \frac{NDVI}{(\beta - NDVI)}\right],\tag{11}$$

where α and β are parameters that determine the shape of the *NDVI-C* curve, and the α-value of 2 and β-value of 1 provide reasonable results of *C* values compared with those estimated as suming a linear relationship [65].

- Support Practice Factor (*P*)

The *P* factor is defined as the ratio of soil loss with a specific support practice to the corresponding loss with upslope and downslope tillage [66]. It is a dimensionless factor with a value between 0 and 1 [44]: 0 means that no soil erosion will occur while 1 means that no soil and water conservation measures have been taken or the measures have completely failed. The *P* factor is closely related to land use type and land use change [67]. Using the maximum likelihood classifier, we classified the land use/cover types in the CLB into six categories, namely farmland, forestland, grassland, water body, construction land, and unused land. The high-resolution satellite images in 2017 provided by Google Earth Pro were used to assess classification accuracy [68]. In total, 500 sample points were randomly generated in the classified image in ArcGIS 10.1 and then imported into Google Earth Pro to retrieve the ground-truthing data. By constructing a confusion matrix, we obtained the overall accuracy (90.1%) and Kappa coefficient (0.86) for this classification. These high values indicate that the classification was well performed and that the classification map could be used for further analysis in this study.

In this study, we used the land use classification of the CLB in 2017 (Figure 3) and assigned the *P* factor values from Zha et al. [56] and Xu et al. [4] for each land use type (Table 2).

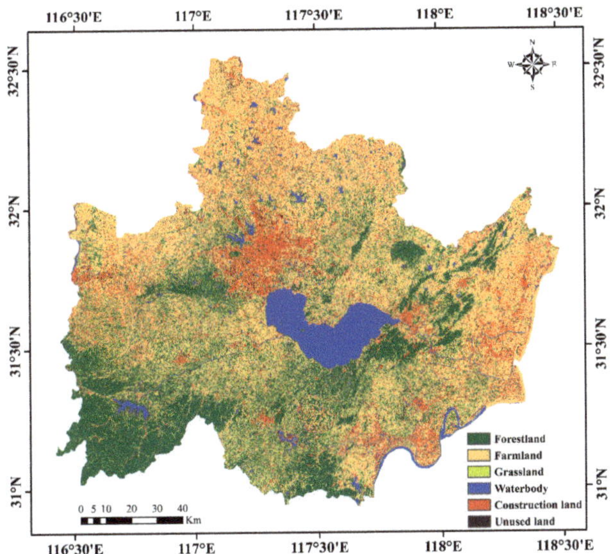

Figure 3. Land use classification map of the CLB in 2017.

Table 2. The *P* factor value of different land use types.

Land Use Type	Farmland	Forestland	Grassland	Waterbody	Construction Land	Unused Land
P factor value	0.35	1	1	0	0	1

2.3.2. Modifying Soil Erodibility (*K*)

In order to modify soil erodibility, gravel content was considered as a key parameter. The piecewise function proposed by Shi [41] was used for the first time to modify soil erodibility, which describes the functional relationship between relative soil erosion and different gravel content ranges. The modification coefficient *M* (relative soil erosion) was determined by the following equation:

$$M = \begin{cases} 0.0781e^{-0.0249R_m} & R_m > 20\% \\ 0.294 - 0.0123R_m & 10\% < R_m \leq 20\%, \\ 1 - 0.0829R_m & R_m \leq 10\% \end{cases} \tag{12}$$

where *M* is the coefficient for modifying the soil erodibility, and R_m is the gravel content. The modified soil erodibility can be obtained by the following equation:

$$K_r = K \times M, \tag{13}$$

where K_r is the modified soil erodibility (Mg·ha·h·ha^{-1}·MJ^{-1}·mm^{-1}), and K is the soil erodibility calculated by the EPIC model (Mg·ha·h·ha^{-1}·MJ^{-1}·mm^{-1}). Using Equations (7), (12), and (13), the soil erodibility K and the modified soil erodibility K_r can be calculated. In order to assess the accuracy of this modified algorithm, five sets of measured soil erodibility data extracted from Zhang et al. [35] were used to compare with our modified soil erodibility.

3. Results

3.1. Fitting Equation of Slope Factor

Using the measured soil erosion dataset, we obtained the fitted equation with high confidence level between slope factor and sine value in the ranges of 10–25° and above 25°, respectively. The regression analysis shown in Figure 4 shows a strong linear relationship between them.

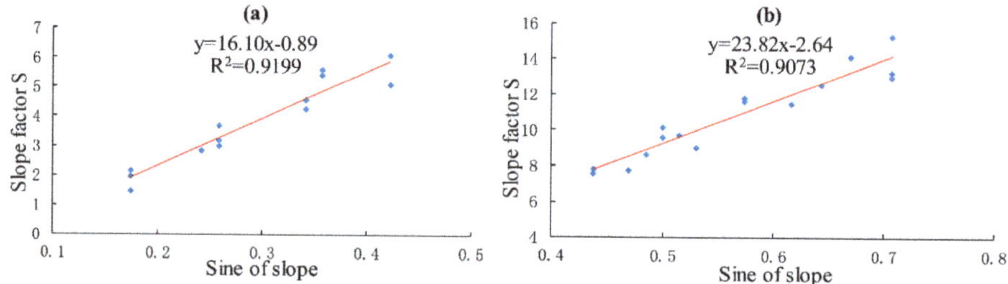

Figure 4. Relationship between the sine of slope and the slope factor: (**a**) slopes between 10° and 25°; and (**b**) slopes above 25°.

As such, based on the algorithm proposed by Liu et al. [33], the estimation of slope factor in the case of slopes higher than 10° can be calculated by the following equation:

$$S = \begin{cases} 10.8\sin\theta + 0.03 & \theta \leq 5° \\ 16.8\sin\theta - 0.50 & 5° < \theta \leq 10° \\ 16.10\sin\theta - 0.89 & 10° < \theta \leq 25° \\ 23.82\sin\theta - 2.64 & \theta > 25° \end{cases} \tag{14}$$

3.2. Accuracy Assessment of Modified Soil Erodibility

The calculated K values and K_r values were shown in Figure 5. We found that the range of the original K values was smaller than that of the modified soil erodibility K_r values. The means of modified soil erodibility K_r were remarkably reduced. This suggests that gravel content has an important effect on the calculation of soil erodibility.

It is clear that the measured K values for the five soil types were all within the range of K_r (Table 3). By comparing the ratios of the calculated K and K_r values to the measured values, we noticed that the calculated K_r values were by far closer to the measured values than the calculated K values. It is therefore believed that this modified algorithm can result in better soil erodibility estimation and has the potential to improve the accuracy of soil erosion prediction.

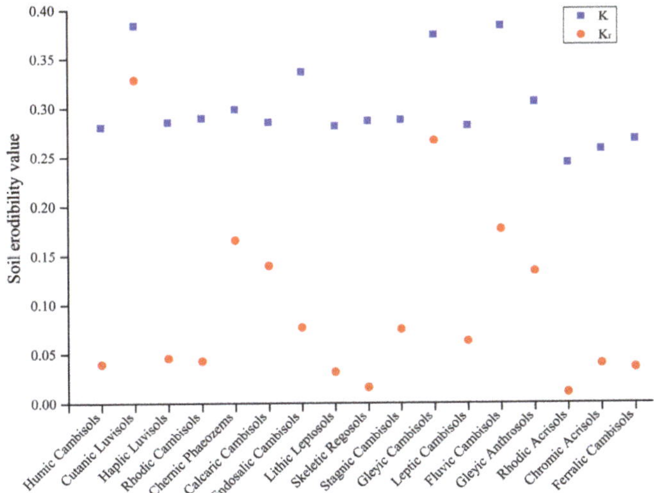

Figure 5. The K values and K_r values for each soil types.

Table 3. Comparison of calculated and measured soil erodibility.

Soil Type	Mean of Gravel Content (%)	Mean of K	Mean of K_r	Measured Value	R_1	R_2
Skeletic Regosols	21.9	0.29	0.02	0.01	20.5	1.1
Chernic Phaeozems	4.6	0.30	0.17	0.29	1.03	0.6
Endosalic Cambisols	9.5	0.31	0.08	0.15	2.3	0.5
Rhodic Acrisols	22.5	0.24	0.01	0.01	17.4	0.8
Haplic Luvisols	10.9	0.29	0.05	0.07	3.9	0.6

Note: [1] R_1 is the ratio of the mean of K to the measured value and R_2 is the ratio of the mean of K_r to the measured value.

With the modified algorithm of soil erodibility factor, the RUSLE model (Equation (1)) can be transformed as follow:

$$A = R \cdot K_r \cdot LS \cdot C \cdot P. \tag{15}$$

3.3. RUSLE Factors

The R values in the CLB varied from 2856.17 to 8985.13 MJ·mm·ha^{-1}·hr^{-1}·year^{-1}, with a mean of 4244.71 MJ·mm·ha^{-1}·hr^{-1}·year^{-1} (Figure 6a). Spatially, the values decreased from southwest to northeast with the highest and smallest values of R observed in the counties of Shucheng and Feixi, respectively.

The K_r value obtained by Equation (13) varied from 0 to 0.38 Mg·ha·h·ha^{-1}·MJ^{-1}·mm^{-1} (Figure 6b). Accounting for 67.9% of the basin's area, gleyic anthrosols was characterized by soil erodibility values ranging between 0.25 and 0.28 Mg·ha·h·ha^{-1}·MJ^{-1}·mm^{-1}. The K_r values for ferralic cambisols, endosalic cambisols, lithic leptosols, skeletic regosols, and stagnic cambisols were nearly equal to 0. The soil types with the highest K_r value were the humic cambisols (0.36 Mg·ha·h·ha^{-1}·MJ^{-1}·mm^{-1}) and calcaric cambisols (0.38 Mg·ha·h·ha^{-1}·MJ^{-1}·mm^{-1}), mainly distributed in the Dabie Mountains in the southwest, the Sangong Mountains in the south, Chaohu east of Chaohu Lake, and mountainous areas in Hanshan.

The LS value in the CLB changed from 0.03 to 40.57, with a mean of 1.04 (Figure 6c). The areas with LS values > 5 were mainly concentrated in the mountainous areas with high elevations and slopes, while the areas with LS values < 0.1 were mostly distributed in flat terrain that was dominated by construction land and lakes.

The C value in the CLB varied from 0 to 1, with a mean of 0.73 (Figure 6d). When the C value = 1, it represents invalid vegetation cover and management measures and a low $NDVI$ value, with a high probability of soil erosion; conversely, when the *C* value = 0, it represents vegetation cover and management measures to inhibit soil erosion with a positive effect. The calculation of *C* value in the CLB shows that the vegetation cover was concentrated in the areas with low *C* value.

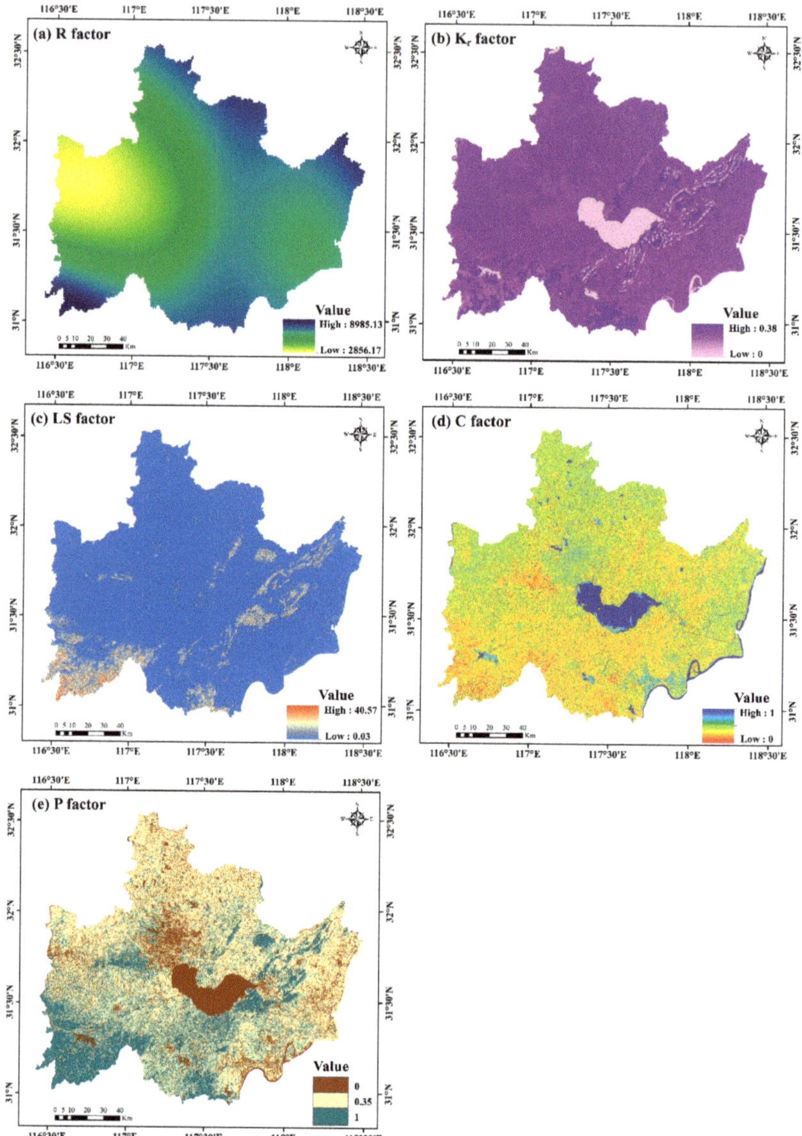

Figure 6. Maps of the Revised Universal Soil Loss Equation (RUSLE) factors: (**a**) *R*, the rainfall–runoff erosivity factor; (**b**) K_r, the soil erodibility factor; (**c**) *LS*, the slope length and steepness factor; (**d**) *C*, the cover fraction factor; (**e**) *P*, the support practice factor.

The results of land use/cover classification (Section 2.3.1) show that the farmland area was 1,196,187 km^2 (accounting for 58.7% of the basin area), which is the largest land use type in the CLB. The forestland area reached 4490.87 km^2 (22%) and the construction land area was 2157.64 km^2 (10.6%).

The areas of water body, grassland, and unused land were relatively small, comprising only 6.9%, 0.1%, and 1.8%, respectively. The P value was as signed according to the land use/cover classification results, varying from 0 to 1 (Figure 6e). It is shown that the areas with large P factor values were concentrated in the mountainous area and its surrounding area. Areas with small P factor values were mainly concentrated in construction land, water body, and unused land, where we as sume that the probability of their soil erosion is low.

3.4. Soil Erosion Estimation

With the factors calculated above, we used the RUSLE in the form of Equations (1) and (15) to estimate the soil erosion of the CLB in 2017. In order to show the differences between soil erosion estimation results clearly, estimated soil erosion was divided into six grades according to the Standards for Classification and Gradation of Soil Erosion issued by the Ministry of Water Resources of the People's Republic of China (SL190-2007) [69]: slight (<5 Mg·ha^{-1}·year^{-1}), light (5–25 Mg·ha^{-1}·year^{-1}), moderate (25–50 Mg·ha^{-1}·year^{-1}), intense (50–80 Mg·ha^{-1}·year^{-1}), extremely intense (80–150 Mg·ha^{-1}·year^{-1}), and severe (>150 Mg·ha^{-1}·year^{-1}). The grading maps are shown in Figure 7.

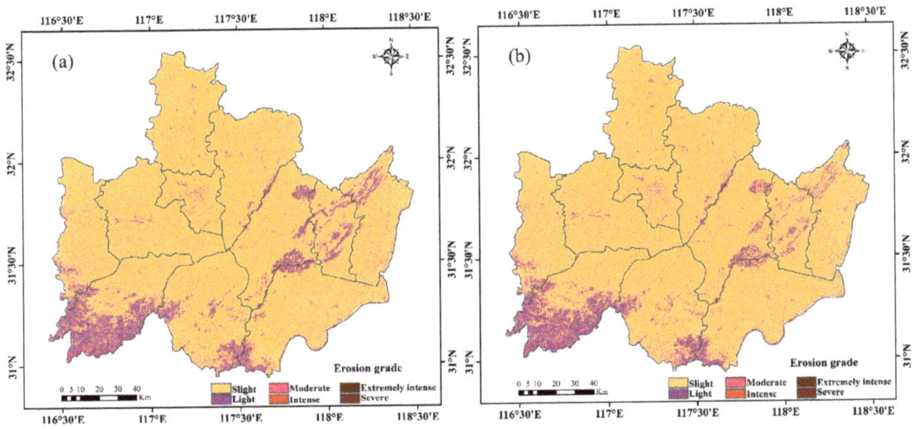

Figure 7. Soil erosion grading maps of the CLB: (**a**) soil erosion modulus estimated from Equation (1); and (**b**) soil erosion modulus estimated from Equation (15).

As seen in Table 4, the two equations resulted in different soil erosion moduli and our estimation (1.78 Mg·ha^{-1}·year^{-1} by Equation (15)) was slightly lower than the original RUSLE (1.92 Mg·ha^{-1}·year^{-1} by Equation (1)). Both calculation results showed that Shucheng was the largest contributor (~5 Mg·ha^{-1}·year^{-1}) to soil loss in the CLB. In contrast, smallest soil erosion moduli were observed in Changfeng, Feidong, and Feixi, all lower than 1 Mg·ha^{-1}·year^{-1}. The comparison also revealed that there was almost no change in the average annual soil erosion modulus for Changfeng, Jin'an, Urban Hefei, Feidong, and Feixi while the estimation from Equation (15) was lower than that from Equation (1) for the rest of the districts and counties.

In addition, we compared the measured data of 11 districts in the CLB provided by Anhui Provincial Water Conservancy Station with the results of Equations (1) and (15) (Figure 8). It can be observed that all the measured data were smaller than the result of Equation (1), where the largest difference was 0.59 Mg·ha^{-1}·year^{-1} in Shucheng and the smallest difference was 0.11 Mg·ha^{-1}·year^{-1} in Changfeng. In general, the result of Equation (15) was closer to the measured data than that of Equation (1). The largest difference between the result of Equation (15) and the measured data was 0.44 Mg·ha^{-1}·year^{-1} in Shucheng and the smallest difference was 0.08 Mg·ha^{-1}·year^{-1} in Hexian, both smaller than those for Equation (1). Figure 8 also shows that the differences between the results of

Equations (1) and (15) with the measured data were the same in Changfeng, Jin'an, Urban Hefei, Feidong, and Feixi. For the entire study area, the differences between the two results and the measured data were 0.26 Mg·ha^{-1}·year^{-1} for the result of Equation (1) and 0.12 Mg·ha^{-1}·year^{-1} for the result of Equation (15).

Table 4. Average soil erosion modulus of the CLB in 2017 (Mg·ha^{-1}·year^{-1}).

Area	Equation (1)	Equation (15)	Differences
CLB	1.92	1.78	−0.14
Wuwei	1.89	1.72	−0.17
Changfeng	0.96	0.96	0.00
Jin'an	1.58	1.57	−0.01
Urban Hefei	1.32	1.32	0.00
Feidong	0.98	0.98	0.00
Feixi	0.96	0.96	0.00
Shucheng	5.10	4.95	−0.15
Lujiang	2.07	1.98	−0.09
Chaohu	1.80	1.47	−0.33
Hanshan	2.44	1.80	−0.64
Hexian	1.67	1.36	−0.31

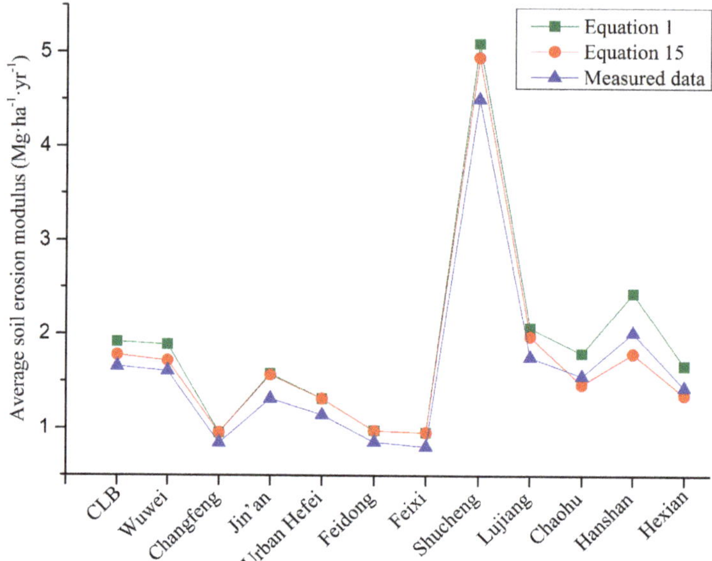

Figure 8. Comparison of the results of Equations (1) and (15) with the measured data.

Table 5 shows the areas and proportions of each soil erosion grade. The area for each grade was quite similar from the two estimations. It is clear that the erosion-affected area (from slight to severe level) in the CLB in 2017 was 19,087.35 km^2, accounting for 93.6% of the total area. The slight level was the largest (~85%) among all the levels, followed the light level (~8%). The areas of intense and extremely intense levels were quite small, both composing less than 1%. Despite a large erosion-affected area, there was no severe-level soil erosion in the CLB.

Table 5. Area for each soil erosion grade.

Grade	Equation (1)		Equation (15)	
	Area (km^2)	Proportion (%)	Area (km^2)	Proportion (%)
No erosion	1303.08	6.39	1303.08	6.39
Slight	17,266.30	84.66	17,445.76	85.55
Light	1686.46	8.27	1531.47	7.51
Moderate	116.24	0.57	97.88	0.48
Intense	14.27	0.07	10.20	0.05
Extremely intense	4.08	0.02	4.08	0.02
Severe	0.00	0.00	0.00	0.00

Figure 9 shows the soil erosion grade for each district and county of the CLB; the slight level was the largest at over 60% for each part of the CLB. Shucheng had more light-erosion (>30%) and moderate-erosion (>2%) areas than the other 10 districts and counties. While Wuwei, Shucheng, and Chaohu had the largest intense erosion areas, no extremely intense erosion was estimated in Changfeng, Jin'an, Urban Hefei, Feidong, and Feixi. Figure 9 also reveals that the two estimations resulted in quite similar distributions of soil erosion grades in these districts, except for the intense level for Hanshan (Figure 9d) and the extremely intense level for Chaohu (Figure 9e).

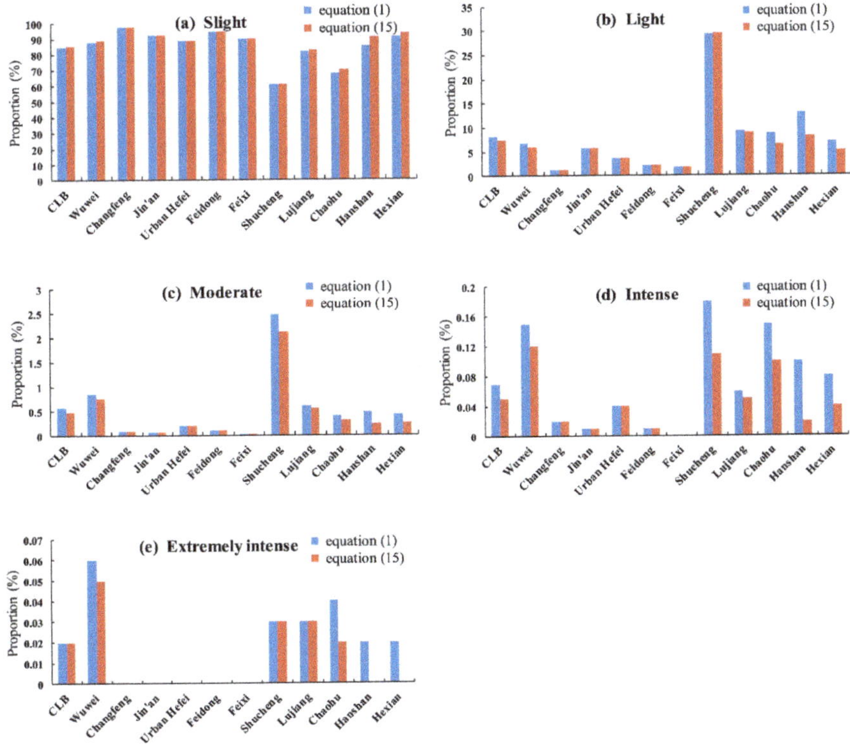

Figure 9. The distribution of soil erosion grade in the CLB and its 11 districts and counties: (**a**) Slight; (**b**) Light; (**c**) Moderate; (**d**) Intense; and (**e**) Extremely intense.

4. Discussion

In this study, we estimated the soil erosion in the CLB in 2017 using the original RUSLE model and the RUSLE model with modified soil erodibility. In addition, the distribution of soil erosion grades in the CLB and its 11 administrative districts were investigated. The interpretation of the results and their implications are given in this section.

4.1. Factors of the RUSLE Model

Soil erosion is a complex process influenced by a variety of natural and human-induced factors [70]. Five factors are considered to estimate soil erosion in the RUSLE model [71,72], namely rainfall erosivity (R), soil erodibility (K), slope length and steepness (LS), cover fraction (C), and support practice (P). The R factor, K factor, and LS factor contribute to greater soil erosion [4] while the C factor and P factor play an important role in preventing soil erosion [27]. By comparing the maps of the R factor, K factor, and LS factor (Figure 6) and the estimated soil erosion (Figure 7), we notice that the three factors are highly consistent with soil erosion in spatial distribution. This helps to explain why soil erosion was higher in Wuwei, Shucheng, Lujiang, Chaohu, Hanshan, and Hexian than the other parts of the CLB. Similarly, Figure 6 shows that the C factor and P factor play an irreplaceable role in the control of soil erosion, especially in the areas with topographic fluctuation. Since it is difficult to change the natural factors such as rainfall, topography, and soil properties, optimizing land use structure and improving vegetation coverage are considered the most effective measures to prevent soil erosion [18].

4.2. Influence of Gravel Content on Soil Erosion Estimation

Based on the comparison of the results of Equations (1) and (15) with the measured data, we found that all the calculated soil erosion modulus combined with the modified soil erodibility were closer to the measured data in Wuwei, Shucheng, Lujiang, Chaohu, Hanshan, and Hexian, which are characterized by mountainous area with high gravel content. However, there were no differences between the two results with the measured data in the districts with small mountainous areas, such as Changfeng, Jin'an, Urban Hefei, Feidong, and Feixi. Therefore, we consider that the accuracy of soil erodibility has an important impact on soil erosion estimation [73,74]. The K values calculated by the method in the RUSLE model were much larger than the measured values. In this study, the gravel content parameter was added into the modified algorithm for soil erodibility, and the estimated K_r values were closer to the measured ones. The accuracy of soil erosion estimation in the CLB was accordingly effectively improved using the modified soil erodibility. Therefore, we believe that this might be because the effect of gravel content on soil erosion was not fully considered.

Many previous studies about the effect of rock fragment and gravel content on soil erosion have also reached similar conclusions. Rodrigo-Comino et al. [37] carried out an investigation with 96 rainfall simulation experiments at the pedon scale and found that the soil losses are inversely proportional to rock fragment cover on the soil surface. Cerdà [38] carried out 20 experiments on bare areas of natural soils and the results showed that water and soil losses were reduced by the rock fragments. Poesen et al. [39] have reported the various effects of rock fragments on soil erosion and the key finding shows that rock fragment cover will offer protection to topsoil and have different efficiencies in different nested spatial scales. The results of two laboratory flume experiments carried out by Jomaa et al. [40] revealed that the rock fragments decreased the sediment transport capacity. These studies provide a reliable support for our views.

4.3. Characteristics of Soil Erosion in the CLB

Overall, the erosion-affected areas of the CLB were mainly distributed along the SW–NE direction. While the slight-level soil erosion was mostly found in the alluvial plains along the middle and lower reaches of the Nanfei River, Hangbu River, and Tianhe River, and the low mountain and hilly areas with high vegetation coverage (Figure 7), the areas with high soil erosion modulus in the CLB concentrated in the northeast of Dabie Mountain, the north of Sangong Mountain, and the mountainous areas of Chaohu and Hanshan. Particularly, the population density of Longhekou Reservoir area in Shucheng was high and inappropriate land use existed in this area, such as steep slope reclamation and excessive vegetation destruction. The intensive interaction between human and nature has caused reservoir siltation, thus serious soil erosion problems [75].

4.4. Limitations

Soil erosion estimation is a key to the understanding and management of the ecological environment, particularly in ecologically vulnerable regions [76]. Although it is considered as a widely used approach to soil erosion estimation [74,77], the application of the RUSLE model might be region-specific due to the complexity of the ecological environment [23]. In the case study of the CLB, the gravel content parameter was used to modify the algorithm of soil erodibility factor in the RUSLE model. Such revision has proved to improve soil erosion estimation for the CLB. Despite the improvement, there are some issues that should be addressed in further research:

(1) the accuracy of soil erodibility obtained by the modified algorithm using the gravel content was assessed based on only five soil types and an exhaustive assessment is required; and

(2) due to the limitation of data acquisition, only 13 meteorological stations could provide rainfall data to estimate rainfall erosivity, which might reduce the accuracy of rainfall erosivity estimation.

5. Conclusions

We estimated and compared the soil erosion of the Chaohu Lake Basin (CLB) in 2017 using the original RUSLE model and the RUSLE model with modified soil erodibility. The average annual soil erosion estimated with the K_r algorithm was 0.14 Mg·ha^{-1}·year^{-1} lower than the estimation result with the original K algorithm in the CLB. In other words, taking gravel content into account helps to improve the calculation of soil erodibility and soil erosion estimation. The overall soil erosion in the CLB was low with a majority of slight erosion (accounting for 85.6%), and the mountainous and hilly areas are more prone to soil erosion. The superposition of inappropriate land use and natural factors (including climate, soil properties, and topography) is the main reason for soil erosion of the CLB and should be optimized for soil erosion prevention in the CLB.

Quantitative analysis of soil erosion is highly beneficial in natural resource management and policy-making to relieve the pressure of soil erosion and land degradation. The findings of this study provide useful insights into the spatial distribution of soil erosion and the driving mechanism in this ecologically important region.

Author Contributions: Conceptualization, S.H. and L.L.; methodology, S.H.; software, L.C. (Liang Cheng) and L.Y.; validation, S.H. and X.H.; formal analysis, S.H.; data curation, X.H. and T.Z.; writing—original draft preparation, S.H.; writing—review and editing, L.L. and L.C. (Longqian Chen); visualization, S.H.; supervision, project administration, and funding acquisition, L.C. (Longqian Chen). All the authors reviewed and approved the final manuscript version.

Funding: This research was funded by the Fundamental Research Funds for the Central Universities under grant number 2018ZDPY07.

Acknowledgments: The authors thank the Geospatial Data Cloud for freely providing the Landsat data, the Cold and Arid Regions Sciences Data Center for the soil data, and the National Meteorological Information Center for the rainfall data. Furthermore, we appreciate the editors and reviewers for their constructive comments and suggestions.

Conflicts of Interest: The authors declare no conflict of interest.

Appendix A

Table A1. Measured soil erosion dataset.

Slope Gradient (°)	Land Use	Soil Loss Rate (Mg·ha^{-1}·yr^{-1})	Reference
15°	arable land	198.2	Yang [78]
15°	arable land	234.64	Yang [78]
21°	arable land	366.12	Yang [78]
21°	arable land	432.14	Yang [78]
29°	arable land	474.04	Yang [78]
42°	arable land	865.88	Yang [78]
45°	arable land	991.48	Yang [78]
45°	arable land	897.58	Yang [78]
45°	arable land	958.22	Yang [78]
10°	bare	83.9	Mu [79]
15°	bare	120.5	Mu [79]
20°	bare	136.21	Mu [79]
25°	bare	210.5	Mu [79]
30°	bare	272.71	Mu [79]
40°	bare	308.08	Mu [79]
10°	arable land	44.21	Bi [80]
20°	arable land	103.26	Bi [80]
25°	shrub	139.87	Bi [80]
28°	shrub	140.7	Bi [80]
32°	bare	103.24	Tang et al. [81]
35°	bare	152.86	Tang et al. [81]
38°	bare	217.74	Tang et al. [81]
30°	bare	83.65	Zhang [82]
35°	bare	100.12	Zhang [82]
10°	grassland	355.19	Xu et al. [83]
14°	arable land	38.4	Wang [84]
26°	fallow	214.62	Lin [85]
26°	arable land	35.39	Wu et al. [86]
31°	fallow	92.66	Liu et al. [87]

References

1. Patowary, S.; Sarma, A.K. GIS-based estimation of soil loss from hilly urban area incorporating hill cut factor into RUSLE. *Water Resour. Manag.* **2018**, *32*, 3535–3547. [CrossRef]
2. Zhao, J.; Vanmaercke, M.; Chen, L.; Govers, G. Vegetation cover and topography rather than human disturbance control gully density and sediment production on the Chinese Loess Plateau. *Geomorphology* **2016**, *274*, 92–105. [CrossRef]
3. Rahman, M.R.; Shi, Z.H.; Chongfa, C. Soil erosion hazard evaluation—An integrated use of remote sensing, GIS and statistical approaches with biophysical parameters towards management strategies. *Ecol. Model.* **2009**, *220*, 1724–1734. [CrossRef]
4. Xu, L.; Xu, X.; Meng, X. Risk assessment of soil erosion in different rainfall scenarios by RUSLE model coupled with Information Diffusion Model: A case study of Bohai Rim, China. *Catena* **2012**, *100*, 74–82. [CrossRef]
5. Pimentel, D. Soil erosion: A food and environmental threat. *Environ. Dev. Sustain.* **2006**, *8*, 119–137. [CrossRef]
6. Park, S.; Jeon, S.; Choi, C.; Oh, C.; Jung, H. Soil erosion risk in Korean watersheds, assessed using the revised universal soil loss equation. *J. Hydrol.* **2011**, *399*, 263–273. [CrossRef]
7. Vrieling, A.; de Jong, S.M.; Sterk, G.; Rodrigues, S.C. Timing of erosion and satellite data: A multi-resolution approach to soil erosion risk mapping. *Int. J. Appl. Earth Obs. Geoinf.* **2008**, *10*, 267–281. [CrossRef]
8. Montanarella, L. Agricultural policy: Govern our soils. *Nature* **2015**, *528*, 32–33. [CrossRef]

9.   Rodrigo-Comino, J.; Keesstra, S.; Cerdà, A. Soil erosion as an environmental concern in vineyards: The case study of Celler del Roure, eastern Spain, by means of rainfall simulation experiments. *Beverages* **2018**, *4*, 31. [CrossRef]

10. Rodrigo-Comino, J.; Davis, J.; Keesstra, S.D.; Cerdà, A. Updated measurements in vineyards improves accuracy of soil erosion rates. *Agron. J.* **2017**, *110*, 411–417. [CrossRef]

11. Cerdà, A.; Ackermann, O.; Terol, E.; Rodrigo-Comino, J. Impact of farmland abandonment on water resources and soil conservation in citrus plantations in eastern Spain. *Water* **2019**, *11*, 824. [CrossRef]

12. Cerdà, A.; Rodrigo-Comino, J.; Giménez-Morera, A.; Novara, A.; Pulido, M.; Kapović-Solomun, M.; Keesstra, S.D. Policies can help to apply successful strategies to control soil and water losses: The case of chipped pruned branches (CPB) in Mediterranean citrus plantations. *Land Use Policy* **2018**, *75*, 734–745. [CrossRef]

13. Zhao, J.; Van Oost, K.; Chen, L.; Govers, G. Moderate topsoil erosion rates constrain the magnitude of the erosion-induced carbon sink and agricultural productivity losses on the Chinese Loess Plateau. *Biogeosciences* **2016**, *13*, 4735–4750. [CrossRef]

14. Rodrigo-Comino, J.; Iserloh, T.; Lassu, T.; Cerdà, A.; Keestra, S.D.; Prosdocimi, M.; Brings, C.; Marzen, M.; Ramos, M.C.; Senciales, J.M.; et al. Quantitative comparison of initial soil erosion processes and runoff generation in Spanish and German vineyards. *Sci. Total Environ.* **2016**, *565*, 1165–1174. [CrossRef] [PubMed]

15. Thomas, J.; Joseph, S.; Thrivikramji, K.P. Assessment of soil erosion in a tropical mountain river basin of the southern Western Ghats, India using RUSLE and GIS. *Geosci. Front.* **2018**, *9*, 893–906. [CrossRef]

16. Keesstra, S.D.; Bouma, J.; Wallinga, J.; Tittonell, P.; Smith, P.; Cerdà, A.; Montanarella, L.; Quinton, J.N.; Pachepsky, Y.; Van Der Putten, W.H.; et al. The significance of soils and soil science towards realization of the United Nations sustainable development goals. *Soil* **2016**, *2*, 111–128. [CrossRef]

17. Keesstra, S.; Mol, G.; de Leeuw, J.; Okx, J.; Molenaar, C.; de Cleen, M.; Visser, S. Soil-related sustainable development goals: Four concepts to make land degradation neutrality and restoration work. *Land* **2018**, *7*, 133. [CrossRef]

18. Wang, X.; Hu, S.; Wen, Q.; Yi, L.; Zuo, L.; Xu, J.; Zhang, Z.; Liu, B.; Zhao, X.; Liu, F. Assessment of soil erosion change and its relationships with land use/cover change in China from the end of the 1980s to 2010. *Catena* **2016**, *137*, 256–268. [CrossRef]

19. Bhattarai, R.; Dutta, D. Estimation of soil erosion and sediment yield using GIS at catchment scale. *Water Resour. Manag.* **2007**, *21*, 1635–1647. [CrossRef]

20. Kinnell, P.I.A. Event soil loss, runoff and the Universal Soil Loss Equation family of models: A review. *J. Hydrol.* **2010**, *385*, 384–397. [CrossRef]

21. Fu, B.J.; Zhao, W.W.; Chen, L.D.; Zhang, Q.J.; Lu, Y.H.; Gulinck, H.; Poesen, J. Assessment of soil erosion using RUSLE and GIS: A case study of the Yangou watershed in the Loess Plateau, China. *Land Degrad. Dev.* **2005**, *16*, 73–85. [CrossRef]

22. Cohen, M.J.; Shepherd, K.D.; Walsh, M.G. Empirical reformulation of the universal soil loss equation for erosion risk assessment in a tropical watershed. *Geoderma* **2005**, *124*, 235–252. [CrossRef]

23. Napoli, M.; Cecchi, S.; Orlandini, S.; Mugnai, G.; Zanchi, C.A. Simulation of field-measured soil loss in Mediterranean hilly areas (Chianti, Italy) with RUSLE. *Catena* **2016**, *145*, 246–256. [CrossRef]

24. Yang, X.; Gray, J.; Chapman, G.; Zhu, Q.; Tulau, M.; McInnes-Clarke, S. Digital mapping of soil erodibility for water erosion in New South Wales, Australia. *Soil Res.* **2017**, *56*, 158–170. [CrossRef]

25. Kijowska-Strugała, M.; Bucała-Hrabia, A.; Demczuk, P. Long-term impact of land use changes on soil erosion in an agricultural catchment (in the Western Polish Carpathians). *Land Degrad. Dev.* **2018**, *29*, 1871–1884. [CrossRef]

26. Kumar, S.; Kushwaha, S.P.S. Modelling soil erosion risk based on RUSLE-3D using GIS in a Shivalik sub-watershed. *J. Earth Syst. Sci.* **2013**, *122*, 389–398. [CrossRef]

27. Farhan, Y.; Nawaiseh, S. Spatial assessment of soil erosion risk using RUSLE and GIS techniques. *Environ. Earth Sci.* **2015**, *74*, 4649–4669. [CrossRef]

28. Zhang, P.; Yao, W.; Liu, G.; Xiao, P. Experimental study on soil erosion prediction model of loess slope based on rill morphology. *Catena* **2019**, *173*, 424–432. [CrossRef]

29. Smith, D.D.; Wischmeier, W.H. Factors affecting sheet and rill erosion. *Trans. Am. Geophys. Union* **1957**, *38*, 889–896. [CrossRef]

30. Wischmeier, W.H.; Smith, D.D. *Predicting Rainfall Erosion Losses: A Guide to Conservation Planning*; Agriculture Handbook No. 537; US Department of Agriculture Science and Education Administration: Washington, DC, USA, 1978; p. 163.

31. McCool, D.K.; Brown, L.C.; Foster, G.R.; Mutchler, C.K. Revised slope steepness factor for the Universal Soil Loss Equation. *Trans. ASAE* **1987**, *30*, 1387–1396. [CrossRef]

32. Liu, B.Y.; Nearing, M.A.; Shi, P.J.; Jia, Z.W. Slope length effects on soil loss for steep slopes. *Soil Sci. Soc. Am. J.* **2000**, *64*, 1759–1763. [CrossRef]

33. Liu, B.; Song, C.; Shi, Z.; Tao, H. Correction algorithm of slope factor in Universal Soil Loss Equation in earth-rocky mountain area of Southwest China. *Soil Water Conserv. China* **2015**, *8*, 49–52. (In Chinese)

34. Williams, J.R.; Jones, C.A.; Dyke, P.T. A modeling approach to determining the relationship between erosion and soil productivity. *Trans. ASAE* **1984**, *27*, 129–144. [CrossRef]

35. Zhang, K.L.; Shu, A.P.; Xu, X.L.; Yang, Q.K.; Yu, B. Soil erodibility and its estimation for agricultural soils in China. *J. Arid Environ.* **2008**, *72*, 1002–1011. [CrossRef]

36. Poesen, J.; De Luna, E.; Franca, A.; Nachtergaele, J.; Govers, G. Concentrated flow erosion rates as affected by rock fragment cover and initial soil moisture content. *Catena* **1999**, *36*, 315–329. [CrossRef]

37. Rodrigo-Comino, J.; García-Díaz, A.; Brevik, E.C.; Keestra, S.D.; Pereira, P.; Novara, A.; Jordán, A.; Cerdà, A. Role of rock fragment cover on runoff generation and sediment yield in tilled vineyards. *Eur. J. Soil Sci.* **2017**, *68*, 864–872. [CrossRef]

38. Cerda, A. Effects of rock fragment cover on soil infiltration, interrill runoff and erosion. *Eur. J. Soil Sci.* **2001**, *52*, 59–68. [CrossRef]

39. Poesen, J.W.; Torri, D.; Bunte, K. Effects of rock fragments on soil erosion by water at different spatial scales: A review. *Catena* **1994**, *23*, 141–166. [CrossRef]

40. Jomaa, S.; Barry, D.A.; Heng, B.C.P.; Brovelli, A.; Sander, G.C.; Parlange, J.Y. Influence of rock fragment coverage on soil erosion and hydrological response: Laboratory flume experiments and modeling. *Water Resour. Res.* **2012**, *48*, W05535. [CrossRef]

41. Shi, C. Effects of gravel content on soil erodibility and the methods of estimating soil erodibility factor K. *Chin. J. Soil Sci.* **2009**, *40*, 1398–1401. (In Chinese)

42. Sun, W.Y.; Shao, Q.Q.; Liu, J.Y.; Zhai, J. Assessing the effects of land use and topography on soil erosion on the Loess Plateau in China. *Catena* **2014**, *121*, 151–163. [CrossRef]

43. Zhang, Z.; Gao, J.; Gao, Y. The influences of land use changes on the value of ecosystem services in Chaohu Lake Basin, China. *Environ. Earth Sci.* **2015**, *74*, 385–395. [CrossRef]

44. Renard, K.G.; Foster, G.R.; Weesies, G.A. *Predicting Soail Erosion by Water: A Guide to Conservation Planning with the Revised Universal Soil Loss Equation (RUSLE)*; Agriculture Handbook No. 703; US Department of Agriculture Science and Education Administration: Washington, DC, USA, 1997; pp. 1–251.

45. Antoneli, V.; Rebinski, E.; Bednarz, J.; Rodrigo-Comino, J.; Keesstra, S.; Cerdà, A.; Pulido Fernández, M. Soil Erosion Induced by the Introduction of New Pasture Species in a Faxinal Farm of Southern Brazil. *Geosciences* **2018**, *8*, 166. [CrossRef]

46. Rodrigo-Comino, J.; Keesstra, S.D.; Cerdà, A. Connectivity assessment in Mediterranean vineyards using improved stock unearthing method, LiDAR and soil erosion field surveys. *Earth Surf. Process. Landf.* **2018**, *43*, 2193–2206. [CrossRef]

47. Mondal, A.; Khare, D.; Kundu, S. International Soil and Water Conservation Research Change in rainfall erosivity in the past and future due to climate change in the central part of India. *Int. Soil Water Conserv. Res.* **2016**, *4*, 186–194. [CrossRef]

48. Amanambu, A.C.; Li, L.; Egbinola, C.N.; Obarein, O.A.; Mupenzi, C.; Chen, D. Spatio-temporal variation in rainfall-runoff erosivity due to climate change in the Lower Niger Basin, West Africa. *Catena* **2019**, *172*, 324–334. [CrossRef]

49. Teng, H.; Shi, Z.; Ma, Z.; Li, Y. Estimating spatially downscaled rainfall by regression kriging using TRMM precipitation and elevation in Zhejiang Province, southeast China. *Int. J. Remote Sens.* **2014**, *35*, 7775–7794. [CrossRef]

50. Liu, B.; Tao, H.; Song, C.; Guo, B.; Shi, Z.; Zhang, C.; Kong, B.; He, B. Temporal and spatial variations of rainfall erosivity in China during 1960 to 2009. *Geogr. Res.* **2013**, *32*, 245–256. (In Chinese)

51. Karydas, C.; Petriolis, M.; Manakos, I. Evaluating alternative methods of soil erodibility mapping in the mediterranean island of Crete. *Agriculture* **2013**, *3*, 362–380. [CrossRef]

52. Lin, B.S.; Chen, C.K.; Thomas, K.; Hsu, C.K.; Ho, H.C. Improvement of the K-Factor of USLE and soil erosion estimation in Shihmen reservoir watershed. *Sustainability* **2019**, *11*, 355. [CrossRef]

53. Wang, M.; Baartman, J.E.M.; Zhang, H.; Yang, Q.; Li, S.; Yang, J.; Cai, C.; Wang, M.; Ritsema, C.J.; Geissen, V. An integrated method for calculating DEM-based RUSLE LS. *Earth Sci. Inform.* **2018**, *11*, 579–590. [CrossRef]

54. Yang, X. Digital mapping of RUSLE slope length and steepness factor across New South Wales, Australia. *Soil Res.* **2015**, *53*, 216–225. [CrossRef]

55. Xue, J.; Lyu, D.; Wang, D.; Wang, Y.; Yin, D.; Zhao, Z.; Mu, Z. Assessment of soil erosion dynamics using the GIS-Based RUSLE Model: A case study of wangjiagou watershed from the Three Gorges Reservoir Region, Southwestern China. *Water* **2018**, *10*, 1817. [CrossRef]

56. Zha, L.; Deng, G.; Gu, J. Dynamic changes of soil erosion in the ChaohuWatershed from 1992 to 2013. *Acta Geogr. Sin.* **2015**, *70*, 1708–1719. (In Chinese)

57. Li, L.; Canters, F.; Solana, C.; Ma, W.; Chen, L.; Kervyn, M. Discriminating lava flows of different age within Nyamuragira's volcanic field using spectral mixture analysis. *Int. J. Appl. Earth Obs. Geoinf.* **2015**, *40*, 1–10. [CrossRef]

58. Li, H.; Li, L.; Chen, L.; Zhou, X.; Cui, Y.; Liu, Y.; Liu, W. Mapping and Characterizing Spatiotemporal Dynamics of Impervious Surfaces Using Landsat Images: A Case Study of Xuzhou, East China from 1995 to 2018. *Sustainability* **2019**, *11*, 1224. [CrossRef]

59. Zhou, X.; Li, L.; Chen, L.; Liu, Y.; Cui, Y. Discriminating Urban Forest Types from Sentinel-2A Image Data through Linear Spectral Mixture Analysis: A Case Study of Xuzhou, East China. *Forests* **2019**, *10*, 478. [CrossRef]

60. Lu, D.; Li, G.; Valladares, G.S.; Batistella, A.M. Mapping soil erosion risk in Rond Nia, Brazilian Amazonia: Using RUSLE, remote sensing and GIS. *Land Degrad. Dev.* **2004**, *15*, 499–512. [CrossRef]

61. Markose, V.J.; Jayappa, K.S. Soil loss estimation and prioritization of sub-watersheds of Kali River basin, Karnataka, India, using RUSLE and GIS. *Environ. Monit. Assess.* **2016**, *188*, 225–240. [CrossRef]

62. Li, L.; Bakelants, L.; Solana, C.; Canters, F.; Kervyn, M. Dating lava flows of tropical volcanoes by means of spatial modeling of vegetation recovery. *Earth Surf. Process. Landf.* **2017**, *43*, 840–856. [CrossRef]

63. Amri, R.; Zribi, M.; Lili-Chabaane, Z.; Duchemin, B.; Gruhier, C.; Chehbouni, A. Analysis of vegetation behavior in a North African semi-arid region, Using SPOT-VEGETATION NDVI data. *Remote Sens.* **2011**, *3*, 2568–2590. [CrossRef]

64. Gutman, G.; Ignatov, A. The derivation of the green vegetation fraction from NOAA/AVHRR data for use in numerical weather prediction models. *Int. J. Remote Sens.* **1998**, *19*, 1533–1543. [CrossRef]

65. Van der Knijff, J.M.; Jones, R.J.A.; Montanarella, L. *Soil Erosion Risk Assessment in Europe*; European Soil Bureau, Joint Research Centre, European Commission and Space Applications Institute: Brussel, Belgium, 2000; pp. 17–19.

66. Panagos, P.; Borrelli, P.; Meusburger, K.; van der Zanden, E.H.; Poesen, J.; Alewell, C. Modelling the effect of support practices (P-factor) on the reduction of soil erosion by water at European scale. *Environ. Sci. Policy* **2015**, *51*, 23–34. [CrossRef]

67. Borrelli, P.; Alewell, C.; Ballabio, C.; Ferro, V.; Robinson, D.A.; Panagos, P.; Lugato, E.; Van Oost, K.; Schütt, B.; Fleischer, L.R.; et al. An assessment of the global impact of 21st century land use change on soil erosion. *Nat. Commun.* **2013**, *8*, 1–13. [CrossRef] [PubMed]

68. Cui, Y.; Li, L.; Chen, L.; Zhang, Y.; Cheng, L.; Zhou, X.; Yang, X. Land-use carbon emissions estimation for the Yangtze River Delta Urban Agglomeration using 1994–2016 Landsat image data. *Remote Sens.* **2018**, *10*, 1334. [CrossRef]

69. Ministry of Water Resources of the People's Republic of China. *Standards for Classification and Gradation of Soil Erosion (SL190-2007)*; China Water&Power Press: Beijing, China, 2007.

70. Hewett, C.J.M.; Simpson, C.; Wainwright, J.; Hudson, S. Communicating risks to infrastructure due to soil erosion: A bottom-up approach. *Land Degrad. Dev.* **2018**, *29*, 1282–1294. [CrossRef]

71. Liu, Y.; Shi, Z.; Chappell, A.; Liang, Z.; Chen, S.; Yu, W.; Teng, H.; Viscarra Rossel, R.A. Current and future assessments of soil erosion by water on the Tibetan Plateau based on RUSLE and CMIP5 climate models. *Sci. Total Environ.* **2018**, *635*, 673–686.

72. Moges, D.M.; Bhat, H.G. Integration of geospatial technologies with RUSLE for analysis of land use/cover change impact on soil erosion: Case study in Rib watershed, north-western highland Ethiopia. *Environ. Earth Sci.* **2017**, *76*, 1–14. [CrossRef]

73. Bryan, R.B. Soil erodibility and processes of water erosion on hillslope. *Geomorphology* **2000**, *32*, 385–415. [CrossRef]

74. Wang, L.; Huang, J.; Du, Y.; Hu, Y.; Han, P. Dynamic assessment of soil erosion risk using landsat TM and HJ satellite data in danjiangkou reservoir area, China. *Remote Sens.* **2013**, *5*, 3826–3848. [CrossRef]

75. Mandal, D.; Sharda, V.N. Appraisal of soil erosion risk in the Eastern Himalayan Region of India for soil conservation planning. *Land Degrad. Dev.* **2013**, *24*, 430–437. [CrossRef]

76. Wynants, M.; Solomon, H.; Ndakidemi, P.; Blake, W.H. Pinpointing areas of increased soil erosion risk following land cover change in the Lake Manyara catchment, Tanzania. *Int. J. Appl. Earth Obs. Geoinf.* **2018**, *71*, 1–8. [CrossRef]

77. Alexakis, D.D.; Hadjimitsis, D.G.; Agapiou, A. Integrated use of remote sensing, GIS and precipitation data for the assessment of soil erosion rate in the catchment area of "Yialias" in Cyprus. *Atmos. Res.* **2013**, *131*, 108–124. [CrossRef]

78. Yang, Z.S. The Topographic factor of soil erosion of sloping cultivated land in the northeast mountain region of Yunnan Province. *J. Mt. Sci.* **1999**, *17*, 16–18.

79. Mu, E.S. Analysis on runoff yield characteristics of slope surface runoff under different control measure conditions in Shiqiao small watershed of Karst Area. *J. Anhui Agric. Sci.* **2012**, *40*, 939–941.

80. Bi, X.G.; Duan, S.H.; Li, Y.G.; Liu, B.Y.; Fu, S.H.; Ye, Z.H.; Yuan, A.P.; Lu, B.J. Study on soil loss equation in Beijing. *Sci. Soil Water Conserv.* **2006**, *4*, 6–13.

81. Tang, K.L.; Xiong, G.S.; Liang, J.Y.; Jing, K.; Zhang, S.L.; Chen, Y.Z.; Li, S.M. *Varieties of erosion and runoff sediment in Yellow River*; Chinese Sciences and Technique Press: Beijing, China, 1993; pp. 155–162.

82. Zhang, L. Study on the Influence of Slope Factor of Longitudinal Ridge Valley on Soil Erosion. Master's Thesis, Kunming University of Science and Technology, Kunming, China, 2008.

83. Xu, J.; Wang, Y.; Sun, J.H.; Chen, Z.J.; Xue, C.K.; Xu, Z.H.; Wang, X.R. Effects of slope factor on soil erosion in Jinan city. *J. Univ. Jinan* **2017**, *31*, 433–437.

84. Wang, X.Y.; Cao, W.H.; Chen, D. Study on relationship between soil erosion and land slope. *J. Sediment Res.* **1998**, *2*, 36–41.

85. Lin, J.L.; Cai, Z.F.; Chen, M.H.; Zhou, F.J.; Huang, Y.H. Relationship between slope degree and soil erosion in the south Fujian. *Fujian J. Agric. Sci.* **2002**, *17*, 86–89.

86. Wu, N.N.; Liang, Y.H.; Zhang, J.M.; Ma, L.S. Soil erosion characteristics of different slope gradients and aspects: Taking low mountain and hilly areas in north branches of Huaihe River as an Example. *Hubei Agric. Sci.* **2014**, *53*, 3755–3759.

87. Liu, S.Y.; Qin, F.C.; Xiang, Y.H.; Lei, F.Y. Study on the relationship between slope and the amount of soil loss based on WEPP. *J. Arid Land Resour. Environ.* **2006**, *20*, 97–101.

Article

Towards an Assessment of the Ephemeral Gully Erosion Potential in Greece Using Google Earth

Christos Karydas [1] and Panos Panagos [2,*]

[1] Remote Sensing Engineer, Mesimeri P.O. Box 413, 57500 Epanomi, Greece; xkarydas@gmail.com
[2] European Commission, Joint Research Centre (JRC), 21027 Ispra, Italy
*   Correspondence: panos.panagos@ec.europa.eu

Received: 10 February 2020; Accepted: 20 February 2020; Published: 23 February 2020

Abstract: Gully erosion may cause considerable soil losses and produce large volumes of sediment. The aim of this study was to perform a preliminary assessment on the presence of ephemeral gullies in Greece by sampling representative cultivated fields in 100 sites randomly distributed throughout the country. The almost 30-ha sampling surfaces were examined with visual interpretation of multi-temporal imagery from the online Google Earth for the period 2002–2019. In parallel, rill and sheet erosion signs, land uses, and presence of terraces and other anti-erosion features, were recorded within every sample. One hundred fifty-three ephemeral gullies were identified in total, inside 22 examined agricultural surfaces. The mean length of the gullies was 55.6 m, with an average slope degree of 9.7%. Vineyards showed the largest proportion of gullies followed by olive groves and arable land, while pastures exhibited limited presence of gullies. Spatial clusters of high gully severity were observed in the north and east of the country. In 77% of the surfaces with gullies, there were no terraces, although most of these surfaces were situated in slopes higher than 8%. It was the first time to use visual interpretation with Google Earth image time-series on a country scale producing a gully erosion inventory. Soil conservation practices such as contour farming and terraces could mitigate the risk of gully erosion in agricultural areas.

Keywords: land degradation; soil erosion; soil conservation; remote sensing

---

1. Introduction

Gully erosion is a key process of land degradation and desertification posing a significant threat to ecosystem services [1]. Gullies are defined as erosional channels deeper than 0.5 m, caused by concentrated water flow during and immediately after a heavy rainfall event [2]. Gullies have dynamic character, affected by topography, soil properties, vegetation cover, climate, and land management. Topography and soil properties are practically constant in time, whereas vegetation cover and land management may vary with time. Erosion-prone conditions include erodible soils, soft subsurface, or instable slopes; though, anthropogenic influences are usually the main driver of gully erosion potential [3]. Understanding the dynamics of this phenomenon in agricultural lands (especially, with regard to climate or land use changes) is important for land managers in order to assess the potential of gully initiation in a specific area of interest [4].

Verstraeten and Poesen (1999) [5] argue that gully erosion contributes significantly to soil degradation in different landscapes by: (1) causing considerable soil losses as they allow massive movement of soil particles by overland flow; (2) producing large sediment volumes (as an immediate result of massive soil loss); and (3) expanding connectivity in the landscape, thus increasing the potential of sediment transfer to watercourses, with respective acceleration of known offsite erosion effects [4]. Gullies are evidence of past intense soil erosion processes causing landscape changes, but also indicators of the impact of environmental change, caused by the geomorphological characteristics of the landscape, land use changes and extreme climatic events [6].

Gullies can be formed in any land use and if they are well-established, they are called 'permanent gullies' or 'classic gullies' [4,7]. They can also be formed in agricultural lands and be removed by tillage operations (farmers can easily refill them), so they are called 'ephemeral gullies' [8]. Usually, ephemeral gullies are less than 0.5 m deep, their formation starts with small erosional channels and then accelerates (or aggravates) with subsequent runoff events [9].

As the soil losses due to gullies are extremely high [10], there is a high interest to investigate gullies as part of environmental change, soil erosion risk, land degradation, and record of the past [11]. Scientists recognize that soil losses due to ephemeral gullies may be significantly greater than those losses attributable to sheet and rill erosion [12]. Many ephemeral gullies that develop within croplands are tillage induced, as farmers tend to redistribute the soil during plowing [13]. In case of an eventual reactivation of the gully during overland flow events, water runoff removes this additional soil material, thus reducing topsoil thickness over the entire tilled portion of the landscape [14]. Once the gullies develop, they form erosional channels, grow larger, facilitate water runoff, and accelerate water erosion rates in a feedback loop [10].

The most known conceptual model specifically developed for ephemeral gully erosion estimation is the ephemeral gully erosion model (EGEM) [15]. Several attempts have contributed to gully erosion detection and mapping using remote sensing and geographic information systems (GIS). For example, Rundquist (2002) [16] determined a ranking schema of fields for potential development of ephemeral gullies by using multi-temporal remote-sensing maps of fractional vegetation cover extracted from 16-day composites of normalized difference vegetation index (NDVI) layers, together with precipitation figures and topographic data. Hessel and Van Asch (2003) [17] studied the rolling hills region of the Chinese Loess Plateau, an area with one of the highest erosion rates on earth using the Limburg soil erosion model (LISEM) [18]. Although LISEM is a mechanistic model originally developed for storm events, the authors adapted it with regard to the positioning of existing gully heads, thus enabling it to assess the amount of material produced by permanent gullies (but only) during runoff events.

More recently, Nwakwasi (2018) [19] predicted gully erosion rates and identified the major factors contributing to gully erosion development in Southeastern Nigeria, using the negative binomial regression model. It was revealed that heavy rainfalls, extractive industries, and excess farming activities were the most influencing factors for gully initiation and formation. Zabihi et al. (2018) [20] used three bivariate statistical predictive models of susceptibility of a site to gully erosion from elevation, slope aspect, slope degree, slope length, topographical wetness index (TWI), plan curvature, profile curvature, land use, lithology, distance from river, drainage density, and distance from a road. All the models achieved prediction by about 80%. Finally, Domazetović et al. (2019) [21] developed a GIS using multicriteria analysis (namely, the GAMA model), allowing at the same time, automation and simplification of multicriteria grouping, weighting, coefficient assignment, and aggregation, towards a generic gully susceptibility modelling. The model has been tested in Pag Island, Croatia, with promising results.

Taking 2010 as a reference year, Panagos et al. (2015) [22] have put Greece among the five European countries with the highest risk for erosion (4.13 t ha^{-1} yr^{-1}), higher than the pan European mean (2.46 t ha^{-1} yr^{-1}). However, quantitative erosion studies in Greece have focused mainly to the sheet erosion form, neglecting so far, the magnitude and distribution of rill and gully formations throughout the country [23–25]. Few studies have also reported coastal erosion problems [26,27], while pan-European assessments addressed the wind erosion problem [28]. According to an outdated (before 1986) rough estimation by A. Vousaros, in Greece, there were over 800 active torrents transporting more than 30 Million m^3 of solid material [29].

Greece was included in a Mediterranean-wide study on determination of channel initiation thresholds for gully erosion according to the geomorphic and power-law equations [30]. The field study sites were located in Lesvos Island and targeted to permanent gullies found in rangelands; grazing is one of the most common contributors to soil erosion in Greece, especially in the islands. The thresholds for Lesvos follow a significantly lower regression slope and a significantly higher intercept than the threshold determined for the other two concurrent Mediterranean studies (Alentejo, Portugal and Sierra de Gata, Spain); this fact indicates less sensitivity in Lesvos than the other sites. It is noted, however, that many of the gullies detected in the Lesvos study were initiated by landslides, as identified from their typical morphology [30].

On the contrary to the lack of enough scientific research on the gully erosion problem in Greece, there is adequate evidence on the significance of gully erosion in the country, including articles in local newspapers and magazines of environmental or societal concern, several items of gray literature, and some reference in legal documents. Most of these items are related either to coastal erosion or permanent gullies. The legal references are associated to private or public works, where evidence of gully formulations is listed among reporting parameters prior to acceptance.

The main goal of this study was to assess ephemeral gully erosion potential in the agricultural lands of Greece. For this purpose, the agricultural land was sampled throughout the country, focusing on ephemeral gullies. The sampled surfaces were examined with visual interpretation of multi-temporal imagery available on Google Earth.

Use of Google Earth has been limited in gully erosion studies up until now. Gilad et al. (2012) [31] mapped gullies within only natural lands contributing sediment to the Great Barrier Reef in Australia and Boardman (2016) [32] studied gully erosion at the agricultural field level in Western Rother valley in the southern part of England. In practice, only the latter focuses on agricultural lands, while it was limited at a small geographic scale (60 km^2). In the current study, we propose a methodology for gully erosion estimation in agricultural lands on a country scale.

2. Materials and Methods

One hundred surfaces were selected throughout the agricultural lands of Greece using a random number generator of x,y coordinates within the country's outline. Then, circular surfaces of 300 m radius were traced and the contained surfaces (each having an extent of 28.3 ha) were examined for signs of ephemeral gullies. It was considered that a 600-m diameter circle was adequate for detecting and identifying an entire gully or part of it.

Provided that the agricultural lands in Greece cover an extent of 5,170,968 ha (source: CORINE Land Cover, 2018), the total sampled agricultural surfaces account for about 0.055%, with all agricultural land use categories included in the sample (Figure 1). According to the CORINE nomenclature, agricultural lands correspond to the 2nd class of the 1st level classification. Random sampling has also been employed by Gilad et al. (2012) [31], which though was dedicated exclusively to natural lands; moreover, it was calibrated with geostatistical analysis of drainage network data in order to exclude areas where gully formation potential was lower.

According to Marzolff et al. (2011) [33], short-term data collection for gullies are not representative of longer-term gully development and demonstrate the necessity for medium- to long-term monitoring. In this research, the detection and recognition of the ephemeral gullies was based on visual interpretation of diachronic Google Earth (GE) imagery, ranging from 2002 to 2019 for most of the samples. The frequency of available GE image scenes was higher in the later years and mostly after 2010. GE comprises a time-series repository of archived, very high-resolution satellite imagery (usually around 0.5 m, such as WorldView, Pleiades, GeoEye, etc.) in visible mode (RGB), available online. The selected samples contained from 4 up to 61 images, acquired on different dates, averaging at about 15 images per sample. GE provides the necessary spatial detail and temporal sequence, to assess gullies' presence and evolution; for example, to indicate channel initiation, to measure gully length, or to distinguish ephemeral from permanent gullies.

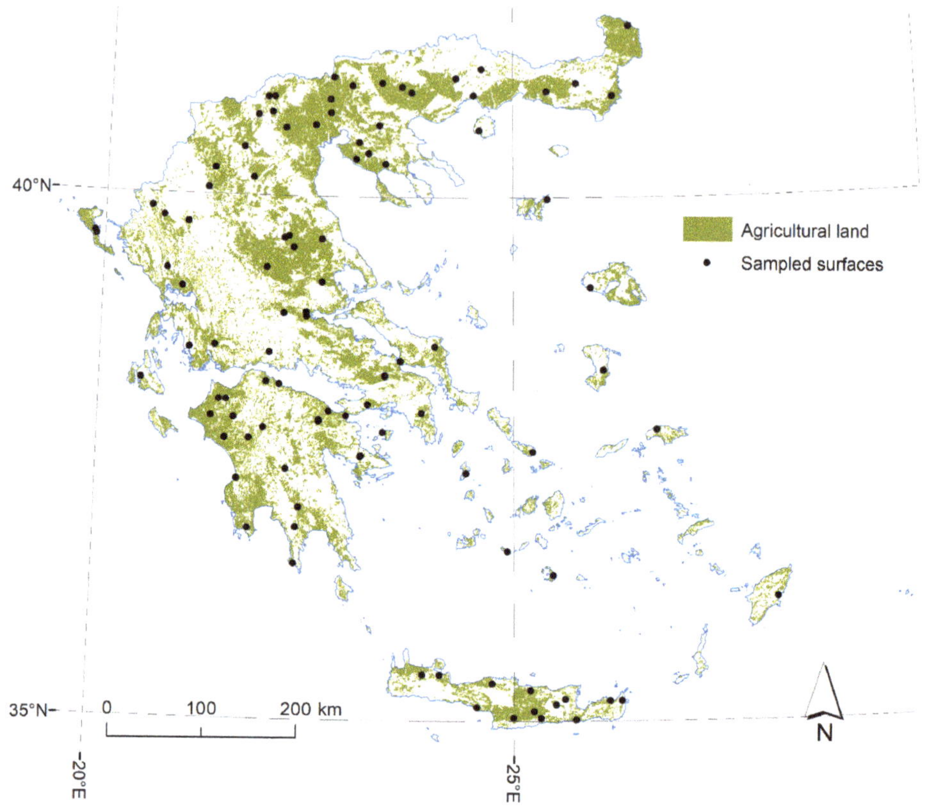

Figure 1. The random sampling scheme within the agricultural land of Greece.

As users of GE know, the time series covering a specific site is never complete, with many images missing due mainly to cloudy days. However, this fact would not affect seriously the high possibility of capturing existing ephemeral gullies, because the latter remain for a long period before they disappear by the farmers' soil tillage operations. This is especially true for the field preparation period, which might be quite long before seeding. In addition, the average number of images per sample (15 images) and the variety in season of image acquisition further empowers the possibility to capture ephemeral gullies before they disappear. Certainly, the possibility of some missed ephemeral gullies due to discontinuity of monitoring renders the true number of ephemeral gullies higher than the detected ones.

The ephemeral gullies were identified only as features within agricultural fields; gullies found between agricultural fields and natural lands or clearly inside natural land patches were not recorded. The identification capacity depended on the clarity of each image, thus rendering the scale of interpretation to vary between 50 and 100 m in terms of an 'eye altitude'. Eye altitude is the estimated altitude at which the observer is supposed to 'fly' over the imaged area. The visual interpretation criteria for indicating possible ephemeral gullies within agricultural fields were like those followed by Boardman (2016) [32]; specifically:

- Curved or straight linear segments inside agricultural fields, darker or lighter from their surroundings.
- Presence of natural vegetation along with linear features, either meeting or not the drainage network.

Every identified ephemeral gully was digitized as a continuous linear feature in a GIS. Ephemeral gullies along the same flow path but interrupted even for a few meters, were digitized as separate features. In order to avoid misinterpretation, we paid attention to traces in the fields created by agricultural machinery, which are usually straight and parallel. On the contrary, gullies are usually curved and at random directions.

Together with the gullies, signs of rill and sheet erosion were also examined and recorded qualitatively, in four distinct grades: no sign, slight, moderate, and strong signs. Bodoque et al. (2011) [34] recognized the necessity of studying gully and sheet erosion types within the same context, due to their linkage in geomorphological terms. Finally, we also recorded other important features of interest, such as terraces and hedges, with erosion control characteristics.

In 20 cases, the original sampling surfaces were found to be out of agricultural land uses according to Google Earth (which is ideal for the recognition and identification of most land uses), although they were selected inside agricultural land use polygons according to the CORINE Land Cover. In all these cases, the sampling surfaces were shifted towards the closest agricultural area.

Using visual interpretation of the GE image time-series, we recorded and computed the following parameters per sampled surface (Figure 2):

- Number of ephemeral gullies
- Exact flow path of ephemeral gullies
- Length of every ephemeral gully (in meters)
- Averaged elevation (in meters)
- Averaged slope degree (in %)
- Signs and grade of rills
- Signs and grade of sheet erosion
- All land use categories (with visual interpretation on Google Earth)
- The number of the GE images covering the sampled surface

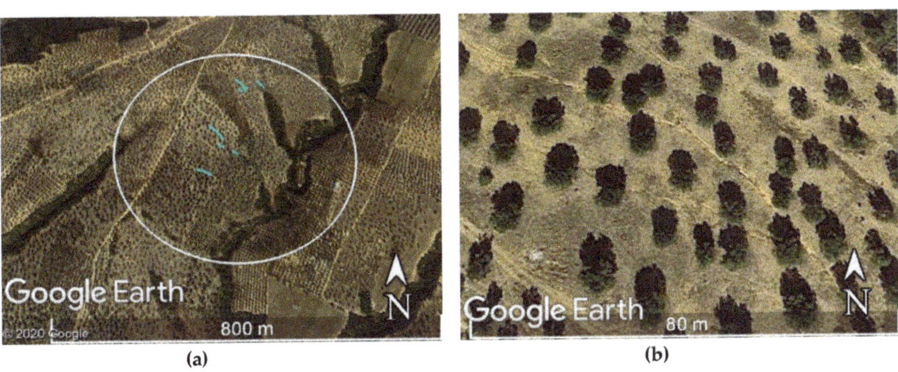

(a)   (b)

Figure 2. *Cont.*

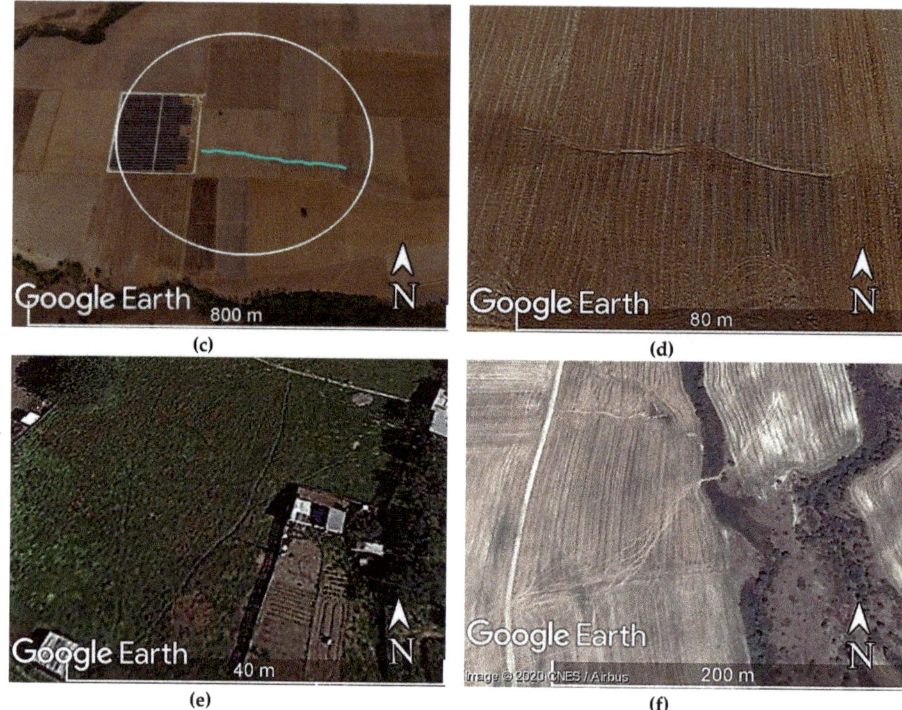

Figure 2. Examples of ephemeral gully detection. The sampled surfaces are denoted by white circles and gullies by cyan lines in the far views (geographic coordinates and image date in brackets) (**a**) sparse gullies formed towards a torrent in sloping olive plantation near Gerakini (40°18′56.41″ N/23°25′45.34″ E, 3 November 2016); (**b**) close view to case (**a**); (**c**) the longest identified gully near Almyros (39°10′55.07″ N/22°40′36.56″ E, 28 October 2013); (**d**) a close view to 0.6-m wide sections of the gully of case (**c**); (**e**) long parallel gullies in an arable field out of Sitia (35°11′42.82″ N/26°6′41.08″ E, 12 April 2013); (**f**) intensive rill and sheet erosion signs close to lake Doirani (41°8′26.07″ N/22°46′18.52″ E, 4 September 2013).

3. Results

The results start with a section with descriptive statistics of the gully inventory (no. of signs, length, elevation, slope, etc.) followed by the geographical distribution. The third section uses advanced indexes (I Moran's, Getis-Ord Gi, and Anselin Local Moran's) to verify the possible biases in sampling and the representativeness of the inventory. The fourth section provides an analysis in relation to land use, topography, and conservation practices followed by possible gully erosion correlations to drainage network.

3.1. Descriptive Statistics

Twenty-two samples out of the 100 examined throughout the country, were detected with clearly formulated ephemeral gullies. The number of gullies identified within every sample varied from 1 to 35, with an average of 6.9 gullies per sample location, or 0.25 gullies per hectare. In total, we identified and mapped 153 distinct ephemeral gullies. The longest gully was found to be 379 m and the shortest 3.5 m, with an average gully length of 55. 6 m. The total length of gullies within every sample varied from 30.6 to 1174 m, while the average total length per sample was 386 m, or 13.6 m per hectare.

The elevation at which the gullies were found ranged from 14 to 568 m, with an average of 201.3 m; while the elevation of the entire sampled agricultural surfaces ranged from 5 to 962 m, with an average of 205.7 m. The slope degree of the sampled surfaces with gullies ranged from 1.8% to 28.7%, with an average of 9.7%; while the slope degree of all the sampled agricultural surfaces ranged from 0.2% to 35.7%, with an average of 9.4%.

Rill signs were identified in 130 cases in total. In most of them (108 samples), rills were classified as slight, in 20 samples as moderate, and in 2 samples as strong. In five samples out of 22 with gullies, there were no signs of rills. Inversely, in 27 cases, rills were found in samples without gullies. Sheet erosion signs were identified in 81 cases in total. In most of them (61 samples), sheet erosion was classified as strong, in 13 samples as moderate, and in 7 samples as slight. In eight samples out of 22 with gullies, there were no signs of sheet erosion. Inversely, in 23 cases, sheet erosion was identified in samples without gullies (Table 1).

Table 1. A summary of the findings.

Parameter	Arithmetic Figures
Samples with gullies	22/100
Number of gullies	153
Length of gullies	Mean = 55.6 m Min = 3.5 m Max = 379 m
Elevation of: Sampled surfaces Samples surfaces with gullies	205.7 m (5 m–962 m) 201.3 m (14 m–568 m)
Slope of: Sampled surfaces Samples surfaces with gullies	9.4% (0.2%–35.7%) 9.7% (1.8%–28.7%)
Signs of rills (total)	130/153 (85%)
Slight (1)	108/153 (70%)
Moderate (2)	20/153 (13%)
Strong (3)	2/153 (1.3%)
Signs of sheet erosion (total)	81/153 (53%)
Slight (1)	7/153 (4.5%)
Moderate (2)	13/153 (8.5%)
Strong (3)	61/153 (39%)

3.2. Geographic Distribution

In geographic terms, most of the random samples where gullies were found, were situated in the north part of the country (Central Macedonia, East Macedonia, and Thrace) and the east part of the country (Thessaly, Attica, Viotia, East Peloponnese, and East Crete), except the small islands of the Aegean. Moderate gully lengths (<100 m) were recorded in south Greece, whereas higher values in north Greece. We recorded the extreme gully length value (335 m long) close to Almyros town in Thessaly, central part of Greece (Figure 3).

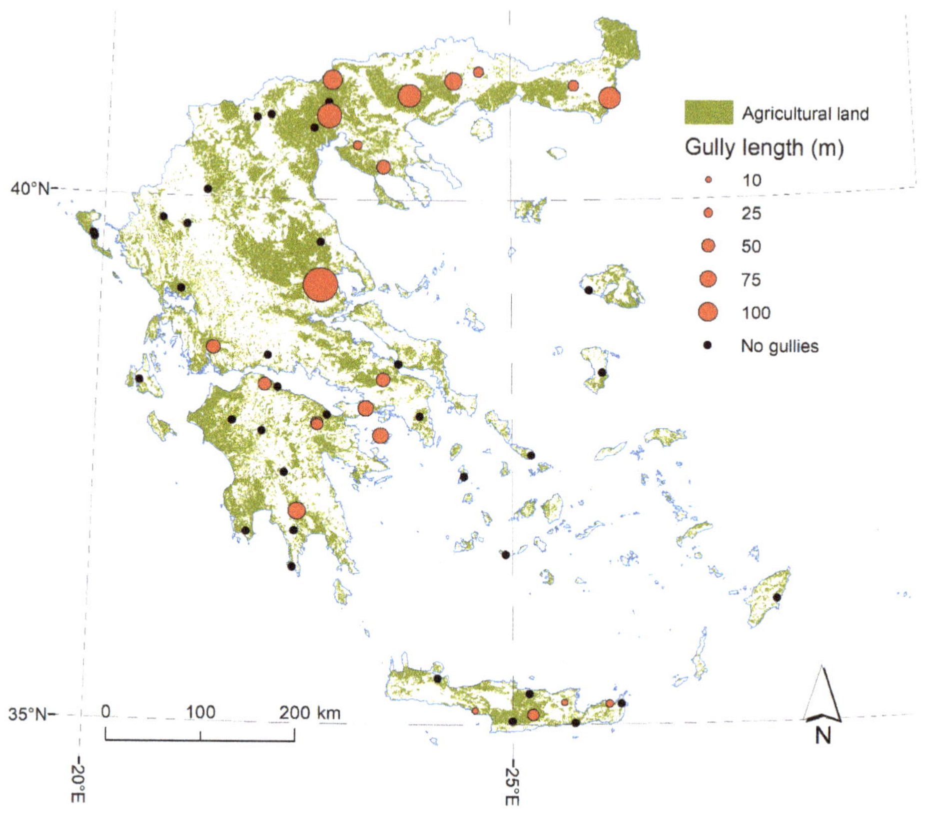

Figure 3. Geographic distribution of the total gully length per randomly sampled site (graduated symbol in five categories of magnitude).

In 66 samples, erosion was present either as gullies, rills, or sheet erosion sings. In 7 samples, gullies were found together with rills and no sheet erosion, in 4 samples gullies were found together with sheet erosion and no rills, and in 10 samples the three forms of erosion were present altogether. The latter category can be found in different sides of the country (Figure 4).

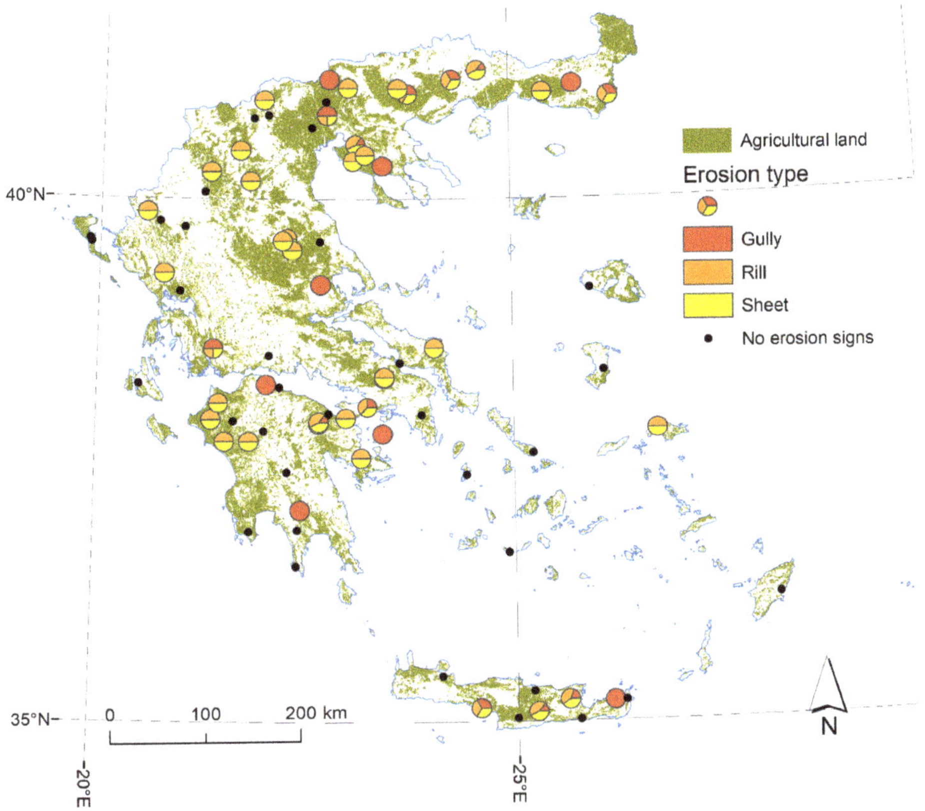

Figure 4. Relative proportion of gullies, rills, and sheet erosion in the randomly sampled surfaces; five categories of severity were considered for gullies, and three for rills and sheet erosion.

3.3. Geostatistical Analysis

The sampling scheme was verified to be unbiased by computing the spatial autocorrelation of the gully length variable. The I Moran's index was used, resulting into −0.0114962, which indicates a random distribution of the samples [35]. This means that the observations were enough far apart, so as not to affect representativeness of the sampling scheme. The independence of the sampling dataset was visualized by two kinds of maps: (a) one identifying statistically significant hot and cold spots at the global level, using the Getis-Ord Gi* statistic; and (b) one identifying the statistically significant hot spots, cold spots, and spatial outliers at the local level, using the Anselin Local Moran's I statistic. As it is shown, at the global level all the samples indicate non-significant differentiations, while at the local level only one high-low spatial outlier was detected (Figure 5).

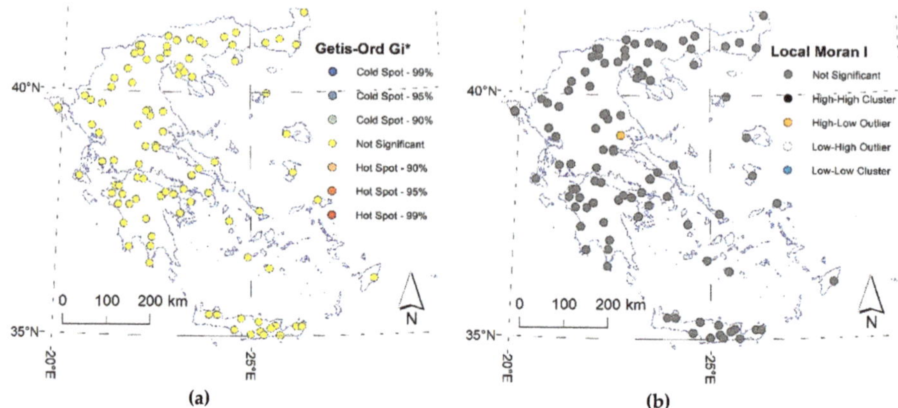

Figure 5. Spatial independence of gully length visualized: (**a**) at the global scale using the Getis-Ord Gi* statistic (percentages indicate level of confidence); and (**b**) at the local scale using Anselin Local Moran's I statistic.

3.4. Gully Erosion Trends in Relation to Land Use and Topography

From the examination of the ephemeral gullies with the concurrent land use identified in Google Earth, it was found that ephemeral gullies were found mainly in arable lands (9 cases), vineyards (4 cases), olive groves (8 cases), and pastures (one case) (Figure 6). Proportionally to each sampled land use class, it was found that 21% of the arable land, 25% of the olive groves, 50% of the vineyards, and 9% of the pastures were found with ephemeral gullies (Figure 7). Inside the arable land, the mean length of gullies was significantly larger than the overall average (90.7 m compared to 55.6 m). In olive groves, the mean gully length was significantly smaller than the overall average (35.8 m compared to 55.6 m). For the vineyards, the mean gully length was close to the average (55.5 m), whereas for the pastures, it was far larger (146 m).

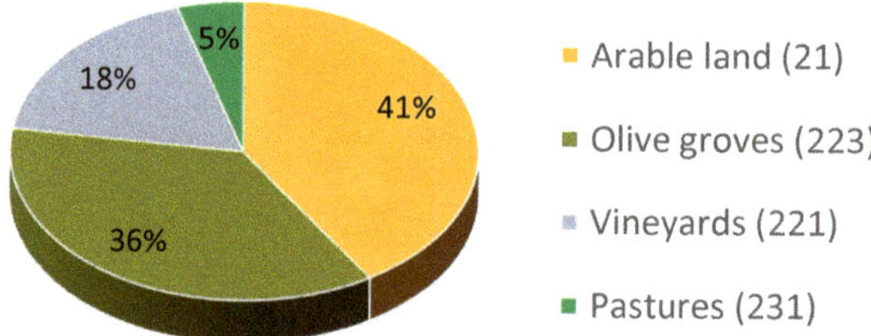

Figure 6. Share of the main agricultural land uses in Greece found to contain ephemeral gullies (CORINE coding in brackets).

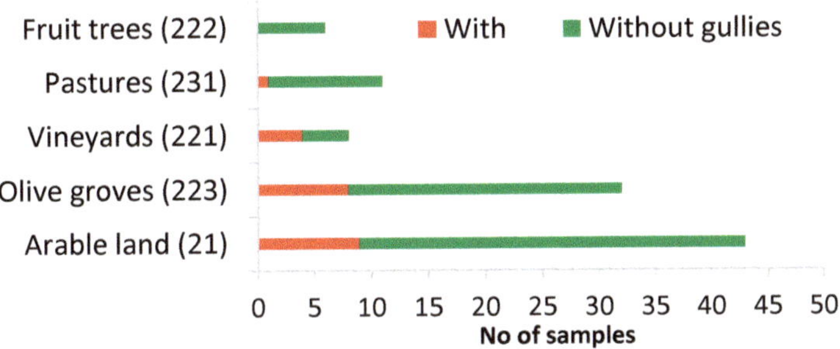

Figure 7. Number of samples found with and without ephemeral gullies (CORINE coding in brackets).

Terraces are among the most significant conservation practices to mitigate soil erosion especially in hot spots [36]. We checked the presence of terraces between or inside the agricultural fields. In this qualitative assessment, 26 samples out of 100 contained terraces. Gullies were present only in 5 locations with terraces presence, thus in 19% of the cases. The field samples with presence of terraces but without gullies have a slope degree ranging from 1% to 34%, with an average of 14.2%. These findings verify in a degree the substantial role of terraces as measures of erosion prevention. However, in most of the samples with gullies, terraces were absent, although having a slope degree ranging from 2% to 26%, with an average of 8.7%. The cases of terraced land with gullies only indicates the possibility that gully formation could be even worse if terraces were absent (Table 2).

Table 2. A summary of the relation of gullies with terraces per land use in Greece.

Land Use	G+T Samples *	Average Slope (%)	G-T Samples *	Average Slope (%)
Arable land	0	-	9	6.5
Olive groves	4	13.9	4	12.3
Vineyards	1	12.6	3	10.8
Pastures	0	-	1	8.3
Fruit trees	0	-	0	-
Overall	5	13.6	17	8.7

* G+T: with gullies and with terraces; G-T: with gullies and without terraces.

We also investigated possible trends of the main numerical parameters, such as the length of the mapped gullies, the number of the gullies, slope degree, and elevation, within the subset of the samples with gullies. Averaged elevation and slope values for each of the samples were derived from a 30-m resolution ASTER GDEM (digital elevation model) [37]. Low to moderate non-linear trends (in terms of coefficient of determination, R^2) appear in the following pairs of variables (Figure 8):

- Length vs. number of gullies (mostly negative)
- Number of gullies vs. slope (mostly positive)
- Total length of gullies vs. slope (mostly negative)
- Severity of sheet erosion sings vs. elevation (positive)

The number or length of the detected gullies were not correlated with signs of rills or sheet erosion by any means.

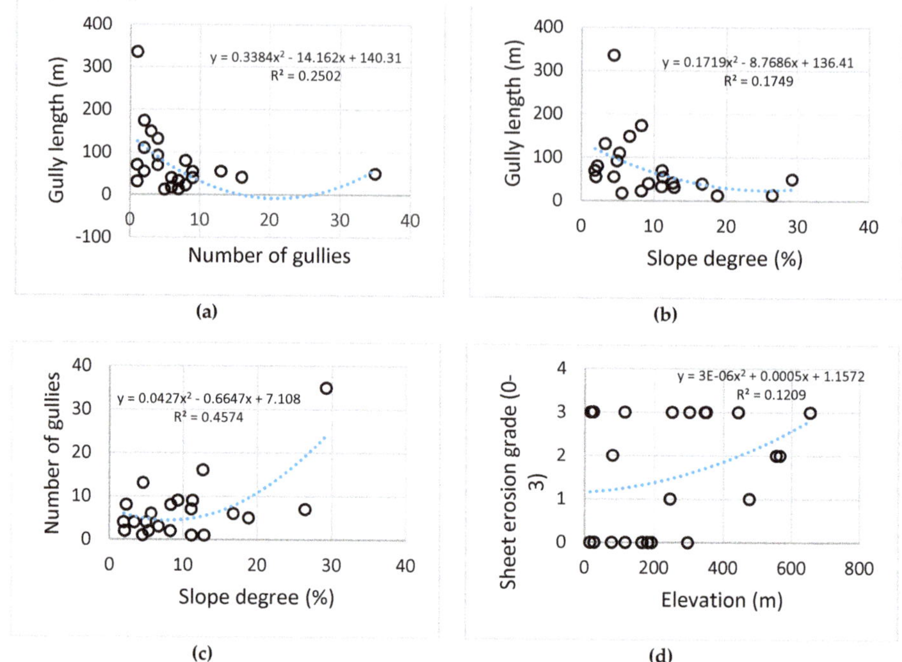

Figure 8. Trend plots: negative trend between total gully length and number of gullies (**a**) and between gully length and slope degree (**b**); positive trend between number of gullies and slope degree (**c**) and sheet erosion grade and elevation (**d**).

3.5. Gully Erosion Correlation with Drainage Network

Finally, we examined possible correlation of the ephemeral gullies with drainage network on the most detailed scale, in the study sites. Drainage network was extracted using the D8 method [38], with the available ASTER GDEM. The path-lines of the mapped ephemeral gullies were overlaid on the drainage network and the portion of the path lines within every stream order of the network was recorded (Figure 9).

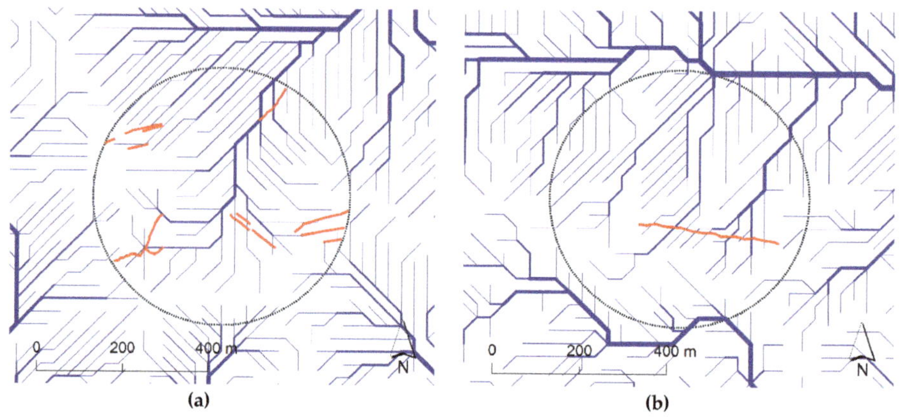

Figure 9. *Cont.*

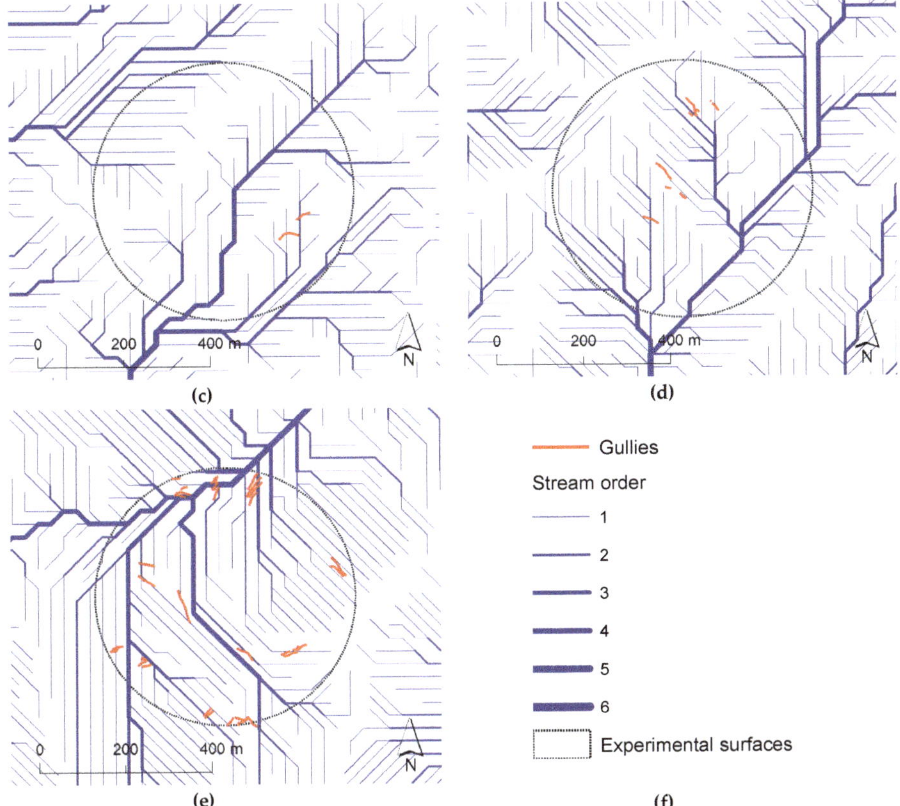

Figure 9. Indicative cases of detected gullies overlaid on the drainage network (**a**) variety of gullies in a 4.5% slope vines - arable land complex, close to Thiva (38°17′8.45″ N/23°27′1.68″ E, 20 February 2014); (**b**) the longest detected gully in a 4.4% slope naked arable soil, close to Almyros (39°10′55.07″ N/22°40′36.56″ E, 28 October 2013); (**c**) short gullies in a 2% slope arable land, close to Agrinio (38°33′59.95″ N/21°23′35.14″ E, 9 September 2009); (**d**) moderate gullies in 11.4% slope olive groves, close to Gerakini (40°18′56.41″ N/23°25′45.34″ E, 3 November 2016); (**e**) highly dense gully pattern in a 29.2% slope olive-vines complex, close to Egio (38°33′59.95″ N/21°23′35.14″ E, 9 July 2009); (**f**) legend.

The results indicate that the detected ephemeral gullies joint or intersected drainage network segments of 1st up to the 5th order in a 6-order drainage network after Shtrahler (1957) [39]. According to the Strahler method, all segments without any tributaries are assigned an order of 1, whereas stream order increases only when segments of the same order intersect. The average order of all drainage network segments which the detected gullies joint or intersected was 1.53. The latter shows that in general, gullies were found mainly close to low-order segments of the drainage network, with very limited exceptions; in seven cases, the gullies were found to join or intersect 5-order segments. Visual inspection showed that in most cases, the detected gullies did not comply absolutely with the direction of the mapped drainage network segments, even in sloping sites.

4. Discussion

The statistical analysis of ephemeral gullies has focused on land cover, land use, topographic features (elevation and slope), drainage, and conservation practices (terraces). Although soil type is an important driver for gully erosion formation [4], we have not investigated this correlation in this study due to the absence of a high spatial resolution soil map.

Considering that the sampling scheme was random and unbiased (no autocorrelation was found), the representativeness of the sample was statistically reliable. The numerical figures indicate a rather densified network of gullies wherever detected. In several cases, the gullies were found in more than one image, or were identified as rills in earlier years, which were evolved into gullies in later years; this notification agrees with the known persistence of gully formation in the same sites over time [40].

Furthermore, the detected gullies were found in any possible location and direction within fields of different crops; in several cases they were parallel between them, while in others they were attributing to the gullies of higher order in a structured drainage network. In few cases, the farmers purposely converted ephemeral channels crossing their fields into permanent drainage channels, thus facilitating water removal. This kind of development was understood by the evolution of naturally curved channels into artificially widened straight ones.

Gully initiation was found to be associated with specific agricultural land uses, mainly vineyards, olive groves, and arable land in sloping sites (8.7% on average). This confirms the hypothesis that soil erosion rates in vineyards are among the highest ones (especially in the Mediterranean basin) due to a combination of bare soil conditions, low vegetation protection, slope areas and anthropogenic factors (tillage, compaction, use of herbicides) [41,42]. Finally, 11% of the entire sampled surfaces were found to contain ephemeral gullies at an average slope degree larger than 5%, but without any terraces. However, the detected ephemeral gullies did not show correlation to specific elevation, nor to specific slope ranges.

Not surprisingly, gullies coexist with rill or sheet erosion in most of the cases. Rills, though, were observed in 85% of the samples, while sheet erosion in 53% of the samples; thus, in many cases rill and sheet erosion forms were present irrespective of ephemeral gullies.

The findings of this study indicate a moderate to high risk for gully erosion in Greece (22% on average), non-uniformly distributed in the country. The Land Use and Land Cover Survey (LUCAS) of 2018 verifies a medium to high gully erosion risk in Greece. According to the LUCAS soil survey 2018, 33 gully points were recorded in a sample of 500 visited sites in Greece (6.6%), thus rendering Greece second in gully density among all European Union (EU) countries after Spain [43]. In 20,000 surveyed points in the entire EU in all land uses, the surveyors noticed gully erosion in 211 points. However, LUCAS followed an a priori systematic point-sampling scheme updated regularly every three years, which can be further improved in order to capture a very localized and temporary phenomenon like gully formulation.

Considering that gully erosion research is very limited and outdated in Greece, a fast-track study was found to be necessary in order to get a rough assessment of the situation, especially regarding ephemeral gullies, which are associated with intensive agricultural land uses. It was showed that wherever gullies have initiated, the density and length of the channels, thus the severity of gully erosion seems to be significant.

The remote sensing method employed in this study (i.e.,; visual interpretation) was able to respond to the second scientific question set by Poesen et al. (2003) [4] in their review on gullies research needs: "What are appropriate measuring techniques for monitoring and experimental studies of the initiation and development of various gully types at various temporal and spatial scales?" Visual interpretation with Google Earth proved to be efficient in detecting and identifying ephemeral gullies and—moreover—be used as an image background for their detailed mapping; in many cases, an estimation of their width could also be provided.

Future work in Greece should combine the current, large scale detection with application of the geomorphic method introduced by Vandaele et al. (1996) [44], to detect potential initiation of gullies in a GIS environment; the empirical constants of the power-law equation could be indicated by pilot studies in different land uses located in preselected indicative sites. In this direction, visual interpretation with Google Earth can be used for verification. Different correlations examined (e.g.,;

the negative correlation between number of gullies and slope) could also contribute to modelling approaches in a geospatial context, together with other remote sensing or field variables. Also, some qualitative correlations, such as coexistence of gullies with sheet erosion signs, which are of permanent nature, could be seen as a possibility factor towards gullies' detection modelling.

5. Concluding Remarks

The presence of gullies at a regional scale can be assessed by using field surveys, visual interpretation, and remote sensing. The increased availability of high-resolution remote sensing images combined with computing capacity and the growing disposal of photographs and data collection will further facilitate the compilation of gully erosion datasets at a regional scale.

The visual interpretation of Google Earth images has proved to be a fruitful technique for gully erosion evolution and inventory. In addition, in terms of costs, the evaluation of 100 points requested about 50 working hours of a remote sensing engineer, including preparatory work sampling design and collection in a geographic information system. Potentially, the evaluation of one or two orders of magnitude (10,000 points) will contribute to a national representative gully erosion dataset. This can be combined with existing field surveys (e.g.,; LUCAS) and other advanced techniques, such as machine learning or semi-automated procedures for gully identification.

The gully inventory can be useful for advising farmers to apply appropriate management practices. Tillage practices and period of plow are key factors influencing gully erosion in hot-dry environments [45], such as Greece. Contour ridge tillage can be an appropriate conservation practice in agricultural lands with ephemeral gully signs.

Author Contributions: Conceptualization, P.P. and C.K.; methodology, C.K.; software, C.K.; validation, C.K. and P.P.; formal analysis, C.K.; investigation, P.P and C.K.; resources, C.K and P.P.; data curation, C.K.; writing—original draft preparation, C.K.; writing—review and editing, P.P.; visualization, C.K.; supervision, P.P.; project administration, P.P.; funding acquisition, P.P. All authors have read and agreed to the published version of the manuscript.

Funding: This research received no external funding. The APC was funded by European Commission, Joint Research Centre (JRC).

Acknowledgments: Many thanks to Google Earth online.

Conflicts of Interest: The authors declare no conflict of interest. The funders had no role in the design of the study; in the collection, analyses, or interpretation of data; in the writing of the manuscript, or in the decision to publish the results.

References

1. Vanmaercke, M.; Poesen, J.; Verstraeten, G.; de Vente, J.; Ocakoglu, F. Sediment yield in Europe: Spatial patterns and scale dependency. *Geomorphology* **2011**, *130*, 142–161. [CrossRef]
2. *Soil Science Society of America, Glossary of Soil Science Terms*; Soil Science Society of America: Madison, WI, USA, 2001. Available online: http://www.soils.org/sssagloss/ (accessed on 8 February 2020).
3. Castillo, C.; Gómez, J.A. A century of gully erosion research: Urgency, complexity and study approaches. *Earth-Sci. Rev.* **2016**, *160*, 300–319. [CrossRef]
4. Poesen, J.; Nachtergaele, J.; Verstraeten, G.; Valentin, C. Gully erosion and environmental change: Importance and research needs. *Catena* **2003**, *50*, 91–133. [CrossRef]
5. Verstraeten, G.; Poesen, J. The nature of small-scale flooding, muddy floods and retention pond sedimentation in central Belgium. *Geomorphology* **1999**, *29*, 275–292. [CrossRef]
6. Ionita, I.; Fullen, M.A.; Zgłobicki, W.; Poesen, J. Gully erosion as a natural and human-induced hazard. *Nat. Hazards* **2015**, *79*, S1–S5. [CrossRef]
7. Foster, G.R. Understanding ephemeral gully erosion. In *Soil Conservation: Assessing the National Resources Inventory*; National Academy Press: Washington, DC, USA, 1986; Volume 2, pp. 90–118.
8. Casali, J.; Bennett, S.J.; Robinson, K.M. Processes of ephemeral gully erosion. *Int. J. Sediment Res.* **2000**, *15*, 31–41.

9.   Bennett, S.J.; Wells, R.R. Gully erosion processes, disciplinary fragmentation, and technological innovation. *Earth Surf. Process. Landf.* **2019**, *44*, 46–53. [CrossRef]

10.   Valentin, C.; Poesen, J.; Li, Y. Gully erosion: Impacts, factors and control. *Catena* **2005**, *63*, 132–153. [CrossRef]

11.   Frank, A.; Nyssen, J.; De Dapper, M.; Haile, M.; Billi, P.; Munro, R.N.; Deckers, J.; Poesen, J. Linking long-term gully and river channel dynamics to environmental change using repeat photography (Northern Ethiopia). *Geomorphology* **2011**, *129*, 238–251. [CrossRef]

12.   Gordon, L.M.; Bennett, S.J.; Alonso, C.V.; Bingner, R.L. Modeling long-term soil losses on agricultural fields due to ephemeral gully erosion. *J. Soil Water Conserv.* **2008**, *63*, 173–181. [CrossRef]

13.   Bingner, R.L.; Wells, R.R.; Momm, H.G.; Rigby, J.R.; Theurer, F.D. Ephemeral gully channel width and erosion simulation technology. *Nat. Hazards* **2016**, *80*, 1949–1966. [CrossRef]

14.   Vanmaercke, M.; Poesen, J.; Van Mele, B.; Demuzere, M.; Bruynseels, A.; Golosov, V.; Bezerra, J.F.R.; Bolysov, S.; Dvinskih, A.; Frankl, A.; et al. How fast do gully headcuts retreat? *Earth-Sci. Rev.* **2016**, *154*, 336–355. [CrossRef]

15.   Capra, A.; Mazzara, L.M.; Scicolone, B. Application of the EGEM model to predict ephemeral gully erosion in Sicily, Italy. *Catena* **2005**, *59*, 133–146. [CrossRef]

16.   Rundquist, B.C. The influence of canopy green vegetation fraction on spectral measurements over native tallgrass prairie. *Remote Sens. Environ.* **2002**, *81*, 129–135. [CrossRef]

17.   Hessel, R.; Van Asch, T.W.J. Modelling gully erosion for a small catchment on the Chinese Loess Plateau. *Catena* **2003**, *54*, 131–146. [CrossRef]

18.   de Roo, A.; Jetten, V.; Wesseling, C.; Ritsema, C. LISEM: A physically-based hydrologic and soil erosion catchment model. In *Modelling Soil Erosion by Water*; Springer: Berlin/Heidelberg, Germany, 1998; pp. 429–440. Available online: https://link.springer.com/chapter/10.1007/978-3-642-58913-3_32 (accessed on 10 February 2020).

19.   Nwakwasi, N.L. Modelling of Gully Erosion Site Data in Southeastern Nigeria, Using Poisson and Negative Binomial Regression Models. *J. Civ. Constr. Environ. Eng.* **2018**, *3*, 111–117. [CrossRef]

20.   Zabihi, M.; Mirchooli, F.; Motevalli, A.; Darvishan, A.K.; Pourghasemi, H.R.; Zakeri, M.A.; Sadighi, F. Spatial modelling of gully erosion in Mazandaran Province, northern Iran. *Catena* **2018**, *161*, 1–13. [CrossRef]

21.   Domazetović, F.; Šiljeg, A.; Lončar, N.; Marić, I. GIS automated multicriteria analysis (GAMA) method for susceptibility modelling. *MethodsX* **2019**, *6*, 2553–2561. [CrossRef]

22.   Panagos, P.; Borrelli, P.; Poesen, J.; Ballabio, C.; Lugato, E.; Meusburger, K.; Montanarella, L.; Alewell, C. The new assessment of soil loss by water erosion in Europe. *Environ. Sci. Policy* **2015**, *54*, 438–447. [CrossRef]

23.   Panagos, P.; Karydas, C.G.; Gitas, I.Z.; Montanarella, L. Monthly soil erosion monitoring based on remotely sensed biophysical parameters: A case study in Strymonas river basin towards a functional pan-European service. *Int. J. Digit. Earth* **2012**, *5*, 461–487. [CrossRef]

24.   Panagos, P.; Karydas, C.; Ballabio, C.; Gitas, I. Seasonal monitoring of soil erosion at regional scale: An application of the G2 model in Crete focusing on agricultural land uses. *Int. J. Appl. Earth Obs. Geoinf.* **2014**, *27*, 147–155. [CrossRef]

25.   Kazamias, A.P.; Sapountzis, M. Spatial and temporal assessment of potential soil erosion over Greece. *Water* **2017**, *59*, 315–321.

26.   Evelpidou, N.; Vassilopoulos, A.; Leonidopoulou, D.; Poulos, S. An investigation of the coastal erosion causes in Samos Island, Eastern Aegean Sea. *J. Landsc. Ecol.* **2008**, *6*, 295–310.

27.   Andredaki, M.; Georgoulas, A.; Hrissanthou, V.; Kotsovinos, N. Assessment of reservoir sedimentation effect on coastal erosion in the case of Nestos River, Greece. *Int. J. Sediment Res.* **2014**, *29*, 34–48. [CrossRef]

28.   Borrelli, P.; Panagos, P.; Ballabio, C.; Lugato, E.; Weynants, M.; Montanarella, L. Towards a pan-European assessment of land susceptibility to wind erosion. *Land Degrad. Dev.* **2016**, *27*, 1093–1105. [CrossRef]

29.   Nemes, I.; Constantinescu, L. The Gully Erosion Effect on the Environment. In Proceedings of the International Conference on Information and Communication Technologies for Sustainable Agri-Production and Environment (HAICTA 2011), Skiathos, Greece, 8–11 September 2011.

30.   Vandekerckhove, L.; Poesen, J.; Oostwoudwijdenes, D.; Nachtergaele, J.; Kosmas, C.; Roxo, M.J.; De Figueiredo, T. Thresholds for gully initiation and sedimentation in Mediterranean Europe. *Earth Surf. Process. Landf.* **2000**, *25*, 1201–1220. [CrossRef]

31. Gilad, U.; Denham, R.; Tindall, D. Gullies, Google Earth and the Great Barrier Reef: A remote sensing methodology for mapping gullies over extensive areas. International Archives of the Photogrammetry. In Proceedings of the International Archives of the Photogrammetry, Remote Sensing and Spatial Information Science, XXII ISPRS Congress, Melbourne, Australia, 25 August–1 September 2012. Available online: https://www.int-arch-photogramm-remote-sens-spatial-inf-sci.net/XXXIX-B8/469/2012/ (accessed on 10 February 2020).

32. Boardman, J. The value of Google Earth ™ for erosion mapping. *Catena* **2016**, *143*, 123–127. [CrossRef]

33. Marzolff, I.; Ries, J.B.; Poesen, J. Short-term versus medium-term monitoring for detecting gully-erosion variability in a Mediterranean environment. *Earth Surf. Process. Landf.* **2011**, *36*, 1604–1623. [CrossRef]

34. Bodoque, J.M.; Lucia, A.; Ballesteros, J.A.; Martin-Duque, J.F.; Rubiales, J.M.; Genova, M. Measuring medium-term sheet erosion in gullies from trees: A case study using dendrogeomorphological analysis of exposed pine roots in central Iberia. *Geomorphology* **2011**, *134*, 417–425. [CrossRef]

35. Getis, A.; Ord, J.K. The analysis of Spatial Association by use of Distance Statistics. *Geogr. Anal.* **1992**, *24*, 189–206. [CrossRef]

36. Panagos, P.; Borrelli, P.; Meusburger, K.; van der Zanden, E.H.; Poesen, J.; Alewell, C. Modelling the effect of support practices (P-factor) on the reduction of soil erosion by water at European scale. *Environ. Sci. Policy* **2015**, *51*, 23–34. [CrossRef]

37. NASA. Jet Propulsion Laboratory ASTER. 2012. Available online: https://asterweb.jpl.nasa.gov/gdem.asp (accessed on 8 February 2020).

38. Jenson, S.K.; Domingue, J.O. Extracting Topographic Structure from Digital Elevation Data for Geographic Information System Analysis. *Photogramm. Eng. Remote Sens.* **1988**, *54*, 1593–1600.

39. Strahler, A.N. Quantitative analysis of watershed geomorphology. *Trans. Am. Geophys. Union* **1957**, *38*, 913–920. [CrossRef]

40. Di Stefano, C.; Ferro, V.; Pampalone, V.; Sanzone, F. Field investigation of rill and ephemeral gully erosion in the Sparacia experimental area, South Italy. *Catena* **2013**, *101*, 226–234. [CrossRef]

41. Rodrigo-Comino, J. Five decades of soil erosion research in "terroir". The State-of-the-Art. *Earth-Sci. Rev.* **2018**, *179*, 436–447. [CrossRef]

42. Rodrigo-Comino, J.; Senciales, J.M.; Sillero-Medina, J.A.; Gyasi-Agyei, Y.; Ruiz-Sinoga, J.D.; Ries, J.B. Analysis of Weather-Type-Induced Soil Erosion in Cultivated and Poorly Managed Abandoned Sloping Vineyards in the Axarquía Region (Málaga, Spain). *Air Soil Water Res.* **2019**, *12*, 1178622119839403. [CrossRef]

43. Orgiazzi, A.; Ballabio, C.; Panagos, P.; Jones, A.; Fernández-Ugalde, O. LUCAS Soil, the largest expandable soil dataset for Europe: A review. *Eur. J. Soil Sci.* **2018**, *69*, 140–153. [CrossRef]

44. Vandaele, K.; Poesen, J.; Govers, G.; van Wesemael, B. Geomorphic threshold conditions for ephemeral gully incision. *Geomorphology* **1996**, *16*, 161–173. [CrossRef]

45. Rong, L.; Duan, X.; Zhang, G.; Gu, Z.; Feng, D. Impacts of tillage practices on ephemeral gully erosion in a dry-hot valley region in southwestern China. *Soil Tillage Res.* **2019**, *187*, 72–84. [CrossRef]

Article

Gully Erosion Susceptibility Mapping Using Multivariate Adaptive Regression Splines—Replications and Sample Size Scenarios

Narges Javidan [1], Ataollah Kavian [1,*], Hamid Reza Pourghasemi [2], Christian Conoscenti [3] and Zeinab Jafarian [4]

[1] Department of Watershed Management, Faculty of Natural Resources, Sari Agricultural Sciences and Natural Resources University (SANRU), Sari 48441-74111, Iran; narges.javidan20@gmail.com
[2] Department of Natural Resources and Environmental Engineering, College of Agriculture, Shiraz University, Shiraz 71441-65186, Iran; hr.pourghasemi@shirazu.ac.ir
[3] Department of Earth and Marine Sciences (DISTEM), University of Palermo, Via Archirafi 22, 90123 Palermo, Italy; christian.conoscenti@unipa.it
[4] Department of Range Management, Sari Agricultural Sciences and Natural Resources University (SANRU), Sari 48441-74111, Iran; z.jafarian@sanru.ac.ir
* Correspondence: a.kavian@sanru.ac.ir

Received: 14 September 2019; Accepted: 29 October 2019; Published: 6 November 2019

Abstract: Soil erosion is a serious problem affecting numerous countries, especially, gully erosion. In the current research, GIS techniques and MARS (Multivariate Adaptive Regression Splines) algorithm were considered to evaluate gully erosion susceptibility mapping among others. The study was conducted in a specific section of the Gorganroud Watershed in Golestan Province (Northern Iran), covering 2142.64 km^2 which is intensely influenced by gully erosion. First, Google Earth images, field surveys, and national reports were used to provide a gully-hedcut evaluation map consisting of 307 gully-hedcut points. Eighteen gully erosion conditioning factors including significant geoenvironmental and morphometric variables were selected as predictors. To model sensitivity of gully erosion, Multivariate Adaptive Regression Splines (MARS) was used while the Area Under the Receiver Operating Characteristic (ROC) Curve (AUC), drawing ROC curves, efficiency percent, Yuden index, and kappa were used to evaluate model efficiency. We used two different scenarios of the combination of the number of replications, and sample size, including 90%/10% and 80%/20% with 10 replications, and 70%/30% with 5, 10, and 15 replications for preparing gully erosion susceptibility mapping (GESM). Each one involves a various subset of both positive (presence), and negative (absence) cases. Absences were extracted as randomly distributed individual cells. Therefore, the predictive competency of the gully erosion susceptibility model and the robustness of the procedure were evaluated through these datasets. Results did not show considerable variation in the accuracy of the model, with altering the percentage of calibration to validation samples and number of model replications. Given the accuracy, the MARS algorithm performed excellently in predictive performance. The combination of 80%/20% using all statistical measures including SST (0.88), SPF (0.83), E (0.79), Kappa (0.58), Robustness (0.01), and AUC (0.84) had the highest performance compared to the other combinations. Consequently, it was found that the performance of MARS for modelling gully erosion susceptibility is quite consistent while changes in the testing and validation specimens are executed. The intense acceptable prediction capability of the MARS model verifies the reliability of the method employed for use of this model elsewhere and gully erosion studies since they are qualified to quickly generating precise and exact GESMs (gully erosion sensitivity maps) to make decisions and management edaphic and hydrologic features.

Keywords: gully erosion susceptibility; GIS; robustness; MARS algorithm

---

1. Introduction

Soil erosion through water is found to be a drastic soil destruction process, which accounts for about one billion hectares worldwide [1], resulting in low plant development, filling reservoirs and valleys, geo-environmental degradation, degradation of a large part of the soil, and siltation of watercourses [2–4]. One of the soil degradation processes is gully erosion and serves as the most intricate erosion phenomena [5], usually stimulated or exacerbated integrating extreme rainstorms and unwise land exploitation [6]. Such erosion contains wide varieties of small processes, including head-cut, fluting, piping, continuous cracking progress, and mass flow [7,8]. Generally, the increasing attention towards analysis of gully erosion indicates the necessity to enhance our awareness regarding its consequences and condition agents that change due to various factors [6]. Gullies involve complex pathways adjusted through the variability of correlated variables including soil texture, lithology, land use and plant canopy, climate, and topography [9]. As gully erosion is a threshold phenomenon [8], various studies have emphasized characterizing the topographic as well as hydraulic conditions to forecast and assess the starting gullies susceptibility mapping [10], to a threshold approach where characterization and positions of erosion processes might as well as be anticipated by using bivariate to multivariate methods. Such techniques provide scholars the ability to describe soil erosion processes, through evaluating space distribution of the gully erosion forms (the consequences) compared to some predictors (geological and environmental factors). As for geomorphology, actuarial methods are enormously used to evaluate landslide susceptibility mapping [11–17].

The large number of studies have at the same time used the probabilistic method to map erosion sensitivity and other hazards. As well as bivariate methods [18–20], various multivariate actuarial approaches were considered to satisfy this end including logistic regression [21,22], classification and regression trees [23,24], and multivariate adaptive regression splines [25–28]. Gutiérrez et al. [27] used the Multivariate Adaptive Regression Splines (MARS) model to predict gully creation locations. The results showed that this model has good performance in geomorphic research.

Gully erosions have the highest sediment production potential. In recent decades, gully erosion has developed in most watersheds of Iran. Gullies are an important sediment source and often cause environmental problems [29]. Due to their damages, such as loss of productive capacity and significant land degradation, high sediment discharge and sediment yields, which can transport both pollutants and nutrients, reducing the water capacity of the reservoirs and damage to the infrastructure and transport routes, prediction of susceptible areas is, therefore, considered essential in management of watersheds [30]. The result of gully erosion study demonstrates the susceptibility to erosion over a country, providing beneficial information for remediation strategies and establishing land use plans [29].

In ongoing research, we adopted the multivariate adaptive regression splines [31] as a multivariate actuarial method for analyzing, assessing, and forecasting the local incidence of gully erosion pathways. The MARS model serves as the most common actuarial method that previously proved to offer credible patterns of gully sensitivity. The reason behinds choosing MARS models for the forecasting gully erosion are as follows: (1) possibility for modeling curvilinear association among the conditioning factors and gully incidence; (2) it allows working with various outcome variables and may manipulate data from different measures; (3) according to previous studies in this area, that any studies have applied this model for evaluating its ability and robustness for gully susceptibility.

This area in Gorganrood has witnessed gully erosion that caused many issues in this area and led organizations to reassess the Weirdness of adopted constant genesis strategies (CONRWMGP 2009). Astonishingly, the major initial place for the erosion phenomenon is cutting down trees positioned at the upper part of Gorganrood watershed and land-use changes (CONRWMGP 2009). From 1990 to 2005 due to the presence of loess soils in the northern of Golestan Province, 430,000 ha of these areas were affected by erosion. Soil erosion in this province is 5–6 tons/ha/year in forest areas (Department of Golestan Natural Resources and Watershed management). Gully erosion in Maravetape and Kalale counties leads to the loss of soil, the imposition of large costs, reduced agricultural potential, and has caused the migration of people in the villages of this region and exacerbated soil erosion pathways influence farming system efficiency. Consequently, to describe a robust model to evaluate the sensitivity of the territory to the development of gully pathways is necessary and Gorganrood watershed as the susceptible area will be the focus of the present study.

The difference between this study and previous studies who used the MARS model is applying two scenarios of the combination of number of replications and sample size, including 90%/10% and 80%/20% with 10 replications, and 70%/30% with 5, 10, and 15 replications for preparing gully erosion susceptibility mapping (GESM) and assessing their performance by MARS algorithm. Each one involves a various subset of both positive and negative cases. Absences are extracted as randomly distributed individual cells. Therefore the predictive competency of the gully erosion susceptibility model and the robustness of the procedure were evaluated through these datasets.

Therefore, the main scope of the current study is gully erosion modeling based on the MARS technique and explores the ability and robustness of the MARS model to forecast the incidence of gully erosion by various data sets and assessment measures. This study enriches the systematic assessment of the MARS model to map gully erosion susceptibility among others. This study tried to investigate gully erosion susceptibility through the MARS model and analyzed the performance and accuracy of this technique for zoning gully erosion.

The result of this study demonstrates ways of offering beneficial insight for remediation strategies and founding land exploitation projects by erosion susceptibility.

2. Materials and Methods

2.1. Study Area

The area under research belongs to Gorganrood basin, related to Golestan Province located in north-eastern Iran. This area accounts for 2142.64 km^2 and is intensely influenced by gully erosion. Its coordinates span over 37°18′–37°52′ N and 55°18′–56°10′ E (Figure 1). Topographically, this region is considered as plain. The mean height ranges between 56 and 2165 m. Its prevalent soil textures are Silty-loamy (approximately 53.7%) and Silty-clay-loamy (nearly 45.3%) soils. The major land uses in this region include pasturelands (40.1%), agriculture (farming) (29.4%), and forest (18.4%). The mean annual rainfall is approximately between 460 and 603 mm. The average minimum and maximum temperatures are 11 and 18.5 °C, respectively [32]. In the last decade, this area has been challenged with natural hazards and has faced intensive gully erosion. Therefore, this study area was selected as a potential gully erosion-prone area. Figure 2 presents some Google Earth images of gully erosion in the research field.

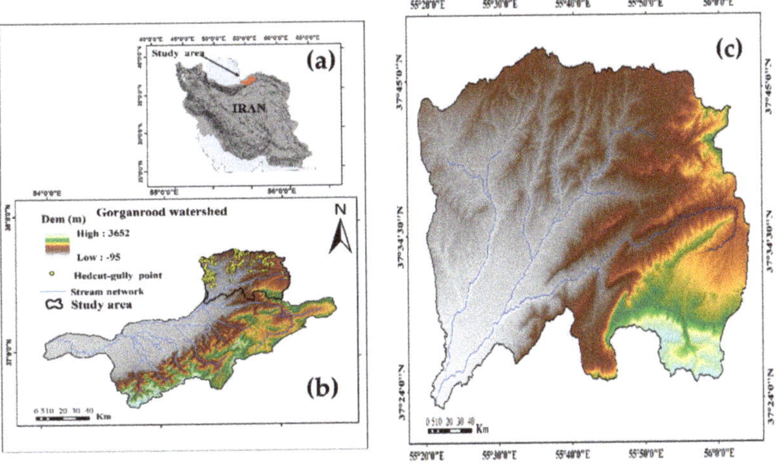

Figure 1. Study area and topographical characteristics. (**a**) Iran, (**b**) Gorganrood Watershed, (**c**) Study area.

Figure 2. Some photographs of the gully erosion zoned on study area.

2.2. Methodology

Figure 3 indicates a flow-diagram for the method applied for the application of multivariate adaptive regression splines model for gully erosion sensitivity zoning developed for this specific research in the north of Iran. As shown, the flowchart consists of five steps: (1) preparing thematic layers or 18 effective conditioning factors; (2) selection of factors using a multi collinearity test; (3) applying two scenarios of the combination of the number of replications, and sample size, including 90%/10% and 80%/20% with 10 replications, and 70%/30% with 5, 10, and 15 replications; (4) gully erosion susceptibility modeling using the MARS technique; (5) validation of the susceptibility maps using the ROC-AUC (Area Under the Receiver Operating Characteristic) curve, efficiency percent, Yuden index, and kappa.

2.2.1. Gully Erosion Inventory Mapping

An essential stage to zoning is to create a hazard evaluation for hazard zones [22]. The gully erosion inventory for the present section of Gorganrood Watershed was provided by Google Earth images, field surveys, and national and regional reports from different organizations. The present map constitutes a set of incidences (307 hedcut points). While designing statistical plans, the training set must differ from sets applied in the validation part [33]. To distinguish training points from the validation points a random dividing algorithm [29,34] was used. In this research, two scenarios were used: these scenarios were selected after altering different sample sizes and the number of replications, including 90%/10% and 80%/20% with 10 replications. To assess the robustness of the model's data sensitivity [8,22,35], 5, 10, and 15 sample data sets, (replicates) for 70%/30% sample size, were prepared through randomly multi-extracting of various data sets in the calibration and validation subsets [36]. Every set was adjusted through addition to positives (i.e., pixels having hedcut points) an equal number of randomly selected negative points, corresponding to pixels without hedcut [37].

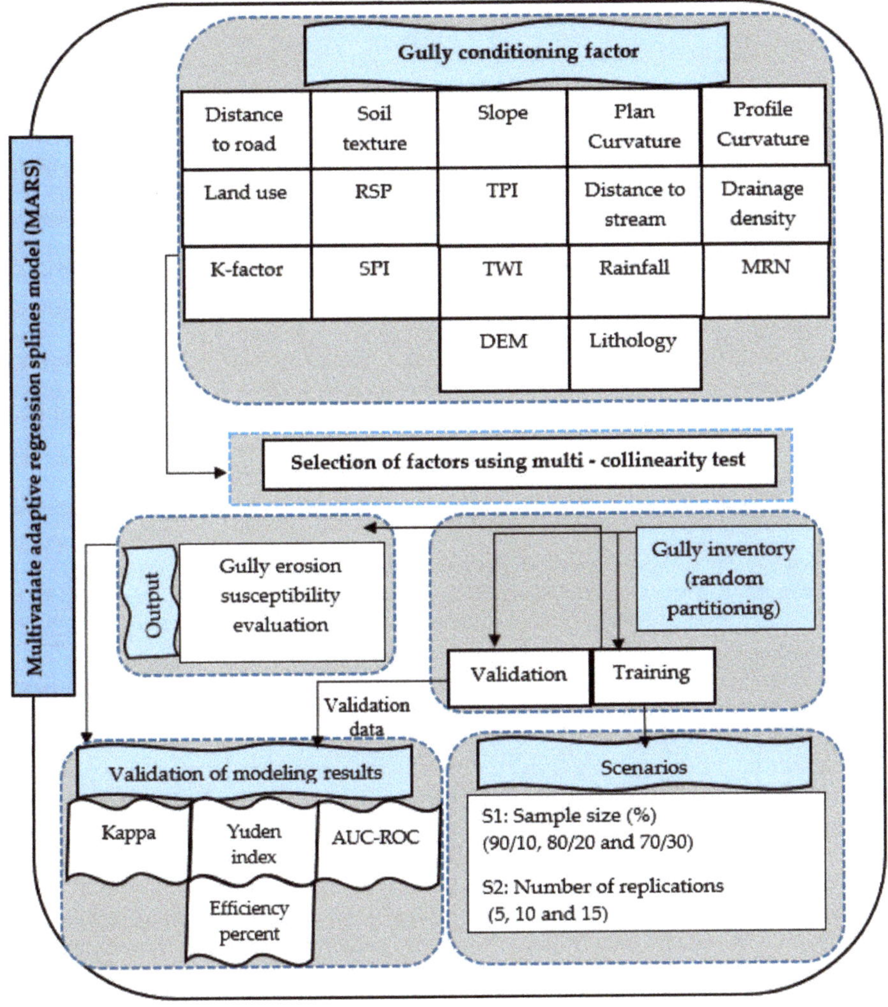

Figure 3. Flow diagram for the research method.

2.2.2. Gully Erosion Predictor Variables (GEPV)

It is essential to determine the effective factors on natural hazards and man-made fatalities in order to performing gully erosion susceptibility maps have great importance [38]. Good knowledge of the main gully erosion-related factors is required to recognize the susceptible areas. Such contributors always apply in studies analyzing gully erosion. Therefore, such factors were chosen from past studies [34,39]. In this study, to generate and exhibit such data grid, ArcGIS 10.5 and a system for automated geoscientific analyses (SAGA) software were used. For the application of the MARS model, all agents were transformed into a raster network with 30×30 m grid pixel. All conditioning factors were primarily continuous, and some of them (litologhy, soil, and land use) were classified within different categories based on expert knowledge and literature review [40–42].

The predicting factors used in this work are (a) digital elevation model (m), (b) aspect map, (c) slope percent, (d) curvature of profile, (e) curvature of plan, (f) land use (LU), (g) soil texture, (h) Topographic wetness index (TWI), (i) distance to streams (m), (j) distance to roads (m), (k) drainage density, (l) annual rainfall (mm), (m) stream power index, (n) relative slope position, (o) lithological formation, (p) K factor, (q) Melton ruggedness number, (r) topographic position index.

Digital contour data obtained from the Department of Natural Resources Management of Iran was applied. A DEM (Digital Elevation Model) (Figure 4a) of the research field characterized with a 30 m pixel was generated. Drawing upon DEM, physiographical and geomorphological grids such as aspect (Figure 4b), slope percent (Figure 4c) as well as curvature layers were extracted using ArcGIS 10.5. The slope percent includes a large section of the intrinsic views and is an important factor as it affects drainage density, surface runoff, influences, vegetation structure the soil erosion, soil moisture, and geomorphological processes [9,43–47]. Slope aspect is another important factor related to precipitation, snow meltwater, land cover, soil moisture patterns, and physiographic trends [48–52]. The suitable geomorphological data may be inferred via curvature assessment [53–55]. Three categories, convex, concave, and flat, were used to develop the slope curvature map. Positive curvature exhibits convex (>+0.1), negative curvature depicts concave (<−0.1), and zero curvature represents flat ((−0.1)–(+0.1)). In addition, profile and plan curvatures (Figure 4d,e) include various negative as well as positive quantities and indicate various description in every measure. Negative as well as positive ones in profile curvature denote convexity (increasing flow velocity) and concavity (reducing flow velocity), respectively. In contrast, negative and positive values in the plan curvature imply concavity (flow convergence) as well as the convexity (flow divergence) [54,56]. Those near to zero denote neutral curvature in both conditions.

LU has a substantial contribution in geomorphological and hydrological pathways through direct or indirect influences on evapotranspiration, infiltration, run-off generation, and sediment dynamics [52,57]. The other hand, agricultural activities have an important impact on gully erosion development as well as genesis [58]. The land-use map in the region in 1:100,000 scale was prepared by the Natural Resources department of Golestan Province and manipulated by Google Earth images. The land-use of the region consists of residential areas (urban), range and farming, forestlands, rangelands, farming, and lake (Figure 4f).

Soil texture (Figure 4g) usually is recognized as a substantial limiting factor in the process of infiltration and overflow and it is effective on gully occurrence [59–61]. Through digitizing the soil texture map of Golestan Province (1:100,000 scale) obtained from the Agriculture Department, Iran the aforementioned layer was prepared. The soil texture in the study area consists of sandy-loam, clay-loam, sandy-clay-loam, silty-clay, silty-clay-loam, as well as silty-loam.

Moore, Grayson, and Ladson [62] and Grabs et al. [63] mentioned that TWI (Topographic wetness index) represents the spatial distribution of wetness conditions, and inclination of gravitation to transport water to downstream. This factor was prepared using Equation (1):

$$\text{TWI} = \ln\left(\frac{\propto}{\tan \beta}\right), \tag{1}$$

where α denotes the aggregated upstream area leaving a point (contour points length) and $\tan \beta$ is point angle. Here, GIS 10.5 software was used for TWI mapping and ranges from 1.20 to 22.92 (Figure 4h).

Distance to streams (Figure 4i) serves as a key determinant because of its important effect on flow magnitude as well as gully erosion [29]. Based on field studies, gully erosions are diffused typically close the linear aspects in particular roads. Undoubtedly, road making imposes an adverse effect on hill sustainability at which flow may be appropriate for gullies [22,64].

Layers of the proximity were produced using the Euclidean metric function in ArcGIS 10.5 and ranged from 0 to 11,720 m for roads (Figure 4j), and 0–15,080 m for streams. The national topographic map at the scale of 1:50,000 was used to extract routes and streams.

The drainage density (Figure 4k) is found to be a great factor which plays an important role in the incidence of many hazards [65]. Several factors such as the structure and nature of the soil characteristics, geological beds, infiltration rate, plant cover condition, and slope degree [66] affect drainage networks. To turn the drainage network model to a reasonable value, the drainage density was specified via an extension of "line density" in ArcGIS 10.5 software.

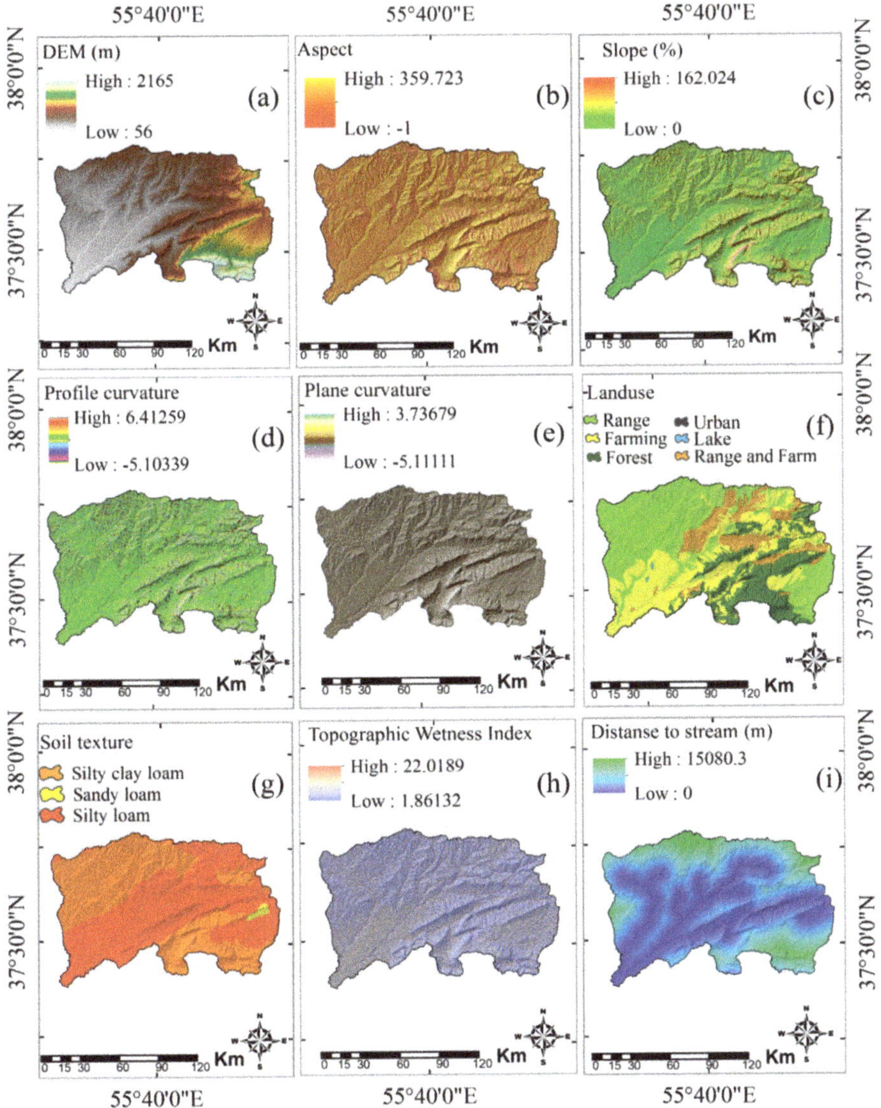

Figure 4. *Cont.*

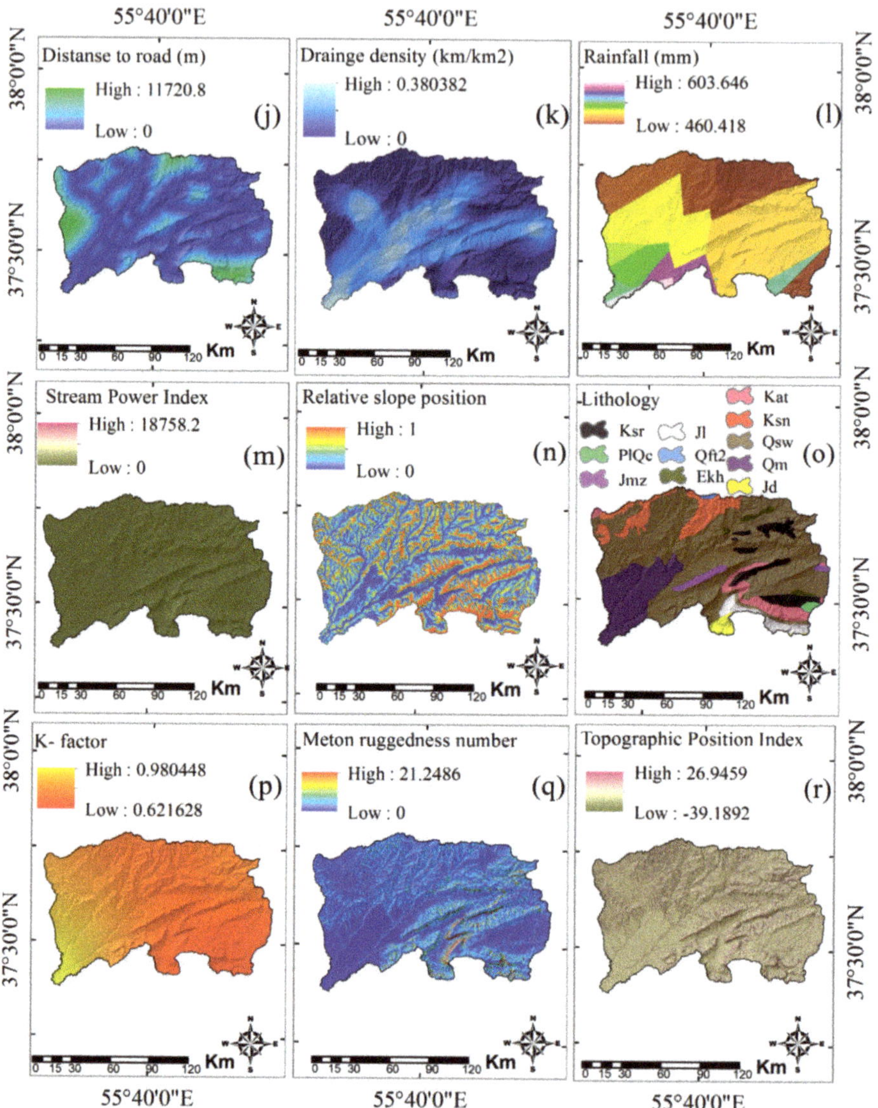

Figure 4. Gully conditioning factor maps. (**a**) Digital Elevation Model (DEM) (m), (**b**) Aspect, (**c**) Slope (%), (**d**) Profile curvature, (**e**) Plan curvature, (**f**) Land use, (**g**) Soil texture, (**h**) Topographic Wetness Index, (**i**) Distance to stream (m), (**j**) Distance to roads (m), (**k**) Drainage density (**l**) Rainfall (mm), (**m**) Stream Power Index, (**n**) Relative slope position (**o**) lithology, (**p**) K factor, (**q**) Melton ruggedness number, (**r**) Topographic Position Index.

The yearly average precipitation map (Figure 4l) of Gorganrood Basin was developed according to precipitation data obtained from the Golestan province Regional Water Organization. The above-mentioned map was developed using the 53 gauges and statistical period of 2009–2016 based on Inverse Distance Weight (IDW) interpolation method [67] (Equation (2)). This map ranges from 460 to 603 mm/year. The rainfall map was developed in 30 × 30 m in ArcGIS 10.5 as an input layer for susceptibility assessment of gully erosion.

$$\lambda i = \frac{Di^{-\alpha}}{\sum_{i=1}^{n} Di^{-\alpha}}, \tag{2}$$

where, λi represents point i weight, Di denotes the space between the point i and the point of unknown, and α implies weighing power [67].

The stream power index (SPI) (Figure 4m) is an index of the water flow erosive power according to the hypothesis that flow is relative to the particular watershed [68].

$$SPI = As \times \tan \sigma, \tag{3}$$

where, As is the particular watershed area per meter and r is the gradient of slope in degrees. The SPI index is the most important factor adjusting slope erosion processes, as erosive power of flow directly affect s river cutting and slope to erosion [68]. The regions with great river power measures have excessive erodibility because it indicates potential energy for limiting sediment [69].

Relative slope position (RSP), (Figure 4n), as a tool, could calculate several terrain indices from the digital elevation model. General information on the computational concept can be found in [70].

Lithology units (Figure 4o), have a dominant contribution in specifying gully erosion sensitivity [9,50,52,71,72] as gully erosion relies on the lithology properties and different lithological units display significant differences in erosion instability. In this research, the lithological map for the region was generated as the present geological maps with a 1:100,000 scale obtained from the Geological Survey Department, Iran. This area of Gorganrood Watershed is full of various types of outcrop formations and divided into 10 classes (Table 1).

Table 1. Lithology of the Gorganrood watershed.

Group	Code	Explanation	Formation
1	Ksr	Shale containing Ammonite with interaction of orbitolin limestone	Sarcheshmeh
2	PlQc	Fluvial conglomerate, Piedmont conglomerate, and sandstone.	-
3	Jmz	Grey thick-fluvial limestone and dolomite	Mozduran
4	Ksn	Brown to block shale and thin layers of siltstone and sandstone	Sanganeh
4	Murm	Light-red to brown marl and gyps marl with sandstone intercalations	-
4	Murmg	Gypsiferous marl	-
4	Elm	Marl, gypsiferous marl and limestone	-
5	Mur	Red marl, gypsiferous marl, sandstone and conglomerate	Dalichai
5	Kad-ab	Usual unit comprising argillaceous limestone, marl and shale	-
5	Jd	Well-bedded to thin-bedded, greenish-grey argillaceous limestone with intercalations of calcareous shale	-
6	Qft1	Concentrated piedmont fan and valley terrace deposits	-
6	Qft2	Low level piedmont fan and valley terrace sedimentation	-
6	Qal	River channel, braided drainage and flood plain sedimentation	-
6	Qs,d	Loose loess sand sedimentation such as dunes	-
7	Jl	Light brown, thin-bedded to massive limestone	Lar
8	Ekh	Olive-green shale and sandstone	Khangiran
9	Kat	Green glauconitic sandstone and shale	Aitamir
10	Qsw	Swamp	-
10	Qm	Swamp and marsh	-

Soil erosion potential (k-factor) (Figure 4p) influences soil persistence to run-off force or rainfall impact [73]. The soil erodibility (K) in Universal Soil Loss Equation (USLE), is measured by the texture, organic matter, infiltration, and soil structure.

A simplified flow accumulation measure, computed as discrepancy among max and min elevation in the watershed divided by the square root of the watershed area is Melton ruggedness number (MRN) (Figure 4q). The measurement is done for every grid pixel, hence minimum elevation equals elevation in the cell's location. Flow measurement is easily performed with Deterministic 8 given the inconsistent nature of a single maximum elevation [74–76].

Topographic position index (TPI) (Figure 4r) serves as an approach widely applied to evaluate topographic slope location, and to zone ordination automation. This function produces single-band raster characterized quantities measured upon elevation. TPI is an abbreviation for Topographic Position Index, in turn, described as the difference between the main pixel and the average of its adjacent pixels [77].

2.3. Multi-Collinearity Test

The above-mentioned factors were used to consider the effect of correlation among them as the independent variable. If both predictor variables are intensely related, it is a problemin the modelling process. The issue is named collinearity. The VIF (variance inflation factor) and Tolerance include both significant measures of multi-collinearity recognition. Indeed, VIF is simply the reciprocal of tolerance, on the other hand, tolerance is 1-R2 for the variable regression versus predictors, deprived of the dependent variable [78]. A VIF of five or 10 and above and/or a tolerance of less than 0.20 or 0.10 indicates a multi-collinearity problem [79,80].

2.4. Multivariate Adaptive Regression Splines (MARS Model)

MARS can adapt complicated, non-linear associations among the independent and dependent measures although providing an explainable model [81]. The MARS algorithm has been applied in geomorphology to forecast and mapping the incidence of gullies [25–27] and landslides [17,38,82].

This technique functions via dividing magnitudes of the explanatory predictors into areas and through the fitting, for every area, a linear regression equation. "Knots" divides quantities among regions, whereas the phrase "basis function" (BF) denotes implying every various condition factor interval (dependents). BFs are functions of the following:

$$\max(0, \, x - k) \text{ or}$$

$$\max(0, \, k - x),$$

where k denotes constant equal to a knot and x is an independent variable. The MARS may be stated generally as below:

$$y = (x) = \alpha + \sum_{n=1}^{N} \beta_n h_n(x), \tag{4}$$

where y is predictor variable anticipated via function f(x), N denotes terms number, who is shaped by a coefficient βn, α is a constant, and h_n (x) represents a separate basis function or multiplication of BFs. MARS analysis was carried out by the Earth version of the R software [83–85]

Evaluation of the Model

The evaluation pathway comprises the fitting degree assessment (goodness of fit), robustness, and prediction skills of the model [22,86]. The goodness of fit demonstrations capability of the approach in forecasting the training subset, whereas the predictive efficiency (prediction skills) is a fundamental step for model precision to forecast a validation set (the percent of gully points that do not use the training process) [39,87]. While changing training and validation points, the precision of the forecasting model is determined as the consistency of outputs of the model in respect to model precision. Here, predictive performance and goodness of fit of the model are subjected to assessment using both threshold-driven and non-threshold-dependent methods. ROC curve, as a threshold-driven method, was drawn for all datasets as well as afterwards goodness of fit (i.e., degree of fitting) and forecasting efficacy of the algorithms were studied, respectively [88].

The Kappa coefficient, Youden index, and efficiency (as threshold dependent performance) were calculated based on the components of the confusion matrix [89]. Furthermore, sensitivity (SST) and specificity (SPF) are common statistical indexes applied for validation of each model performance [90]. Comparison of the model observed data and results are demonstrated through a contingency matrix (Table 2). According to Table 2, the true negative (TN) and true positive (TP) are the numbers of pixels that are correctly and appropriately classified, whereas false negative (FN) and false positive (FP) are the numbers of pixels fallaciously classified.

$$SST = \frac{TP}{TP + FN}, \tag{5}$$

$$PF = \frac{TN}{TN + FP}. \tag{6}$$

Efficiency (E) is the ratio of gully points and non-gully cells which output model accurately divided:

$$E = \frac{TP + TN}{TP + TN + FP + FN}. \tag{7}$$

The Kappa factor (K) describes the potential of the employed model to categorize the gully point cells [1], and can be expressed as the ratio of given consistency beyond that expected by chance:

$$K = \frac{P_{obs} - P_{exp}}{1 - P_{exp}}, \tag{8}$$

where P_{obs} represents the ratio of cells that are properly divided as gully occurrence or non-gully and P_{exp} denotes the ratio of pixels whose consistency is random [91]. P_{obs} and P_{exp} can be measured as below:

$$P_{obs} = TP + TN, \tag{9}$$

$$P_{exp} = [(TP + FN)(TP + FP) + (FP + TN)(FN + TN)]. \tag{10}$$

The performance of model given the Kappa factor may be categorized as ≤ 0 (weak), 0–0.2 (slight), 0.2–0.4 (fair), 0.4–0.6 (moderate), 0.6–0.8 (high), and 0.8–1 (almost perfect) [1,2].

Table 2. Contingency matrix applied for the Multivariate Adaptive Regression Splines (MARS) model assessment.

Observed	Predicted	
	%Gully (+)	%Non-Gully (−)
Gully (+)	(+|+) True positive (TP)	(−|+) False negative (FN)
Non-gully (−)	(+|−) False positive (FP)	(−|−) True negative (TN)

The ROC (receiver operating characteristics) curve, as a non-threshold-driven method, is widely used to quantify the measure classification [92–94]. The receiver operating characteristics is the area under the curve (AUC-ROC) that describes the performance of a model to precisely forecast non-incidence or an incidence [95,96]. The forecasting precision of the model according to the AUC value may be categorized as three classes of accuracy following the classification proposed by [97]: 0.7, 0.8, and 0.9. AUC thresholds were used for acceptable, superior, and substantial performance, respectively [22,37].

Since the admission of a forecasting model entails, for assessment of its precision, trifle variations of the input data (i.e., data susceptibility), a gully erosion sensitivity model was developed on 10 various samples of mapping measures. Therefore, robustness in forecasting models and their consistency were furthermore evaluated when the training and validation samples are altered (i.e., a replicate method) [8,22,41,96,98]. The model was employed to given data, and subsequently was tested by the validation datasets. As for the ROC curve, the calculated event map by the validation dataset were compared.

Through subtracting the maximum and minimum accuracy values according to each assessment measures precision of the model was calculated [22,96]:

$$R_{AUC-ROC} = AUC - ROC_{max} - AUC - ROC_{min,} \tag{11}$$

where RAUC-ROC is model precision as per AUC-ROC measures, and AUC-ROC$_{max}$ includes maximum precisions within all sets. As well, least accuracy values are represented as AUC-ROC min within whole datasets. In addition, in this research, ROC curves were applied to designate the optimum benchmark point of the model rate, via considering Youden's index (J) [99], with J relative to the maximum perpendicular space between the ROC and the former bisector as follows:

$$J = Maximum\ (sensitiviity + specificity - 1). \tag{12}$$

3. Results

3.1. Gully Erosion Susceptibility Model

First, Google Earth images, field surveys, and national reports were used to provide a gully-hedcut evaluation map consisting of 307 gully-hedcut points. Here, the MARS model was employed on balanced given sets (positives/negatives), and each one comprises all the positive (gully point) pixels and same randomly inferred negative (non-gully spots) cells, that was for two processes of integrating some repeats and samples, involving 90%/10% and 80%/20% with 10 replications. To assess the robustness of the model's data sensitivity, 5, 10 and 15 sample datasets (replicates) for 70%/30% sample size, were prepared through random selection of different data sets in the calibration and validation subsets. Figure 5 indicates the relative diffusion of the mean of gully erosion susceptibility categorizes for 10 groups in the sample data sets (70%/30%).

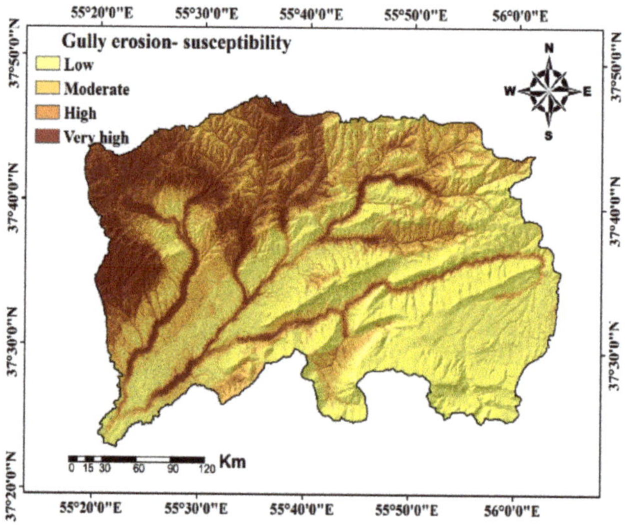

Figure 5. Sensitivity maps of gully erosion of the study area by the MARS model for 70%/30%, 10 replications.

The results of the relative distribution of the average of gully erosion susceptibility classes for other models are presented in Table 3. However, in the other combinations, the results were almost similar together. As well, the actuarial features of the probabilistic forecasting of the gully erosion of 10 sample data sets and replicates are shown in Table 4.

Table 3. Relative distributions of the gully susceptibility classes.

| MARS Model | Relative Distributions of the Gully Susceptibility Classes | | | | |
| | 70%/30% | | | 80%/20% | 90%/10% |
	5 rep *	10 rep	15 rep	10 rep	10 rep
Low	47.14	44.72	45.86	47.55	47.65
Medium	22.83	23.63	22.85	22.16	21.94
High	15.70	17.10	16.27	15.20	16.17
Very high	14.34	14.55	15.02	15.09	14.25

* rep: Replicate.

Table 4. Actuarial features of the probability values inferred from the MARS model.

| MARS Model | Probabilistic Prediction Values | | | | |
| | 70%/30% | | | 80%/20% | 90%/10% |
	5 rep	10 rep	15 rep	10 rep	10 rep
Mean	0.277	0.279	0.283	0.277	0.275
SD	0.281	0.270	0.280	0.285	0.273
Minimum	0.000	0.000	0.000	0.000	0.000
Maximum	0.999	0.997	0.996	0.999	0.998

SD: Standard deviation.

3.2. Evaluation of the Susceptibility in Gully Erosion

The results of the goodness-of-fit are shown in Table 5. The results did not show considerable variation in the accuracy of the model, with altering the percentage of calibration to validation samples and number of model replications.

Table 5. Forecasting efficiency of the model given 10 data sets.

| MARS Model | 70%/30% | | | 80%/20% | 90%/10% |
	5 rep	10 rep	15 rep	10 rep	10 rep
Sensitivity	0.86	0.79	0.85	0.88	0.86
Specificity	0.72	0.81	0.66	0.83	0.72
(Negative predictive value)	0.70	0.78	0.72	0.74	0.75
(Positive predictive value)	0.83	0.73	0.82	0.85	0.84
Efficiency (%)	79.0	76.0	76.0	79.0	77.9
Kappa	0.58	0.51	0.52	0.58	0.58
AUC Mean	0.80	0.82	0.83	0.84	0.83
Robustness	0.03	0.08	0.11	0.01	0.15

The MARS algorithm performed excellently both in predictive performance. It can be observed that the MARS model for combination of 80%/20% had the highest performance in terms of SST (0.88), SPF (0.83), E% (0.79), Kappa (0.58), Robustness (0.01) and AUC (0.84) compared to the other combinations. The combination of 70%/30% with five replicates only had the highest performance in terms of E% (0.79). Figure 6 illustrates the mean ROC curves for the MARS model through 10 replicates. The results of AUC as a threshold-driven method for other scenarios are as follows: (70%/30% with 5, 10 and 15 replicates: 0.80, 0.82, 0.83 respectively; 90%/10%: 0.83). The MARS model revealed from acceptable to excellent performances, with AUC values above the 0.77 and 0.8 thresholds. By adopting the J index, 0.386 probability cut-off values were generated for the susceptibility gully erosion model (Figure 6). Based on these cut-off values, the probability of gully erosion occurrence for each cell was adapted into a binary (positive/negative) prediction to achieve the spatial distribution of cases correctly categorized (true positives and negatives (TP, TN)) and incorrectly classified (false positives and negatives (FP, FN)) for the MARS susceptibility model (Table 6). Figure 7 shows the distribution of true positive (TP), false positive (FP), true negative (TN), and false-negative (FN) cases within the study area. A larger true positive prediction is produced by the model, indicating that the conditions for gully erosion are widespread.

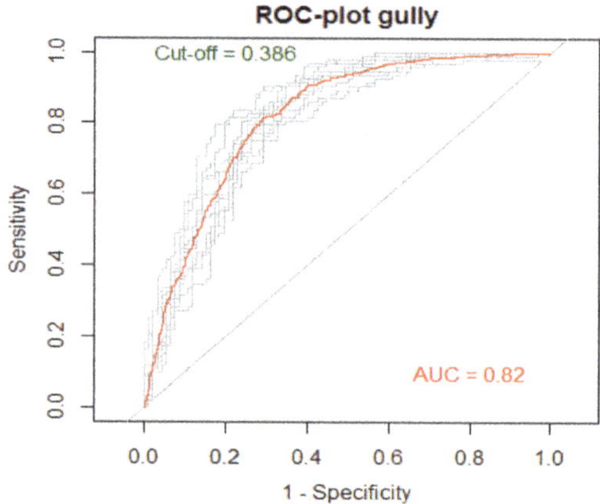

Figure 6. Average Receiver Operating Characteristic (ROC) curves for the MARS model through 10 replicates.

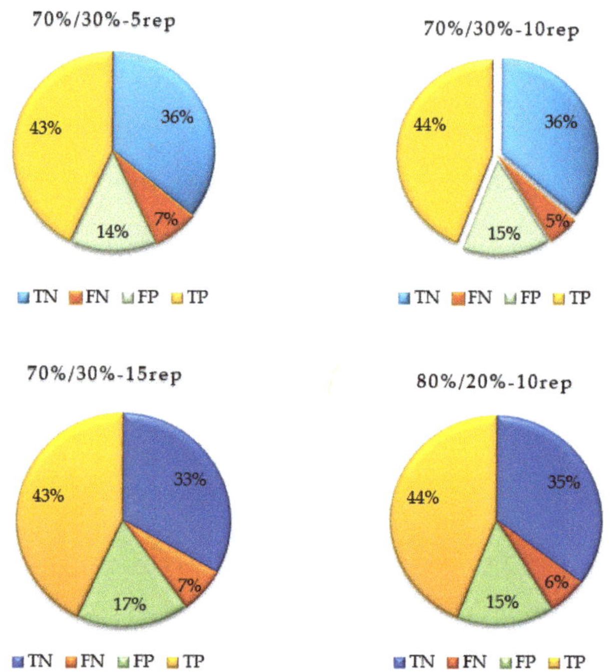

Figure 7. Contingency matrix applied for assessing the MARS model for all scenarios.

Table 6. Contingency matrix applied for assessment of the MARS model.

Observed	Predicted	
	%Gully (+)	**%Non-gully (−)**
Gully (+)	(+|+) 40% (TP)	(−|+) 10% (FN)
Non-gully (−)	(+|−) 14% (FP)	(−|−) 36% (TN)

4. Discussion

The results of the model (based on all sample data sets) show different ranges of susceptibility values of gullying. The output map was divided into four classes of poor, high, and very high for each model by the Natural breaks classification approach [96,100,101]. As the regions of the area were categorized as high and very high, gully erosion susceptibility maps were complementary with the sections of the watershed with low slope and near to roads. For an average of gully susceptibility, more than 31% of the study area has a high (HGES) and very high sensitivity (VHGES). Approximately 23.63% of the research field was classified as moderate classes (MGES). A total of 44.72% of the cells in the study region classified into low susceptibility groups (LGES). On the other hand, the low density of gullies was observed in forest areas with high slopes. In the forest areas, roughness due to vegetation covering may lead to medium runoff factors on this area as well, hence, low degradation force of centralized flow [96]. Areas nearer to roads and streams, with sparse vegetation and higher drainage density than other areas, have more potential for gully occurrence. These findings are in line with [4]. The parts of the basin with low slope and near to roads classified as high and very high gully erosion susceptibility maps (HGES and VHGES), and high density of gully erosions occurred in these parts of the catchment. The distribution of gullies over lithological units by other studies observed that poorly sorted materials are favorable conditions for gully erosion development [102–104].

The MARS model precision for gully erosion susceptibility was assessed by both threshold-driven and non-threshold-driven methods as pointed out in the methodology section. To assess the robustness of the model's data sensitivity similar with references [8,22,35], 5, 10 and 15 sample data sets, (replicates) for 70%/30% sample size, were prepared through random selection of different data sets in the calibration and validation subsets. This method is in agreement with studies of Rotigliano et al. [36]. The MARS model reproduces non-linear relationships using several linear regressions. Hence, this allows MARS to generate models with a better fit to the training data set while maintaining high predictive power [5].

Given the accuracy, the MARS algorithm performed excellently in predictive performance. It can be observed that the MARS model for combination of 80%/20% had the highest performance in terms of SST (0.88), SPF (0.83), E% (0.79), Kappa (0.58), Robustness (0.01), and AUC (0.84) compared to the other combinations. Since the accuracy values are very similar for the all sample data sets, the MARS model was robust when the validation group changed [2]. As well, in the research of Rahmati et al. [29], three various classes of training samples were applied to accompany different machine learning models for predicting the susceptibility of gully erosion. Three different sample data sets (S1, S2, and S3), were randomly prepared to evaluate the robustness of the models. Their results illustrated accurate predictions. Additionally, it was found that performance of RF and RBF-SVM for modelling gully erosion occurrence is quite stable when the learning and validation samples are changed. Regarding forecasting efficiency, the results of AUC as a threshold-driven method for scenarios are as follows: (70%/30% with 5, 10 and 15 replicates: 0.80, 0.82, 0.83 respectively; 90%/10%: 0.83, 80%/20%: 0.84). The MARS model revealed from acceptable to excellent performances, with AUC values above the 0.77 and 0.8 thresholds [97]. This result demonstrated a strong agreement between the distribution of the existing gully erosion points and the final predictive susceptibility map. Additionally, the AUC values for all data sets are approximately similar and the modelling method can be considered as robust to changes in learning points. Our results similar and agree with studies of Gómez-Gutiérrez et al. [26], who also applied the MARS model to predict gully erosion occurrence, and obtained AUC in the range of 0.75–0.98. The other MARS application to gully erosion susceptibility evaluation was made by Gómez-Gutiérrez et al. [27], who achieved a mean AUC of 0.826 and 0.859 in Spain and Sicily respectively. Accordingly, Conoscenti et al. [17] pointed out that even for the worst validation AUC value inferred from five datasets, MARS is much more than the best calibration AUC value calculated compared with the logistic regression (LR) model. In addition, Conoscenti et al. [102] evaluated gully erosion sensitivity in both surrounding farmed basins of Sicily (Italy) by using multivariable adaptive regression splines. Model assessment on the whole basins indicates the outstanding predictive performance of models. This finding supports our results. Based on cut-off values, the probability of gully erosion occurrence for each cell was adapted into a binary (positive/negative) prediction to achieve the spatial distribution of cases correctly categorized (true positives and negatives (TP, TN)) and incorrectly classified (false positives and negatives (FP, FN)) for the MARS susceptibility model (Table 6). A larger true positive prediction is produced by the model, indicating that the conditions for gully erosion are widespread. As Rahmati et al. [96] pointed out, random ordination of datasets is the main origin of uncertainty in spatial modelling. From the validation result, it is clear that the MARS model provided acceptable to excellent performance in predicting the probability of gully erosion occurrence based on independent and dependent assessment measures. In light of abovementioned results, it is obvious that the MARS model can be used as an efficient statistical model for the prediction gully erosion susceptibility map. This is in line with the other studies that applied this model to landslides and gully erosion susceptibility mapping [9,25–27,102]. This is a relevant issue to achieve sustainable land management where gully erosions must be restored when they have developed as a result of human mismanagement, and for this it is necessary to use nature-based solutions [6]. To achieve success in gully erosion control, the strategies must find a way to reduce the connectivity of the flows [7].

5. Conclusions

Recognition of a precise and calibrated model to alleviate errors in modelling gully erosion susceptibility and determining gully erosion susceptible areas is of great importance. The present study enriches the systematic assessment of the multivariate adaptive regression splines model (MARS) to model gully erosion susceptibility among others. The major concluding remarks might be written as below: the aforementioned MARS model not just confirmed superiority for either based on limits-independent and limits-driven approaches, at the same time led to precise forecasting while changing the sample dataset. Hence, it can be inferred that this model is superior to evaluate gully erosion susceptibility while research aims to generate an exact gully erosion susceptibility map (GESM) and to pave the way to offering information on the outstanding potential of predictors. Reconnaissance and gully alleviation controls are not cost-effective and at the same time are not time-effective, and hence, to develop comprehensive susceptibility forecasting system as per modelling seems to be a remarkable possibility. In the current research based on past studies [34] and multicollinearity tests, digital elevation model, aspect map, slope percent, curvature of profile, curvature of plan, land use (LU), soil texture, TWI, distance to streams, distance to roads, drainage density, annual rainfall, stream power index, relative slope position, lithological formation, K factor, melton ruggedness number, and topographic position index are significant factors that influence gullying in the study area. Additionally, in this study we tried to investigate gully erosion susceptibility through the MARS model and analyzed the performance and accuracy of this technique for zoning gully erosion. While changing training and validation spots, the precision of the forecasting model was determined as the consistency of outputs of the model in respect to model precision. Here, predictive performance and goodness of fit of the model was subjected to assessment using both threshold-driven and non-threshold-dependent methods. This validates the robustness as well as the effectiveness of this model. Zonation of the gully erosion susceptibility for that section of Gorganrood display regions with high hazard susceptibility as well as demonstrates a valid map, in which the result is useful to managers and stakeholders to recognize the most prone area gully erosion and dedicate inputs for soil protection actions in the best manner.

Author Contributions: N.J., A.K., Z.J., H.R.P. and C.C. designed the experiments, performed the experiments, analyzed the data, and contributed reagents/materials/analysis tools. N.J. wrote the paper, Writing—review and editing was done by A.K. and H.R.P.

Funding: This research received no external funding.

Acknowledgments: We are grateful to the Sari Agricultural Sciences and Natural Resources University for financial support (SANRU) and Department of Earth and Sea Sciences (DISTEM), University of Palermo.

Conflicts of Interest: The authors declare no conflict of interest.

References

1. Lal, R. Offsetting global CO_2 emissions by restoration of degraded soils and intensification of world agriculture and forestry. *Land Degrad. Dev.* **2003**, *14*, 309–322. [CrossRef]

2. Kosmas, C.; Danalatos, N.; Cammeraat, L.H.; Chabart, M.; Diamantopoulos, J.; Farand, R.; Gutierrez, L.; Jacob, A.; Marques, H.; Martinez-Fernandez, J.; et al. The effect of land use on runoff and soil erosion rates under Mediterranean conditions. *Catena* **1997**, *29*, 45–59. [CrossRef]

3. Vandekerckhove, L.; Poesen, J.; Oostwoud Wijdenes, D.; Gyssels, G.; Beuselinck, L.; de Luna, E. Characteristics and controlling factors of bank gullies in two semi-arid mediterranean environments. *Geomorphology* **2000**, *33*, 37–58. [CrossRef]

4. Vanwalleghem, T.; Poesen, J.; Nachtergaele, J.; Verstraeten, G. Characteristics, controlling factors and importance of deep gullies under cropland on loess-derived soils. *Geomorphology* **2005**, *69*, 76–91. [CrossRef]

5. Chaplot, V.; Giboire, G.; Marchand, P.; Valentin, C. Dynamic modelling for linear erosion initiation and development under climate and land-use changes in northern Laos. *Catena* **2005**, *63*, 318–328. [CrossRef]

6. Li, Y. Gully erosion: Impacts, factors and control. *Catena* **2005**, *63*, 132–153.

7. Imeson, A.C.; Kwaad, F. Gully types and gully prediction. *Geogr. Tydschr.* **1980**, *14*, 430–441.

8. Angileri, S.E.; Conoscenti, C.; Hochschild, V.; Märker, M.; Rotigliano, E.; Agnesi, V. *Water Erosion Susceptibility Mapping by Applying Stochastic Gradient Treeboost to the Imera Meridionale River Basin (Sicily, Italy)*; Elsevier: Amsterdam, The Netherlands, 2016; Volume 262, ISBN 3909123864643.

9. Conforti, M.; Aucelli, P.P.C.; Robustelli, G.; Scarciglia, F. Geomorphology and GIS analysis for mapping gully erosion susceptibility in the Turbolo stream catchment (Northern Calabria, Italy). *Nat. Hazards* **2011**, *56*, 881–898. [CrossRef]

10. Govers, G.; Giménez, R.; Van Oost, K. Rill erosion: Exploring the relationship between experiments, modelling and field observations. *Earth Sci. Rev.* **2007**, *84*, 87–102. [CrossRef]

11. Guzzetti, F.; Reichenbach, P.; Ardizzone, F.; Cardinali, M.; Galli, M. Estimating the quality of landslide susceptibility models. *Geomorphology* **2006**, *81*, 166–184. [CrossRef]

12. Carrara, A.; Pike, R.J. GIS technology and models for assessing landslide hazard and risk. *Geomorphology* **2008**, *3*, 257–260. [CrossRef]

13. Vergari, F.; Della Seta, M.; Del Monte, M.; Fredi, P.; Palmieri, E.L. Landslide susceptibility assessment in the Upper Orcia Valley (Southern Tuscany, Italy) through conditional analysis: A contribution to the unbiased selection of causal factors. *Nat. Hazards Earth Syst. Sci.* **2011**, *11*, 1475. [CrossRef]

14. Pozdnoukhov, A.; Matasci, G.; Kanevski, M.; Purves, R.S. Spatio-temporal avalanche forecasting with Support Vector Machines. *Nat. Hazards Earth Syst. Sci.* **2011**, *11*, 367–382. [CrossRef]

15. Costanzo, D.; Cappadonia, C.; Conoscenti, C.; Rotigliano, E. Exporting a Google Earth™ aided earth-flow susceptibility model: A test in central Sicily. *Nat. Hazards* **2012**, *61*, 103–114. [CrossRef]

16. Ballabio, C.; Sterlacchini, S. Support vector machines for landslide susceptibility mapping: The Staffora River Basin case study, Italy. *Math. Geosci.* **2012**, *44*, 47–70. [CrossRef]

17. Conoscenti, C.; Ciaccio, M.; Caraballo-Arias, N.A.; Gómez-Gutiérrez, Á.; Rotigliano, E.; Agnesi, V. Assessment of susceptibility to earth-flow landslide using logistic regression and multivariate adaptive regression splines: A case of the Belice River basin (western Sicily, Italy). *Geomorphology* **2015**, *242*, 49–64. [CrossRef]

18. Magliulo, P. Soil erosion susceptibility maps of the Janare Torrent Basin (Southern Italy). *J. Maps* **2010**, *6*, 435–447. [CrossRef]

19. Magliulo, P. Assessing the susceptibility to water-induced soil erosion using a geomorphological, bivariate statistics-based approach. *Environ. Earth Sci.* **2012**, *67*, 1801–1820. [CrossRef]

20. Conoscenti, C.; Agnesi, V.; Angileri, S.; Cappadonia, C.; Rotigliano, E.; Märker, M. A GIS-based approach for gully erosion susceptibility modelling: A test in Sicily, Italy. *Environ. Earth Sci.* **2013**, *70*, 1179–1195. [CrossRef]
21. Lucà, F.; Conforti, M.; Robustelli, G. Comparison of GIS-based gullying susceptibility mapping using bivariate and multivariate statistics: Northern Calabria, South Italy. *Geomorphology* **2011**, *134*, 297–308. [CrossRef]
22. Conoscenti, C.; Angileri, S.; Cappadonia, C.; Rotigliano, E.; Agnesi, V.; Märker, M. Gully erosion susceptibility assessment by means of GIS-based logistic regression: A case of Sicily (Italy). *Geomorphology* **2014**, *204*, 399–411. [CrossRef]
23. Geissen, V.; Kampichler, C.; López-de Llergo-Juárez, J.J.; Galindo-Acántara, A. Superficial and subterranean soil erosion in Tabasco, tropical Mexico: Development of a decision tree modeling approach. *Geoderma* **2007**, *139*, 277–287. [CrossRef]
24. Märker, M.; Pelacani, S.; Schröder, B. A functional entity approach to predict soil erosion processes in a small Plio-Pleistocene Mediterranean catchment in Northern Chianti, Italy. *Geomorphology* **2011**, *125*, 530–540. [CrossRef]
25. Gutiérrez, Á.G.; Schnabel, S.; Contador, J.F.L. Using and comparing two nonparametric methods (CART and MARS) to model the potential distribution of gullies. *Ecol. Model.* **2009**, *220*, 3630–3637. [CrossRef]
26. Gomez Gutierrez, A.; Schnabel, S.; Felicísimo, Á.M. Modelling the occurrence of gullies in rangelands of southwest Spain. *Earth Surf. Process. Landf. J. Br. Geomorphol. Res. Gr.* **2009**, *34*, 1894–1902. [CrossRef]
27. Gómez-Gutiérrez, Á.; Conoscenti, C.; Angileri, S.E.; Rotigliano, E.; Schnabel, S. Using topographical attributes to evaluate gully erosion proneness (susceptibility) in two mediterranean basins: Advantages and limitations. *Nat. Hazards* **2015**, *79*, 291–314. [CrossRef]
28. Zabihi, M.; Pourghasemi, H.R.; Motevalli, A.; Zakeri, M.A. Gully erosion modeling using GIS-based data mining techniques in Northern Iran: A comparison Between Boosted Regression Tree and Multivariate Adaptive Regression Spline. In *Natural Hazards GIS-Based Spatial Modeling Using Data Mining Techniques*; Springer: Cham, Switzerland, 2019; pp. 1–26.
29. Rahmati, O.; Tahmasebipour, N.; Haghizadeh, A.; Pourghasemi, H.R.; Feizizadeh, B. Evaluation of different machine learning models for predicting and mapping the susceptibility of gully erosion. *Geomorphology* **2017**, *298*, 118–137. [CrossRef]
30. Nazari Samani, A.; Ahmadi, H.; Mohammadi, A.; Ghoddousi, J.; Salajegheh, A.; Boggs, G.; Pishyar, R. Factors controlling gully advancement and models evaluation (Hableh Rood Basin, Iran). *Water Resour. Manag.* **2010**, *24*, 1531–1549. [CrossRef]
31. Friedman, J.H. Multivariate adaptive regression splines. *Ann. Stat.* **1991**, *19*, 1–67. [CrossRef]
32. Water Resources Company of Golestan (WRCG). Precipitation and Temperature Reports. Available online: http://www.gsrw.ir/Default.aspx (accessed on 15 May 2013).
33. Lee, S.; Ryu, J.-H.; Kim, I.-S. Landslide susceptibility analysis and its verification using likelihood ratio, logistic regression, and artificial neural network models: Case study of Youngin, Korea. *Landslides* **2007**, *4*, 327–338. [CrossRef]
34. Pradhan, B. Flood Susceptible Mapping and Risk Area Delineation Using Logistic Regression, GIS and Remote Sensing. *J. Spat. Hydrol.* **2009**, *9*, 1–18.
35. Cama, M.; Lombardo, L.; Conoscenti, C.; Rotigliano, E. Improving transferability strategies for debris flow susceptibility assessment: Application to the Saponara and Itala catchments (Messina, Italy). *Geomorphology* **2017**, *288*, 52–65. [CrossRef]
36. Rotigliano, E.; Martinello, C.; Agnesi, V.; Conoscenti, C. Evaluation of debris flow susceptibility in El Salvador (CA): A comparison between Multivariate Adaptive Regression Splines (MARS) and Binary Logistic Regression (BLR). *Hung. Geogr. Bull.* **2018**, *67*, 361–373. [CrossRef]
37. Conoscenti, C.; Rotigliano, E.; Cama, M.; Caraballo-Arias, N.A.; Lombardo, L.; Agnesi, V. *Exploring the Effect of Absence Selection on Landslide Susceptibility Models: A Case Study in Sicily, Italy*; Elsevier: Amsterdam, The Netherlands, 2016; Volume 261, ISBN 3909123864.
38. Kia, M.B.; Pirasteh, S.; Pradhan, B.; Mahmud, A.R.; Sulaiman, W.N.A.; Moradi, A. An artificial neural network model for flood simulation using GIS: Johor River Basin, Malaysia. *Environ. Earth Sci.* **2012**, *67*, 251–264. [CrossRef]

39. Lee, M.J.; Kang, J.; Jeon, S. Application of Frequency Ratio Model and Validation for Predictive Flooded Area Susceptibility Mapping Using GIS. In Proceedings of the 2012 IEEE international Geoscience and Remote Sensing Symposium; IEEE: Piscataway, NJ, USA, 2012; pp. 895–898.

40. Youssef, A.M.; Pourghasemi, H.R.; Pourtaghi, Z.S.; Al-Katheeri, M.M. Erratum to: Landslide susceptibility mapping using random forest, boosted regression tree, classification and regression tree, and general linear models and comparison of their performance at Wadi Tayyah Basin, Asir Region, Saudi Arabia (Landslides, 10.10. *Landslides* **2016**, *13*, 1315–1318. [CrossRef]

41. Jaafari, A.; Najafi, A.; Pourghasemi, H.R.; Rezaeian, J.; Sattarian, A. GIS-based frequency ratio and index of entropy models for landslide susceptibility assessment in the Caspian forest, northern Iran. *Int. J. Environ. Sci. Technol.* **2014**, *11*, 909–926. [CrossRef]

42. Jiménez-Perálvarez, J.D.; Irigaray, C.; El Hamdouni, R.; Chacón, J. Landslide-susceptibility mapping in a semi-arid mountain environment: An example from the southern slopes of Sierra Nevada (Granada, Spain). *Bull. Eng. Geol. Environ.* **2011**, *70*, 265–277. [CrossRef]

43. Nagarajan, R.; Roy, A.; Kumar, R.V.; Mukherjee, A.; Khire, M. V Landslide hazard susceptibility mapping based on terrain and climatic factors for tropical monsoon regions. *Bull. Eng. Geol. Environ.* **2000**, *58*, 275–287. [CrossRef]

44. Gallardo-Cruz, J.A.; Pérez-García, E.A.; Meave, J.A. β-Diversity and vegetation structure as influenced by slope aspect and altitude in a seasonally dry tropical landscape. *Landsc. Ecol.* **2009**, *24*, 473–482. [CrossRef]

45. Geroy, I.J.; Gribb, M.M.; Marshall, H.-P.; Chandler, D.G.; Benner, S.G.; McNamara, J.P. Aspect influences on soil water retention and storage. *Hydrol. Process.* **2011**, *25*, 3836–3842. [CrossRef]

46. Kornejady, A.; Ownegh, M.; Bahremand, A. Landslide susceptibility assessment using maximum entropy model with two different data sampling methods. *Catena* **2017**, *152*, 144–162. [CrossRef]

47. Kornejady, A.; Ownegh, M.; Rahmati, O.; Bahremand, A. Landslide susceptibility assessment using three bivariate models considering the new topo-hydrological factor: HAND. *Geocarto Int.* **2018**, *33*, 1155–1185. [CrossRef]

48. Ercanoglu, M.; Gokceoglu, C. Assessment of landslide susceptibility for a landslide-prone area (north of Yenice, NW Turkey) by fuzzy approach. *Environ. Geol.* **2002**, *41*, 720–730.

49. Sidle, R.C.; Ochiai, H. Landslides: Processes, Prediction, and Land Use. In *Water Resources Monograph*; American Geophysical Union: Washington, DC, USA, 2006; Volume 18, p. 307. [CrossRef]

50. Yalcin, A. GIS-based landslide susceptibility mapping using analytical hierarchy process and bivariate statistics in Ardesen (Turkey): Comparisons of results and confirmations. *Catena* **2008**, *72*, 1–12. [CrossRef]

51. Vahidnia, M.H.; Alesheikh, A.A.; Alimohammadi, A.; Hosseinali, F. A GIS-based neuro-fuzzy procedure for integrating knowledge and data in landslide susceptibility mapping. *Comput. Geosci.* **2010**, *36*, 1101–1114. [CrossRef]

52. Meinhardt, M.; Fink, M.; Tünschel, H. Landslide susceptibility analysis in central Vietnam based on an incomplete landslide inventory: Comparison of a new method to calculate weighting factors by means of bivariate statistics. *Geomorphology* **2015**, *234*, 80–97. [CrossRef]

53. Tehrany, M.S.; Pradhan, B.; Jebur, M.N. Flood susceptibility mapping using a novel ensemble weights-of-evidence and support vector machine models in GIS. *J. Hydrol.* **2014**, *512*, 332–343. [CrossRef]

54. Tehrany, M.S.; Lee, M.-J.; Pradhan, B.; Jebur, M.N.; Lee, S. Flood susceptibility mapping using integrated bivariate and multivariate statistical models. *Environ. Earth Sci.* **2014**, *72*, 4001–4015. [CrossRef]

55. Khosravi, K.; Nohani, E.; Maroufinia, E.; Pourghasemi, H.R. A GIS-based flood susceptibility assessment and its mapping in Iran: A comparison between frequency ratio and weights-of-evidence bivariate statistical models with multi-criteria decision-making technique. *Nat. Hazards* **2016**, *83*, 947–987. [CrossRef]

56. Jenness, J. *DEM Surface Tools*; Jenness Enterp.: Flagstaff, AZ, USA, 2013. Available online: http://www.jennessent.com/arcgis/surface_area.htm (accessed on 20 May 2013).

57. Maestre, F.T.; Cortina, J. Spatial patterns of surface soil properties and vegetation in a Mediterranean semi-arid steppe. *Plant Soil* **2002**, *241*, 279–291. [CrossRef]

58. Zucca, C.; Canu, A.; Della Peruta, R. Effects of land use and landscape on spatial distribution and morphological features of gullies in an agropastoral area in Sardinia (Italy). *Catena* **2006**, *68*, 87–95. [CrossRef]

59. Cosby, B.J.; Hornberger, G.M.; Clapp, R.B.; Ginn, T. A statistical exploration of the relationships of soil moisture characteristics to the physical properties of soils. *Water Resour. Res.* **1984**, *20*, 682–690. [CrossRef]

60. Gyssels, G.; Poesen, J.; Nachtergaele, J.; Govers, G. The impact of sowing density of small grains on rill and ephemeral gully erosion in concentrated flow zones. *Soil Tillage Res.* **2002**, *64*, 189–201. [CrossRef]
61. Vandekerckhove, L.; Poesen, J.; Govers, G. Medium-term gully headcut retreat rates in Southeast Spain determined from aerial photographs and ground measurements. *Catena* **2003**, *50*, 329–352. [CrossRef]
62. Moore, I.D.; Grayson, R.B.; Ladson, A.R. Digital terrain modelling: A review of hydrological, geomorphological, and biological applications. *Hydrol. Process.* **1991**, *5*, 3–30. [CrossRef]
63. Grabs, T.; Seibert, J.; Bishop, K.; Laudon, H. Modeling spatial patterns of saturated areas: A comparison of the topographic wetness index and a dynamic distributed model. *J. Hydrol.* **2009**, *373*, 15–23. [CrossRef]
64. Jungerius, P.D.; Matundura, J.; Van De Ancker, J.A.M. Road construction and gully erosion in West Pokot, Kenya. *Earth Surf. Process. Landf.* **2002**, *27*, 1237–1247. [CrossRef]
65. Pourghasemi, H.R.; Moradi, H.R.; Fatemi Aghda, S.M. Landslide susceptibility mapping by binary logistic regression, analytical hierarchy process, and statistical index models and assessment of their performances. *Nat. Hazards* **2013**, *69*, 749–779. [CrossRef]
66. Pourtaghi, Z.S.; Pourghasemi, H.R. GIS-based groundwater spring potential assessment and mapping in the Birjand Township, southern Khorasan Province, Iran. *Hydrogeol. J.* **2014**, *22*, 643–662. [CrossRef]
67. Bui, D.T.; Pradhan, B.; Lofman, O.; Revhaug, I.; Dick, O.B. Spatial prediction of landslide hazards in Hoa Binh province (Vietnam): A comparative assessment of the efficacy of evidential belief functions and fuzzy logic models. *Catena* **2012**, *96*, 28–40.
68. Nefeslioglu, H.A.; Gokceoglu, C.; Sonmez, H. An assessment on the use of logistic regression and artificial neural networks with different sampling strategies for the preparation of landslide susceptibility maps. *Eng. Geol.* **2008**, *97*, 171–191. [CrossRef]
69. Kakembo, V.; Xanga, W.W.; Rowntree, K. Topographic thresholds in gully development on the hillslopes of communal areas in Ngqushwa Local Municipality, Eastern Cape, South Africa. *Geomorphology* **2009**, *110*, 188–194. [CrossRef]
70. Böhner, J.; Selige, T. Spatial Prediction of Soil Attributes Using Terrain Analysis and Climate Regionalisation. In *SAGA—Analysis and Modelling Applications*; Verlag Erich Goltze GmbH: Göttingen, Germany, 2006; Volume 115.
71. Song, Y.; Gong, J.; Gao, S.; Wang, D.; Cui, T.; Li, Y.; Wei, B. Susceptibility assessment of earthquake-induced landslides using Bayesian network: A case study in Beichuan, China. *Comput. Geosci.* **2012**, *42*, 189–199. [CrossRef]
72. Zhu, A.-X.; Wang, R.; Qiao, J.; Qin, C.-Z.; Chen, Y.; Liu, J.; Du, F.; Lin, Y.; Zhu, T. An expert knowledge-based approach to landslide susceptibility mapping using GIS and fuzzy logic. *Geomorphology* **2014**, *214*, 128–138. [CrossRef]
73. Vaezi, A.R.; Sadeghi, S.H.R.; Bahrami, H.A.; Mahdian, M.H. Modeling the USLE K-factor for calcareous soils in northwestern Iran. *Geomorphology* **2008**, *97*, 414–423. [CrossRef]
74. O'Callaghan, J.F.; Mark, D.M. The extraction of drainage networks from digital elevation data. *Comput. Vis. Graph. Image Process.* **1984**, *28*, 323–344. [CrossRef]
75. Melton, M.A. The geomorphic and paleoclimatic significance of alluvial deposits in southern Arizona. *J. Geol.* **1965**, *73*, 1–38. [CrossRef]
76. Marchi, L.; Dalla Fontana, G. GIS morphometric indicators for the analysis of sediment dynamics in mountain basins. *Environ. Geol.* **2005**, *48*, 218–228. [CrossRef]
77. De Reu, J.; Bourgeois, J.; Bats, M.; Zwertvaegher, A.; Gelorini, V.; De Smedt, P.; Chu, W.; Antrop, M.; De Maeyer, P.; Finke, P. Application of the topographic position index to heterogeneous landscapes. *Geomorphology* **2013**, *186*, 39–49. [CrossRef]
78. Daoud, J.I. Multicollinearity and Regression Analysis. *J. Phys. Confer. Ser.* **2017**, *949*, 12009. [CrossRef]
79. O'brien, R.M. A caution regarding rules of thumb for variance inflation factors. *Qual. Quant.* **2007**, *41*, 673–690. [CrossRef]
80. Ozdemir, A. Using a binary logistic regression method and GIS for evaluating and mapping the groundwater spring potential in the Sultan Mountains (Akshir, Turkey). *J. Hydrol.* **2011**, *405*, 123–136. [CrossRef]
81. Leathwick, J.R.; Rowe, D.; Richardson, J.; Elith, J.; Hastie, T. Using multivariate adaptive regression splines to predict the distributions of New Zealand's freshwater diadromous fish. *Freshw. Biol.* **2005**, *50*, 2034–2052. [CrossRef]

82. Felicísimo, Á.M.; Cuartero, A.; Remondo, J.; Quirós, E. Mapping landslide susceptibility with logistic regression, multiple adaptive regression splines, classification and regression trees, and maximum entropy methods: A comparative study. *Landslides* **2013**, *10*, 175–189. [CrossRef]

83. Milborrow, S.; Hastie, T.; Tibshirani, R. Earth: Multivariate Adaptive Regression Spline Models; R Software Package. 2019. Available online: http://www.milbo.users.sonic.net/earth (accessed on 12 April 2019).

84. Demšar, J.; Curk, T.; Erjavec, A.; Gorup, Č.; Hočevar, T.; Milutinovič, M.; Možina, M.; Polajnar, M.; Toplak, M.; Starič, A. Orange: Data mining toolbox in Python. *J. Mach. Learn. Res.* **2013**, *14*, 2349–2353.

85. Galili, T. dendextend: An R package for visualizing, adjusting and comparing trees of hierarchical clustering. *Bioinformatics* **2015**, *31*, 3718–3720. [CrossRef]

86. Lombardo, L.; Cama, M.; Conoscenti, C.; Märker, M.; Rotigliano, E. Binary logistic regression versus stochastic gradient boosted decision trees in assessing landslide susceptibility for multiple-occurring landslide events: Application to the 2009 storm event in Messina (Sicily, southern Italy). *Nat. Hazards* **2015**, *79*, 1621–1648. [CrossRef]

87. Umar, Z.; Pradhan, B.; Ahmad, A.; Jebur, M.N.; Tehrany, M.S. Earthquake induced landslide susceptibility mapping using an integrated ensemble frequency ratio and logistic regression models in West Sumatera Province, Indonesia. *Catena* **2014**, *118*, 124–135. [CrossRef]

88. Oh, H.-J.; Pradhan, B. Application of a neuro-fuzzy model to landslide-susceptibility mapping for shallow landslides in a tropical hilly area. *Comput. Geosci.* **2011**, *37*, 1264–1276. [CrossRef]

89. Mouton, A.M.; De Baets, B.; Goethals, P.L.M. Ecological relevance of performance criteria for species distribution models. *Ecol. Model.* **2010**, *221*, 1995–2002. [CrossRef]

90. Shirzadi, A.; Soliamani, K.; Habibnejhad, M.; Kavian, A.; Chapi, K.; Shahabi, H.; Chen, W.; Khosravi, K.; Thai Pham, B.; Pradhan, B. Novel GIS based machine learning algorithms for shallow landslide susceptibility mapping. *Sensors* **2018**, *18*, 3777. [CrossRef] [PubMed]

91. Saito, H.; Nakayama, D.; Matsuyama, H. Comparison of landslide susceptibility based on a decision-tree model and actual landslide occurrence: The Akaishi Mountains, Japan. *Geomorphology* **2009**, *109*, 108–121. [CrossRef]

92. Kontijevskis, A.; Wikberg, J.E.S.; Komorowski, J. Computational proteomics analysis of HIV-1 protease interactome. *Proteins Struct. Funct. Bioinform.* **2007**, *68*, 305–312. [CrossRef] [PubMed]

93. Dai, Q.; Yang, Y.; Wang, T. Markov model plus k-word distributions: A synergy that produces novel statistical measures for sequence comparison. *Bioinformatics* **2008**, *24*, 2296–2302. [CrossRef] [PubMed]

94. Park, S.; Choi, C.; Kim, B.; Kim, J. Landslide susceptibility mapping using frequency ratio, analytic hierarchy process, logistic regression, and artificial neural network methods at the Inje area, Korea. *Environ. Earth Sci.* **2013**, *68*, 1443–1464. [CrossRef]

95. Naghibi, S.A.; Dashtpagerdi, M.M. Evaluation of four supervised learning methods for groundwater spring potential mapping in Khalkhal region (Iran) using GIS-based features. *Hydrogeol. J.* **2017**, *25*, 169–189. [CrossRef]

96. Rahmati, O.; Naghibi, S.A.; Shahabi, H.; Bui, D.T.; Pradhan, B.; Azareh, A.; Rafiei-Sardooi, E.; Samani, A.N.; Melesse, A.M. Groundwater spring potential modelling: Comprising the capability and robustness of three different modeling approaches. *J. Hydrol.* **2018**, *565*, 248–261. [CrossRef]

97. Hosmer, D.W.; Lemeshow, S. Wiley Series in Probability and Statistics. Applied Logistic Regression. 1989. Available online: https://www.worldcat.org/title/applied-logistic-regression/oclc/19514573 (accessed on 8 June 1989).

98. Pourghasemi, H.R.; Yousefi, S.; Kornejady, A.; Cerdà, A. Performance assessment of individual and ensemble data-mining techniques for gully erosion modeling. *Sci. Total Environ.* **2017**, *609*, 764–775. [CrossRef]

99. Youden, W.J. Index for rating diagnostic tests. *Cancer* **1950**, *3*, 32–35. [CrossRef]

100. Ayalew, L.; Yamagishi, H. The application of GIS-based logistic regression for landslide susceptibility mapping in the Kakuda-Yahiko Mountains, Central Japan. *Geomorphology* **2005**, *65*, 15–31. [CrossRef]

101. Akgun, A.; Dag, S.; Bulut, F. Landslide susceptibility mapping for a landslide-prone area (Findikli, NE of Turkey) by likelihood-frequency ratio and weighted linear combination models. *Environ. Geol.* **2008**, *54*, 1127–1143. [CrossRef]

102. Conoscenti, C.; Agnesi, V.; Cama, M.; Caraballo-Arias, N.A.; Rotigliano, E. Assessment of Gully Erosion Susceptibility Using Multivariate Adaptive Regression Splines and Accounting for Terrain Connectivity. *Land Degrad. Dev.* **2018**, *29*, 724–736. [CrossRef]

103. Poesen, J.; Nachtergaele, J.; Verstraeten, G.; Valentin, C. Gully erosion and environmental change: Importance and research needs. *Catena* **2003**, *50*, 91–133. [CrossRef]
104. Dewitte, O.; Daoudi, M.; Bosco, C.; Van Den Eeckhaut, M. Predicting the susceptibility to gully initiation in data-poor regions. *Geomorphology* **2015**, *228*, 101–115. [CrossRef]

Article

Spatial Pattern Analysis and Prediction of Gully Erosion Using Novel Hybrid Model of Entropy-Weight of Evidence

Alireza Arabameri [1,*]**, Artemi Cerda** [2] **and John P. Tiefenbacher** [3]

[1]   Department of Geomorphology, Tarbiat Modares University, Tehran 36581-17994, Iran
[2]   Soil Erosion and Degradation Research Group, Departament de Geografia, Universitat de València, Blasco Ibàñez, 28, 46010-Valencia, Spain; artemio.cerda@uv.es
[3]   Department of Geography, Texas State University, San Marcos, TX 78666, USA; tief@txstate.edu
*   Correspondence: Alireza.ameri91@yahoo.com; Tel.: +98-9191735198

Received: 24 April 2019; Accepted: 24 May 2019; Published: 29 May 2019

Abstract: Gully erosion is an environmental problem in arid and semi-arid areas. Gullies threaten the soil and water resources and cause off- and on-site problems. In this research, a new hybrid model combines the index-of-entropy (IoE) model with the weight-of-evidence (WoE) model. Remote sensing and GIS techniques are used to map gully-erosion susceptibility in the watershed of the Bastam district of Semnan Province in northern Iran. The performance of the hybrid model is assessed by comparing the results with from models that use only IoE or WoE. Three hundred and three gullies were mapped in the study area and were randomly classified into two groups for training (70% or 212 gullies) and validation (30% or 91 gullies). Eighteen topographical, hydrological, geological, and environmental conditioning factors were considered in the modeling process. Prediction-rate curves (PRCs) and success-rate curves (SRCs) were used for validation. Results from the IoE model indicate that drainage density, slope, and rainfall factors are the most important factors promoting gullying in the study area. Validation results indicate that the ensemble model performed better than either the IoE or WoE models. The hybrid model predicted that 38.02 percent of the study area has either high or very high susceptible to gullying. Given the high accuracy of the novel hybrid model, this scientific methodology may be very useful for land use management decisions and for land use planning in gully-prone regions. Our research contributes to achieve Land Degradation Neutrality as will help to design remediation programs to control non-sustainable soil erosion rates.

Keywords: soil; natural resources; modeling; hybrid model; Bastam watershed

---

1. Introduction

Soil erosion in arid and semi-arid regions is one of the important factors that should be taken into consideration in land use planning [1–3]. Soil erosion is a major consequence of environmental and ecological change [4–6]. Water soil erosion has long been of interest to soil conservation researchers [1]. Topography, types of landforms, and resistance to erosion are the main determinants of erosion type and severity, and this is especially the case with linear erosion processes such as gullies [7].

Among the types of erosion, gully erosion most threatens numerous environmental resources and land use sustainability. This threat is not limited to soil degradation, changes in the landscape and/or landcover, limitations of agricultural activities, or the economic exploitation of natural resources. Gully erosion also promotes the initiation and expansion of badlands, promotes floods, lowering of water tables, desertification, and production and transportation of significant volumes of sediments along the watersheds to the coastal and lowlands [8–10].

A gully is a drainage channel with steeply sloping sidewalls and an active, steep eroding head-cut caused by fluctuating surface flows (i.e., during or after severe rainfall) [11]. Most scholars regard the destruction of natural ecosystems, non-sustainable land uses, degradation of vegetative cover, overgrazing, climate change, geological conditions, and human disturbance of natural systems to be the main causes of gullying [12–19].

Some studies have found that the range of sediment production attributable to gully erosion is 10% to 94% in different ecosystems and that spatial and temporal factors, soil types, land uses, topography, climate, and others can also influence sediment production [11]. A study of 22 watersheds in Spain determined that annual sediment production in areas without gullying is 0.74 Mg/ha, but in similar environments where gullying is occurring the annual production is 2.97 Mg/ha [20]. A single gully can generate sediments annually at a rate of up to 93,750 Mg/km [21].

Because the erosion is so fast at the head of a gully, there are no measures that can be taken to control or stop head-cut development; the measures that are taken are likely to only slow down their growth [22]. To prevent the rapid growth of gullies or to minimize the damage they cause, it is vital to understand the morphology of a gully, the manner in which it forms, and the causes for its development. Studying the amount of development of gully head-cut erosion in a specific timeframe can enable the prediction of the rate of expansion and growth, and this can also improve damage estimates. Identification of areas that are more susceptible to gully erosion is therefore possible. Accurate and adequate information about the type and extent of headward head-cut erosion can enable better land use decisions and more effective environmental management [23].

For more than a half-century, prediction and modeling of soil erosion have been a valuable component of soil conservation and engineering planning [24]. Mathematical simulations can be used to estimate erosion that is caused by an array of independent factors. Geomorphologic models are empirically based and must not include theoretical assumptions of relationships. They also must avoid scenarios and should not ignore multicollinearity among conditioning factors for the phenomenon [25]. Without a clear explanation of a geomorphological conceptual model, it is impossible to identify the appropriate sub-models that can numerically express the phenomenon [25]. The solution is to define the modeling goals and to identify the unknowns about the phenomenon; only then can a conceptual model for the desired phenomenon be developed, tested, and validated [26]. Modeling of soil erosion requires many empirical measurements of a landscape. Remote sensing techniques that produce aerial photographs, satellite data, or radar data, have eased the difficulty of this otherwise laborious problem [27].

In the last decades, several gully-erosion models have been developed to aid gully- erosion susceptibility mapping (GESM). These models can be classified into three groups: knowledge based models like the analytic hierarchy process (AHP) [16]; bivariate and multivariate statistical models like conditional probability (CP) [28], information value (IV) [29], frequency ratio (FR) [13], index of entropy (IoE) [30], evidential belief function (EBF) [14], weights-of-evidence (WOE) [31], certainty factor (CF) [32], and logistic regression (LR) [33]; and machine-learning models like maximum entropy (ME) [19], multivariate adaptive regression spline (MARS) [15], artificial neural network (ANN) [34], boosted regression tree (BRT) [35], linear discriminant analysis (ADA) [17], bagging best-first decision tree (Bag-BFTree) [36], random forest (RF) [37], flexible discriminant analysis (FDA) [38], support vector machine (SVM) [39], and classification and regression trees (CART) [15].

Each model has disadvantages and advantages. Therefore, improve performance of gully-erosion models; this study uses a hybrid approach to identify gully-erosion susceptible areas. A region that is heavily impacted by gully erosion is the Bastam district watershed in Semnan Province, of northern Iran. Gully erosion occurs in this area due to the highly erodible soils of these lands and misuse of soil and water resources. This has led to the destruction and transformation of agricultural lands into wastelands, the destruction of communication infrastructure, and damage to residential areas. In this study, two models of WoE and IoE are combined to map gully-erosion susceptibility in the Bastam district watershed. These two models have been previously used in studies of other fields of destructive geomorphological and hydrological phenomena, such as landslides [40–42], floods [43], and groundwater depletion [44,45].

2. Materials and Methods

2.1. Study Area

The Bastam watershed covers a 1329 km^2 area (36°27′02″ to 36°47′13″ N, and 54°24′23″ to 55°11′08″ E (Figure 1). Elevation in the study area ranges from 1357 m a.s.l. to 3893 m a.s.l. The mean elevation is 278.34 m a.s.l.). The minimum and maximum slopes are 0° and 70.66°; the mean is 13.55°. The central and south parts of the watershed have relatively smooth topography with gentle slopes. The rest of the study area is mountainous; is part of the Alborz Mountains. The watershed experiences a mean annual rainfall amount of 262 mm. The mean annual temperature is 12.8 °C [46]. More than 25% of the land is poor rangeland. Another 22.84% is covered by a relatively sparse forest, while 0.01% is covered in dense forest and 0.01% is land on which dryland-farming is practiced. The main lithographic units include high elevation piedmont fans and valley terrace deposits, stream channels, braided channels, floodplain deposits, and low elevation piedmont fans, and valley terrace deposits [47]. Rock outcrops, entisols, entisols/inceptisols, and mollisols are the most common geological and pedological surfaces in the study area [48]. Morphometric analysis of gullies in the study area shows that approximately 9.3% of the study area is affected by gully erosion. This indicates that the study area is very susceptible to gully erosion. There are more gullies in the south-central portion of the study area where slopes are lower than in the northern parts of the study area where slopes are steep and rocky outcrops are abundant and few gullies form. Gullies range in length from several meters to several hundred meters and their depths can be several meters. Gully width also varies, ranging from several centimeters to several meters. Gullies in the northern parts of the study area have V-shaped cross-sections, whereas in the central and southern parts of the region, due to more erodible soils, concentrated runoff because of low slope, and the more resistant sediments on the floor of the watershed prevent erosion, so the valleys are U-shaped.

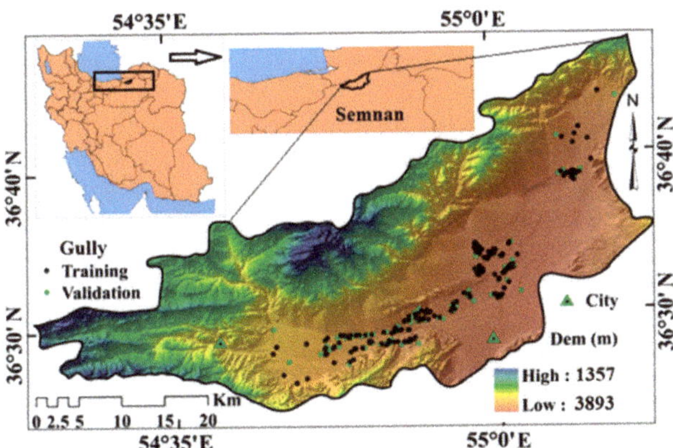

Figure 1. Location of the study area in Iran and Semnan province and locations of gullies used for training (black dots) and validation (green dots).

2.2. Methodology

This study involved six steps (Figure 2): data preparation, including the creation of a GEIM, compilation of thematic layers depicting the spatial distribution of gully-erosion conditioning factors (GECFs), and division of GEIM into training and validation groups; multi-collinearity analyses of the GECFs using indices of tolerance (TOL) and variance inflation factor (VIF); application of the WoE model to analyze the spatial relationship between GECFs and gully locations; determination of the importance of the GECFs using the IoE model; preparation of the GESM using the WoE and IoE models separately; and integration of the hybrid WoE-IoE model; and validation using the area-under-the-curve receiver-operating characteristic (AUROC).

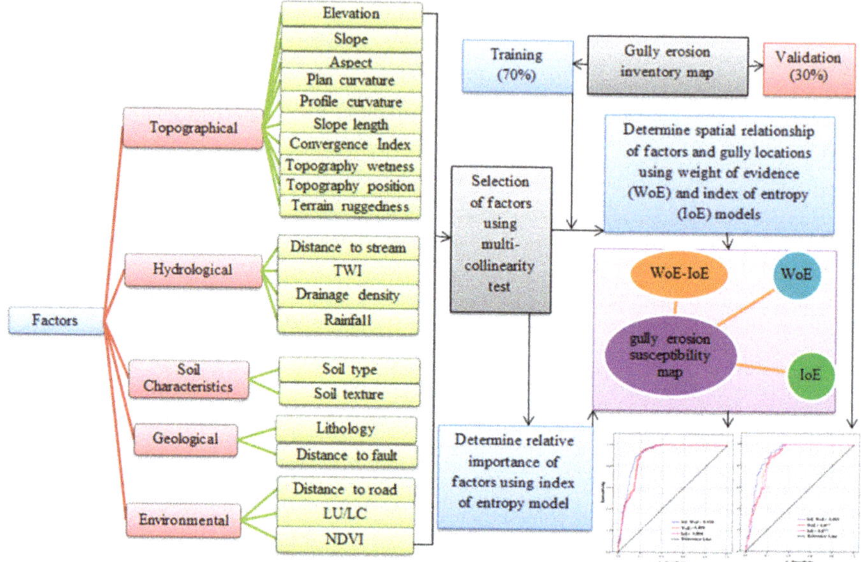

Figure 2. Flowchart of research in the present study for gully erosion susceptibility mapping.

2.2.1. Gully-erosion Inventory Map (GEIM)

A GEIM shows the spatial distribution of gullies and this is necessary to determine the spatial distribution of susceptibility to gullying [15]. Data acquired from the Agricultural and Natural Resources Research Center of Semnan Province were combined with extensive field surveys (Figure 3) and interpretation of satellite imagery to map gully erosion in the study area. Three hundred and three gullies were identified in the study area (Figure 1) and this set was randomly separated into two groups for modeling (70% or 212 gullies) and validation (30% or 91 gullies). In addition, an equivalent number (303) of locations without gullies were randomly selected using the random-point tool in ArcGIS to support calibration and validation [49]. The maximum and minimum lengths of the 303 gullies are 346 m and 0.74 m, respectively, whereas, the maximum depth is 7.2 m and the minimum depth is 0.65 m. The maximum width is 16.6 m and the minimum width is 0.74 m.

Figure 3. Sample of mapped gullies in the study area.

2.2.2. Gully Erosion Conditioning Factors (GECFs)

Gullying results from the interaction of several environmental factors that include hydrology, topography, land use/land cover, soil characteristics, human activities, and rainfall [50]. In this study, 18 GECFs were selected and mapped: elevation (m), slope (°), plan curvature (100/m), aspect, convergence index (100/m), slope length (LS), profile curvature (100/m), topography position index (TPI), terrain ruggedness index (TRI), topography wetness index (TWI), stream power index (SPI), rainfall (mm), distance to road (m), distance to stream (m), drainage density (km/km^2), land use/land cover (LU/LC), soil type, and lithology (Figure 4a–r). These factors were obtained from several sources. The geological map at the scale of 1:100,000 and made by the Geological Society of Iran (GSI) (http://www.gsi.ir/) was used to produce the lithology map. Ten lithological units in the study area were created based upon formation and susceptibility to gully erosion (Table 1). The topographic map at the scale of 1:50,000 were acquired from the National Geographic Organization of Iran (www.ngo-org.ir), and Google Earth images were used to digitize the road network on the topographic map. The ALOS DEM with 12.5 m resolution, downloaded from the Alaska Satellite Facility (ASF) Distributed Active Archive Center (DAAC) was used to extract topographic and hydrological data to map elevation, slope, aspect, plan curvature, profile curvature, slope length (LS), TWI, SPI, TPI, TRI, CI, distance to stream, and drainage density [29,51]. The reproduction of complex morphology and features depends both on accuracy and gridding techniques [52,53]. The quality of the reproduction influence the value of some of the topographical and hydrological gully-erosion conditioning factors, Therefore, in this research, ALOS DEM with a vertical accuracy of 0.3 m was used. Similar to the accuracy assessment procedures implemented by [54], vertical accuracies of the ALOS DEM were assessed by comparing the ALOS DEM elevations with those of the ground control points (GCPs). At each point, the DEM elevations were extracted using ArcGIS 10.5 software. Then, the differences in elevation were computed by subtracting the GCP elevation from its corresponding DEM elevations, and these differences are the measured errors in the ALOS DEM. For a particular DEM, positive errors represent locations where the DEM was above the GCP elevation, and negative errors occur at locations where the DEM was below the control point elevation. From these measured errors, the mean error and RMSE for each DEM were calculated, including standard deviations of the mean errors. The mean error (or bias) indicates if a DEM has an overall vertical offset (either positive or negative) from true ground level [54]. Finally, accuracy assessment results were analyzed. Details on how the ALOS DEM was produced using interferometric synthetic aperture radar (InSAR) are discussed in the papers of References [55,56]. The most important step in InSAR for DEM generation is the phase measurement, then the transformation of phase to height [55].

LU/LC was extracted from a land-use map prepared by the Iranian Soil Conservation and Watershed Management Research Institute (https://www.scwmri.ac.ir). The map of soil types was prepared by the Agricultural and Natural Resources Research Center of Semnan Province. To map annual rainfall, 30 years of annual data (1986 to 2016) were acquired for the meteorological stations at Mojen, Farahzad, Bastam, Abr, Karkhaneh, Semnan, Shahroud, Tarzeh, and Shah Kuh-e Bala. Kriging was used to prepare the final annual rainfall map in ArcGIS 10.5. ArcGIS v10.5, Arc Hydro v10.4, Google Earth Pro 7.8, and ENVI v4.8 toolboxes were used to prepare the conditioning factors for gully-erosion susceptibility assessment. Microsoft Excel v2016 and SPSS v24 were also used for exploratory, statistical, and validation analyses. Calculation of the GECFs is explained elsewhere [57–61].

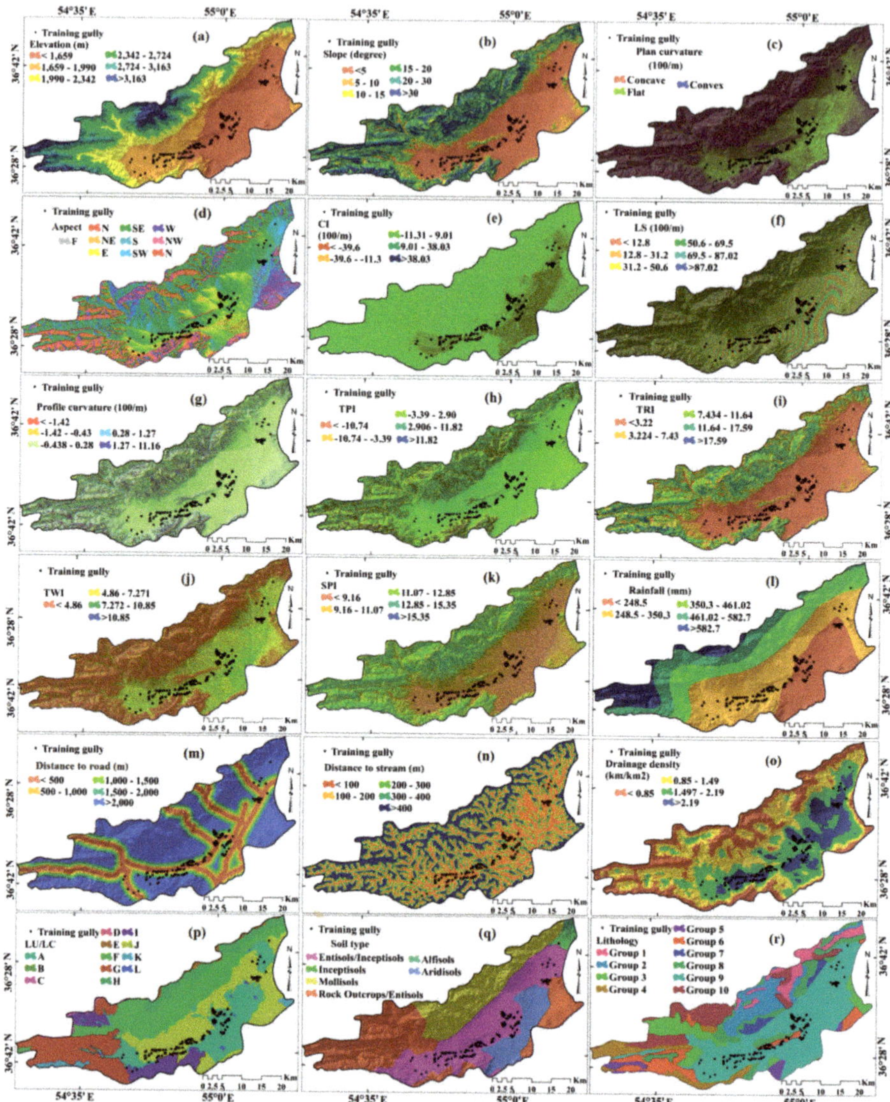

Figure 4. Gully erosion conditioning factors. (**a**) elevation, (**b**) slope, (**c**) plan curvature, (**d**) slope aspect, (**e**) convergence index (CI), (**f**) slope length (LS), (**g**) profile curvature, (**h**) topography position index (TPI), (**i**) terrain ruggedness index (TRI), (**j**) topography wetness index (TWI), (**k**) stream power index (SPI), (**l**) rainfall, (**m**) distance to road, (**n**) distance to stream, (**o**) drainage density, (**p**) land use/land cover, (**q**) soil type, and (**r**) lithology.

Table 1. The lithology of the study area.

Group	Unit	Description	Age
1	Cm	Dark grey to black fossiliferous limestone with subordinate black shale (MOBARAK FM)	Carboniferous
2	DCkh	Yellowish, thin to thick-bedded, fossileferous argillaceous limestone, dark grey limestone, greenish marl and shale, locally including gypsum	Devonian
3	Ea.bvs	Andesitic to basaltic volcano sediment	Eocene
	Ek	Well bedded green tuff and tuffaceous shale (KARAJ FM)	Eocene
	Ea.bv	Andesitic and basaltic volcanic	Eocene
4	Jl	Light grey, thin-bedded to massive limestone (LAR FM)	Jurassic-Cretaceous
	Jd	Well-bedded to thin-bedded, greenish-grey argillaceous limestone with intercalations of calcareous shale (DALICHAI FM)	Jurassic
5	Ku	Upper cretaceous, undifferentiated rocks	Late.Cretaceous
	K2m	Marl, shale and dendritic limestone	Late.Cretaceous
6	Murc	Red conglomerate and sandstone	Miocene
	Murmg	Gypsiferous marl	Miocene
	Mc	Red conglomerate and sandstone	Miocene
7	Osh	Greenish-grey siltstone and shale with intercalations of flaggy limestone (SHIRGESHT FM)	Ordovician
8	Pd	Red sandstone and shale with subordinate sandy limestone (DORUD FM)	Permian
	Plc	Polymictic conglomerate and sandstone	Pliocene
	Pz1a.bv	Andesitic basaltic volcanic	Paleozoic
	P	Undifferentiated Permian rocks	Permian
	PeEz	Reef-type limestone and gypsiferous marl (ZIARAT FM)	Paleocene-Eocene
	PlQc	Fluvial conglomerate, piedmont conglomerate and sandstone.	Pliocene-Quaternary
9	Qft1	High-level piedmont fan and valley terrace deposits	Quaternary
	Qal	Stream channel, braided channel and flood plain deposits	Quaternary
	Qft2	Low-level piedmont fan and valley terrace deposits	Quaternary
10	TRe2	Thick bedded dolomite	Early-Middle.Triassic
	TRJs	Dark grey shale and sandstone (SHEMSHAK FM)	Triassic-Jurassic

2.3. Multi-collinearity Analysis

Collinearity indicates that an independent variable is the linear function of another independent variable. If collinearity in a regression equation is high, there is a high correlation between independent variables and that reduces the accuracy of the model [15]. In this study, the indicators tolerance (TOL) and variance inflation factor (VIF) were used to assess collinearity among the independent variables. These indices do not have standardized thresholds, however, the literature reports that the following intervals have been widely used by several researchers: VIF $\leq$ 5 or 10 and TOL $\leq$ 0.1 or 0.2, imply that no collinearity is present and the gully conditioning factors are independent [15].

2.4. Models

2.4.1. Index of Entropy (IoE)

Entropy is a measurement of the instability, imbalance, disorder, and uncertainty of a system [62]. The value of entropy of a system has a one-to-one relationship with the degree of disorder. This relationship, called the Boltzmann principle, has been used to describe the thermodynamic status of a system [62]. Shannon improved upon the Boltzmann principle and established an entropy model for information theory. The IoE distinguishes the most important factors from less significant effective factors of a target; it identifies the variables that have the greatest impact on the occurrence of an event. While gully erosion is affected by several factors and gully-erosion susceptibility can be determined with a bivariate statistical model like the WoE, where all factors are weighted the same. Factors having a greater impact might be ignored. Therefore, the IoE can provide an important understanding of the conditioning factors and their impacts and the results can inform appropriate management [63]. The entropy of gully erosion refers to the extent that various factors influence the development of a gully. Several important factors provide additional entropy into the index system.

As a result, the entropy value can be used to calculate the objective weights of the index system. To determine the relative importance of GECFs in the gully occurrence and GESM using IoE model, following Equations (1)–(6) have been used [63]:

$$P_{ij} = \frac{b}{a} \tag{1}$$

$$\left(P_{ij}\right) = \frac{P_{ij}}{\sum_{j=1}^{Sj} P_{ij}} \tag{2}$$

$$H_j = -\sum_{i=1}^{Sj} \left(P_{ij}\right) \log_2 \left(P_{ij}\right), \qquad j = 1, \ldots, n \tag{3}$$

$$H_{j\,max} = \log_2 S_j \tag{4}$$

$$I_j = \frac{H_{j\,max} - H_j}{H_{j\,max}}, \, I = (0,1), \qquad j = 1, \ldots, n \tag{5}$$

$$W_{ij} = I_j \times P_{ij} \tag{6}$$

where a and b are the domain and gully erosion percentages, respectively, $\left(P_{ij}\right)$ is the probability density, H_j and $H_{j\,max}$ indicate the entropy values and maximum entropy, respectively, I_j is the information value, S_j is the number of classes and W_j indicates the resulting weight of each factor. W_{ij}'s values range from 0 to 1. After calculation of the final weight of each GECF and of their classes, these values were added to each related thematic layer and then each was weighted. Finally, they were summed to produce the final gully-erosion susceptibility map using Equation (7) and the weighted-sum tool in ArcGIS [43].

$$GESM = \sum_{I=1}^{n} \left(W_J \times P_j\right) \tag{7}$$

where W_j and P_j are the final weight and the probability density for the jth feature.

2.4.2. Weight of Evidence (WoE) Model

WoE is a bivariate statistical method, based on the Bayesian probability framework, to statistically estimate the relative importance of conditioning factors [41]. By overlaying gully erosion locations on each gully-related conditioning factor, the spatial relationship between them can be determined and the explanatory significance of the effective variable for past gullying can be evaluated. The WoE model calculates the positive (W^+) and negative (W^-) weights for each gully conditioning factor (A) based on the presence or absence of gully sites (B). This model measurement the conditioning factors within the study area as follows [64]:

$$W_i^+ = \text{In}\left(\frac{p\{B|AL\}}{P\left\{B|\overline{A}\right\}}\right) \tag{8}$$

$$W_i^- = \text{In}\left(\frac{P\left\{\overline{B}|A\right\}}{P\left\{\overline{B}|\overline{A}\right\}}\right) \tag{9}$$

P is the probability and In is the natural log function. B and $\overline{B}$ indicate the presence and absence of the gully conditioning factors, respectively. A is the presence of gully, and $\overline{A}$ is the absence of a gully. A positive weight (W^+) designates the fact that the conditioning factor is present at the gully locations and its value is an indication of the positive correlation between the presence of the gully conditioning factor and the gullies. Similarly, a negative weight (W^-) explains the absence of the gully conditioning factor and reflects the level of negative correlation. In GESM, the weight contrast (C) measures and specifies the spatial association between the effective factors and gully erosion occurrences. C is negative for a negative spatial relationship and positive for a positive relationship. The standard deviation S(C) of W is calculated by Equation (10):

$$S(C) = \sqrt{S^2(W^+) + S^2(W^-)} \tag{10}$$

where $S^2(W^+)$ is the variance of the W^+ and $S^2(W^-)$ is the variance of W^-. The variances of the weights can be determined as follows:

$$S^2(W^+) = \frac{1}{N\{B \cap L\}} + \frac{1}{\{B \cap L\}} \tag{11}$$

$$S^2(W^-) = \frac{1}{\{\overline{B} \cap L\}} + \frac{1}{\{\overline{B} \cap \overline{L}\}} \tag{12}$$

The studentized contrast (G_{final}) is a measure of confidence and is calculated, using the following equation:

$$G_{final} = \left(\frac{C}{S(C)}\right) \tag{13}$$

C indicates the overall association between a geo-environmental factor and gully occurrence. S(C) is the standard deviation of the contrast and W is the final weight.

After determining the relative weights of the GECFs using IoE model and the spatial relationships between GECFs and gullies in the study area using the WoE model, the two models are integrated to improve performance and decrease the disadvantages of each so that, relative weight of GECFs obtained by IoE multiple with weight of GECF classes obtained by WoE using Equation (14):

$$\begin{aligned}
GESM_{WoE-IoE} = &\left(WoE_{Elevation} \times Elevation_{IoE} + WoE_{Slope} \times Slope_{IoE} + \right. \\
WoE_{Aspect} \times Aspect_{IoE} &+ WoE_{SPI} \times SPI_{IoE} + WoE_{TWI} \times TWI_{IoE} + WoE_{TPI} \times TPI_{IoE} + \\
&\left. \ldots + WoE_{Lithology} \times Lithology_{IoE}\right).
\end{aligned} \tag{14}$$

2.5. Validation Method

Validation of the results is an important step in GESM [65]. The AUROC (area under the receiver operating characteristic) method is an efficient and accurate validation tool of GESMs [66]. This approach helps to visualize the quality of models by showing the incremental percentage of the model's prediction accuracy [67]. In the AUROC approach, the number of pixels correctly predicted by the model is plotted against the number of pixels incorrectly predicted. The AUROC is usually between 0.5 and 1; as the value approaches 1, the higher is the performance of the model [68]. AUROC values can be classified as performance ratings: 0.5–0.6 is poor, 0.6–0.7 is average, 0.7–0.8 is good, 0.8–0.9 is very good, and 0.9–1 is excellent [69].

3. Results

3.1. Multi-collinearity Analysis Among GECFs

Among the 21 GECFs studied, three factors (NDVI with tolerance (or TOL) = 0.095 and variance inflation factor (or VIF) = 10.47), distance to fault (TOL = 0.022 and VIF = 45.04), and soil texture (TOL = 0.03 and VIF = 33.76) have collinearity and cannot be used for modeling (Table 2). The values of TOL and VIF for the other nineteen factors ranged from 0.271 to 0.975 and 1.02 to 4.17, respectively, which indicates non-linearity between them. Therefore, nineteen factors were included in the modeling.

Table 2. Multi-collinearity test among gully erosion conditioning factors.

Factors	Collinearity Statistics		Factors	Collinearity Statistics	
	Tolerance	VIF		Tolerance	VIF
Aspect	0.902	1.109	SPI	0.584	1.919
CI	0.635	1.574	Soil texture	0.030	33.769
NDVI	0.095	10.479	TPI	0.392	2.553
Drainage density	0.322	3.107	TWI	0.374	3.509
Distance to road	0.685	1.460	Lithology	0.271	3.688
Distance to stream	0.604	1.655	SOIL	0.471	2.124
LS	0.975	1.025	LU	0.265	3.778
Plan curvature	0.494	2.025	Elevation	0.643	1.276
Profile curvature	0.508	1.968	TRI	0.426	2.265
Rainfall	0.289	4.179	Slope	0.543	1.498
Distance to fault	0.022	45.046	SPI	0.584	1.919

3.2. Determine the Spatial Relationship between GECFs and Occurred Gullies

The analysis of the spatial relationship between GECFs and the gullies occurred in the study area was conducted first using the IoE and WoE models (Table 3 and Figure 5). Values less than 1 in the IoE model indicate low correlations and values larger than 1 indicate higher correlations. In general, high values of IoE indicate higher probability of gully occurrence and lower values indicate less likelihood that gullying will occur [16]. Negative values generated by the WoE model indicate negative spatial relationships and positive values reflect positive relationships. Zero indicates that the specific class of conditioning factors is not useful in the analysis [70]. Classes of the factors elevation, slope, plan curvature, aspect, and CI have strong relationships with gullying in the IoE model: classes of <1659 m = 2.495, <5° = 2.499, flat curvature = 1.565, east-facing = 1.941, and CI < −39.6 = 2.831. In the WoE model, these same classes also show strong relationships with sites of gullying: <1659 m = 12.43, <5° = 5.746, flat curvature = 4.156, east-facing = 5.914, and CI < −39.6 = 6.463. With regard to the factors in the IoE model, LS, profile curvature, TPI, TRI, and TWI have strongly predictive classes: 50.6 to 69.5 m = 1.490, −0.43 to 0.28 = 1.425, −3.3 to 2.9 = 1.607, <3.22 = 2.134, and > 10.85 = 3.009. These same factors and class are similarly strongly correlated with gullying in the WoE model: 50.6 to 69.5 m = 2.880, −0.43 to 0.28 = 7.096, −3.3 to 2.9 = 4.849, <3.22 = 11.011, and >10.85 = 6.226. In the case of the IoE model, the factors SPI, rainfall, distance to road, distance to stream, and drainage density have classes that strongly indicate locations of highest susceptibility to gully erosion: > 15.35 = 1.910, <248.5 mm = 2.461, <500 m=2.214, <100 m = 1.627, and >2.19 km/km^2 = 3.385. In the WoE model these same classes indicated high susceptibility as well: > 15.35 = 2.636, <248.5 mm = 9.641, < 500 m = 7.139, <100 m = 5.500, and >2.19 km/km^2 = 15.005). The IoE model aligns classes of LU/LC, soil type, and lithology with higher susceptibility: poor rangeland = 2.583, entisols/inceptisols = 2.933, and Group 9 (consisting of quaternary formations that include high elevation piedmont fan and valley terrace deposits, stream channel, braided channel, and floodplain deposits, low elevation piedmont fan and valley terrace deposits = 2.237. The same classes in the WoE model are highly predictive of gullying, as well: poor rangeland = 11.992, entisols/inceptisols = 11.522, and group 9 = 11.297.

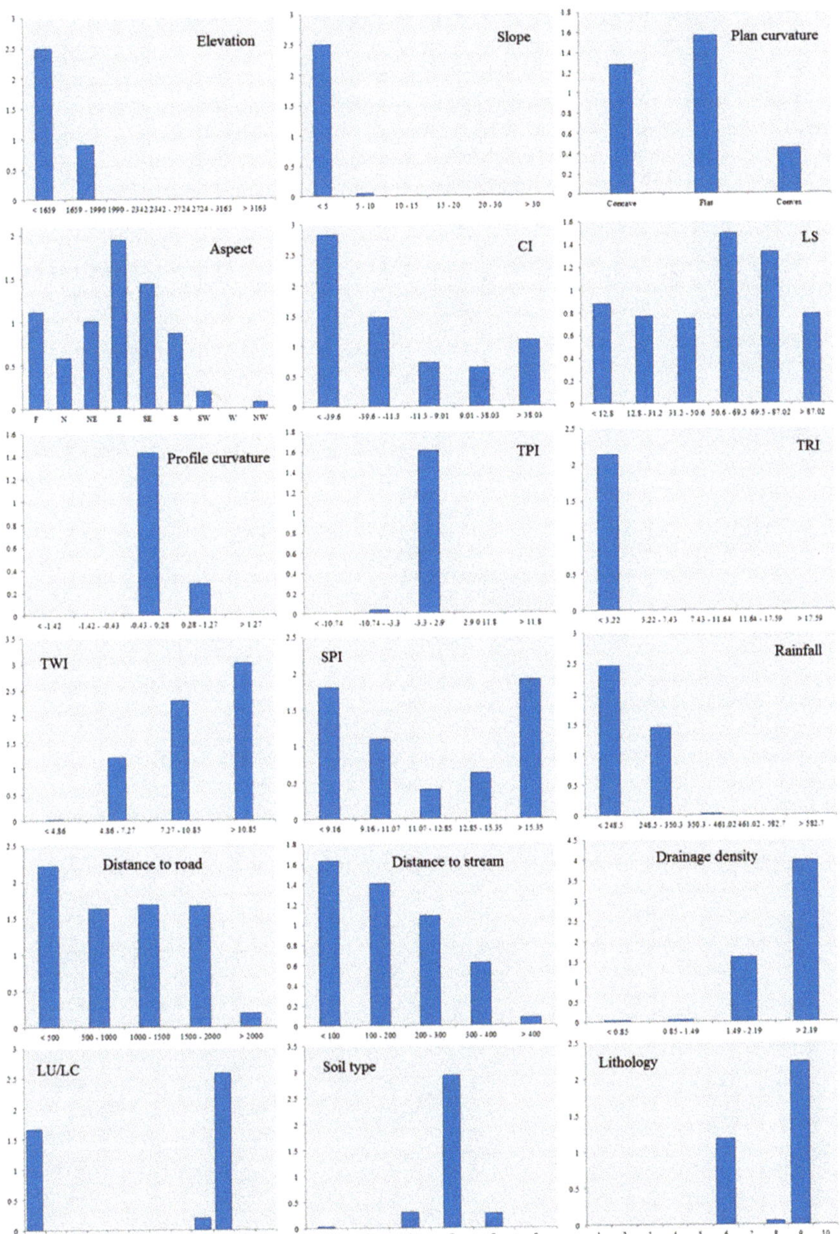

Figure 5. The spatial relationship between conditioning factors and gully locations.

Table 3. The spatial relationship between gully erosion conditioning factors and gully locations.

Factor	Value	Total Pixels		Total Gullies		Weight of Evidence		
		No	%	No	%	C	S©	C/S©
Elevation (m)	<1659	479814	32.482	171	81.043	2.185	0.176	12.439
	1659–1990	309671	20.964	40	18.957	−0.126	0.176	−0.716
	1990–2342	239854	16.238	0	0.000	−0.177	0.069	−2.574
	2342–2724	209109	14.156	0	0.000	−0.153	0.069	−2.218
	2724–3163	146912	9.946	0	0.000	−0.105	0.069	−1.522
	>3163	91792	6.214	0	0.000	−0.064	0.069	−0.932
Slope (°)	<5	588363	39.831	210	99.526	5.760	1.002	5.746
	5–10	149259	10.105	1	0.474	−3.162	1.002	−3.154
	10–15	124159	8.405	0	0.000	−0.088	0.069	−1.276
	15–20	143874	9.740	0	0.000	−0.102	0.069	−1.489
	20–30	280859	19.014	0	0.000	−0.211	0.069	−3.064
	>30	190637	12.906	0	0.000	−0.138	0.069	−2.007
PC	Concave	571885	38.715	104	49.289	0.431	0.138	3.129
	Flat	308591	20.891	69	32.701	0.610	0.147	4.156
	Convex	596675	40.394	38	18.009	−1.127	0.179	−6.289
Aspect	F	6262	0.424	1	0.474	0.112	1.002	0.112
	N	132304	8.957	11	5.213	−0.311	0.360	−0.862
	NE	152509	10.325	22	10.427	0.011	0.225	0.049
	E	245198	16.599	68	32.227	0.871	0.147	5.914
	SE	341071	23.090	70	33.175	0.503	0.146	3.441
	S	273880	18.541	34	16.114	−0.170	0.187	−0.906
	SW	144941	9.812	4	1.896	−1.728	0.505	−3.424
	W	89718	6.074	0	0.000	−0.063	0.069	−0.910
	NW	91269	6.179	1	0.474	−2.627	1.002	−2.621
CI	<−39.6	91462	6.223	37	17.619	1.171	0.181	6.463
	−39.6–−11.3	289807	19.717	61	29.048	0.511	0.152	3.362
	−11.3–9.01	663255	45.125	68	32.381	−0.541	0.147	−3.667
	9.01–38.03	341553	23.238	31	14.762	−0.559	0.195	−2.871
	>38.03	83751	5.698	13	6.190	0.088	0.286	0.308
LS (m)	<12.8	352071	23.834	44	20.853	−0.172	0.169	−1.015
	12.8–31.2	220640	14.937	24	11.374	−0.314	0.217	−1.446
	31.2–50.6	197564	13.375	21	9.953	−0.334	0.230	−1.454
	50.6–69.5	211467	14.316	45	21.327	0.484	0.168	2.880
	69.5–87.02	279653	18.932	53	25.118	0.362	0.159	2.282
	>87.02	215757	14.606	24	11.374	−0.287	0.217	−1.325
Profile curvature	<−1.42	39576	2.679	0	0.000	−0.027	0.069	−0.395
	−1.42–−0.43	150920	10.217	0	0.000	−0.108	0.069	−1.566
	−0.43–0.28	987589	66.858	201	95.261	2.299	0.324	7.096
	0.28–1.27	249999	16.924	10	4.739	−1.410	0.324	−4.351
	>1.27	49067	3.322	0	0.000	−0.034	0.069	−0.491
TPI	<−10.74	69120	4.679	0	0.000	−0.048	0.069	−0.696
	−10.74–−3.3	233435	15.803	1	0.474	−3.674	1.002	−3.666
	−3.3–2.9	914956	61.941	210	99.526	4.860	1.002	4.849
	2.9–11.8	201144	13.617	0	0.000	−0.146	0.069	−2.127
	>11.8	58497	3.960	0	0.000	−0.040	0.069	−0.587
TRI	<3.22	692266	46.865	211	100	0.758	0.069	11.011
	3.22–7.43	286708	19.410	0	0.000	−0.216	0.069	−3.135
	7.43–11.64	285453	19.325	0	0.000	−0.215	0.069	−3.120
	11.64–17.59	174864	11.838	0	0.000	−0.126	0.069	−1.830
	>17.59	37860	2.563	0	0.000	−0.026	0.069	−0.377
TWI	<4.86	582464	39.432	1	0.474	−4.918	1.002	−4.906
	4.86–7.27	585130	39.612	101	47.867	0.336	0.138	2.441
	7.27–10.85	237436	16.074	78	36.967	1.119	0.143	7.848
	>10.85	72122	4.883	31	14.692	1.211	0.194	6.226
SPI	<9.16	322599	21.839	83	39.336	0.842	0.141	5.975
	9.16–11.07	437271	29.602	68	32.227	0.123	0.147	0.835
	11.07–12.85	455113	30.810	26	12.322	−1.153	0.209	−5.507
	12.85–15.35	203521	13.778	18	8.531	−0.539	0.246	−2.185
	>15.35	58648	3.970	16	7.583	0.686	0.260	2.636

Table 3. *Cont.*

Factor	Value	Total Pixels		Total Gullies		Weight of Evidence		
		No	%	No	%	C	S©	C/S©
Rainfall (mm)	<248.5	284466	19.258	100	47.393	1.329	0.138	9.641
	248.5–350.3	535319	36.240	110	52.133	0.650	0.138	4.720
	350.3–461.02	321491	21.764	1	0.474	−4.068	1.002	−4.058
	461.02–582.7	208018	14.082	0	0.000	−0.152	0.069	−2.205
	>582.7	127857	8.656	0	0.000	−0.091	0.069	−1.315
Dis to road (m)	<500	224545	15.201	71	33.649	1.040	0.146	7.139
	500–1000	193768	13.118	45	21.327	0.585	0.168	3.483
	1000–1500	166957	11.303	40	18.957	0.608	0.176	3.459
	1500–2000	147431	9.981	35	16.588	0.584	0.185	3.157
	>2000	744451	50.398	20	9.479	−2.273	0.235	−9.670
Dis to stream (m)	<100	408797	27.675	95	45.024	0.761	0.138	5.500
	100–200	298966	20.239	60	28.436	0.449	0.153	2.939
	200–300	245504	16.620	38	18.009	0.097	0.179	0.542
	300–400	157383	10.654	14	6.635	−0.518	0.277	−1.872
	>400	366502	24.811	4	1.896	−2.838	0.505	−5.622
Drainage density (km/km2)	<0.85	398195	26.957	2	0.939	−3.662	0.710	−5.155
	0.85–1.49	497684	33.692	4	1.878	−3.279	0.505	−6.497
	1.49–2.19	368973	24.979	85	39.906	0.690	0.140	4.935
	>2.19	212300	14.372	122	57.277	2.078	0.139	15.005
LU/LC	Agriculture (A)	281159	19.034	67	31.754	0.683	0.148	4.617
	Dense-forest (B)	155	0.010	0	0.000	0.000	0.069	−0.002
	Good-range (C)	8660	0.586	0	0.000	−0.006	0.069	−0.085
	Agri-dryfarming (D)	243	0.016	0	0.000	0.000	0.069	−0.002
	Dryfarming (E)	16235	1.099	0	0.000	−0.011	0.069	−0.161
	Low-forest (F)	337471	22.846	0	0.000	−0.259	0.069	−3.768
	Woodland (G)	233619	15.816	0	0.000	−0.172	0.069	−2.501
	Mod-forest (H)	83527	5.655	0	0.000	−0.058	0.069	−0.846
	Mod-range (I)	105592	7.148	3	1.422	−1.675	0.581	−2.880
	Poor-range (J)	382114	25.868	141	66.825	1.753	0.146	11.992
	Rock (K)	23154	1.567	0	0.000	−0.016	0.069	−0.230
	Urban (L)	5220	0.353	0	0.000	−0.004	0.069	−0.051
Soil type	Rock Outcrop/Entisols (A)	486282	32.920	2	0.948	−3.938	0.710	−5.542
	Alfisols (B)	7086	0.480	0	0.000	−0.005	0.069	−0.070
	Aridisols (C)	161526	10.935	7	3.318	−1.275	0.384	−3.317
	Entisols/Inceptisols (D)	479698	32.475	201	95.261	3.733	0.324	11.522
	Inceptisols (E)	24777	1.677	1	0.474	−1.276	1.002	−1.273
	Mollisols (F)	317783	21.513	0	0.000	−0.242	0.069	−3.519
Lithology	Group 1	76475	5.177	0	0.000	−0.053	0.069	−0.772
	Group 2	131673	8.914	0	0.000	−0.093	0.069	−1.356
	Group 3	114748	7.768	0	0.000	−0.081	0.069	−1.175
	Group 4	94149	6.374	0	0.000	−0.066	0.069	−0.957
	Group 5	33722	2.283	0	0.000	−0.023	0.069	−0.336
	Group 6	117907	7.982	20	9.479	0.188	0.235	0.801
	Group 7	31564	2.137	0	0.000	−0.022	0.069	−0.314
	Group 8	134059	9.076	1	0.474	−3.043	1.002	−3.036
	Group 9	594531	40.248	190	90.047	2.598	0.230	11.297
	Group 10	148324	10.041	0	0.000	−0.106	0.069	−1.537

3.3. Determining the Relative Importance of GECFs in the IoE Model

The order of importance of the GECFs in the IoE model (Figure 6) is: drainage density (0.564), slope (0.48), rainfall (0.448), TRI (0.427), soil type (0.391), elevation (0.383), TWI (0.373), TPI (0.309), LU/LC (0.248), profile curvature (0.246), lithology (0.24), distance to road (0.136), CI (0.132), distance to stream (0.13), slope aspect (0.117), plan curvature (0.107), SPI (0.105), and LS (0.024).

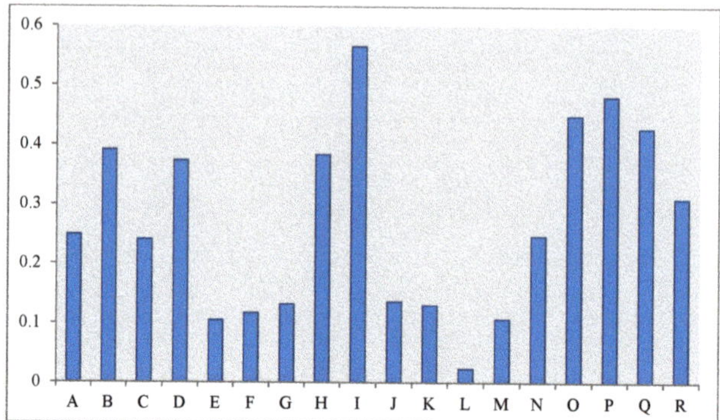

Figure 6. The relative importance of conditioning factors in the IoE model. A: landuse/land cover, B: Soil type, C: lithology, D: topography wetness index (TWI), E: stream power index (SPI), F: aspect, G: convergence index (CI), H: elevation, I: drainage density, J: distance to stream, K: distance to stream, L: slope length (LS), M: plan curvature, N: profile curvature, O: rainfall, P: slope, Q: terrain ruggedness index (TRI), R) topography position index (TPI).

3.4. Gully Erosion Susceptibility Mapping (GESM)

After the computation of the spatial relationship between gully erosion and conditioning factors, and determination of the relative importance of GECFs in the IoE model, GESM calculations based on the WoE model Equation (15), the IoE model Equation (16), and the hybrid WoE- IoE model Equation (17) were constructed.

$$
\begin{aligned}
\text{GESM}_{\text{WoE}} = \ & (W_{\text{WoE}}\text{Elevation}) + (W_{\text{WoE}}\text{Slope }) + (W_{\text{WoE}}\text{Plan curvature}) \\
& + (W_{\text{WoE}}\text{Slope aspect}) + (W_{\text{WoE}}\text{CI}) + (W_{\text{WoE}}\text{LS}) \\
& + (W_{\text{WoE}}\text{Profile curvature}) + (W_{\text{WoE}}\text{TPI}) + (W_{\text{WoE}}\text{TRI}) \\
& + (W_{\text{WoE}}\text{TWI}) + (W_{\text{WoE}}\text{SPI}) + (W_{\text{WoE}}\text{Rainfall}) \\
& + (W_{\text{WoE}}\text{Distance to road}) + (W_{\text{WoE}}\text{Distance to stream}) \\
& + (W_{\text{WoE}}\text{Drinage density}) + (W_{\text{WoE}}\text{LU/LC}) + (W_{\text{WoE}}\text{Soil type}) \\
& + (W_{\text{WoE}}\text{Lithology})
\end{aligned}
\tag{15}
$$

$$
\begin{aligned}
\text{GESM}_{\text{IoE}} = \ & (W_{\text{IoE}}\text{Elevation} \times 0.383) + (W_{\text{IoE}}\text{Slope } \times 0.48) \\
& + (W_{\text{IoE}}\text{Plan curvature} \times 0.107) + (W_{\text{IoE}}\text{Slope aspect} \times 0.117) \\
& + (W_{\text{IoE}}\text{CI} \times 0.132) + (W_{\text{IoE}}\text{LS} \times 0.024) \\
& + (W_{\text{IoE}}\text{Profile curvature} \times 0.246) + (W_{\text{IoE}}\text{TPI} \times 0.309) \\
& + (W_{\text{IoE}}\text{TRI} \times 0.427) + (W_{\text{IoE}}\text{TWI} \times 0.373) + (W_{\text{IoE}}\text{SPI} \times 0.105) \\
& + (W_{\text{IoE}}\text{Rainfall} \times 0.448) + (W_{\text{IoE}}\text{Distance to road} \times 0.136) \\
& + (W_{\text{IoE}}\text{Distance to stream} \times 0.13) + (W_{\text{IoE}}\text{Drinage density} \times 0.565) \\
& + (W_{\text{IoE}}\text{LU/LC} \times 0.248) + (W_{\text{IoE}}\text{Soil type} \times 0.391) \\
& + (W_{\text{IoE}}\text{Lithology} \times 0.24)
\end{aligned}
\tag{16}
$$

$$
\begin{aligned}
\text{GESM}_{\text{WoE-IoE}} = \ & (W_{\text{WoE}}\text{Elevation} \times 0.383) + (W_{\text{WoE}}\text{Slope } \times 0.48) \\
& + (W_{\text{WoE}}\text{Plan curvature} \times 0.107) + (W_{\text{WoE}}\text{Slope aspect} \times 0.117) \\
& + (W_{\text{WoE}}\text{CI} \times 0.132) + (W_{\text{WoE}}\text{LS} \times 0.024) \\
& + (W_{\text{WoE}}\text{Profile curvature} \times 0.246) + (W_{\text{WoE}}\text{TPI} \times 0.309) \\
& + (W_{\text{WoE}}\text{TRI} \times 0.427) + (W_{\text{WoE}}\text{TWI} \times 0.373) \\
& + (W_{\text{WoE}}\text{SPI} \times 0.105) + (W_{\text{WoE}}\text{Rainfall} \times 0.448) \\
& + (W_{\text{WoE}}\text{Distance to road} \times 0.136) + (W_{\text{WoE}}\text{Distance to stream} \times 0.13) \\
& + (W_{\text{WoE}}\text{Drinage density} \times 0.565) + (W_{\text{WoE}}\text{LU/LC} \times 0.248) \\
& + (W_{\text{WoE}}\text{Soil type} \times 0.391) + (W_{\text{WoE}}\text{Lithology} \times 0.24).
\end{aligned}
\tag{17}
$$

Values of gully erosion susceptibility for the WoE model, the IoE model, and the hybrid WoE-IoE model varied for the minimums (−77.558, 0.265, and −23.3362) and the maximums (143.404, 13.810, and 51.6535) of each, respectively. These values were classified into five classes using the natural breaks method: very low, low, moderate, high, and very high susceptibility classes (Figure 7a–c). The results for each category using the WoE model were 41.45%, 14.59%, 10.91%, 18.5%, and 14.52%, respectively (Figure 8). The results for each category using the IoE model were 40.03%, 15.43%, 11.10%, 17.18%, and 16.22%, respectively (Figure 8). And using WoE-IoE integrated model, the classes of susceptibility were 27.95%, 23.78%, 10.22%, 16.60%, and 21.42%, respectively.

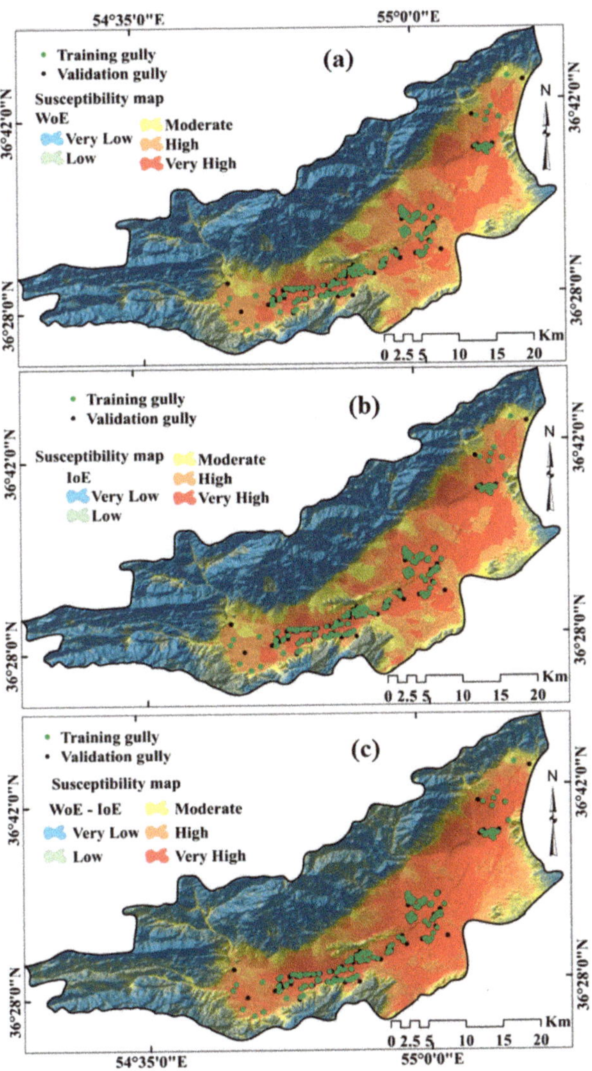

Figure 7. Gully erosion susceptibility map. (**a**) weight-of-evidence (WoE) model, (**b**) index-of-entropy (IoE) model, and (**c**) WoE-IoE integrated model.

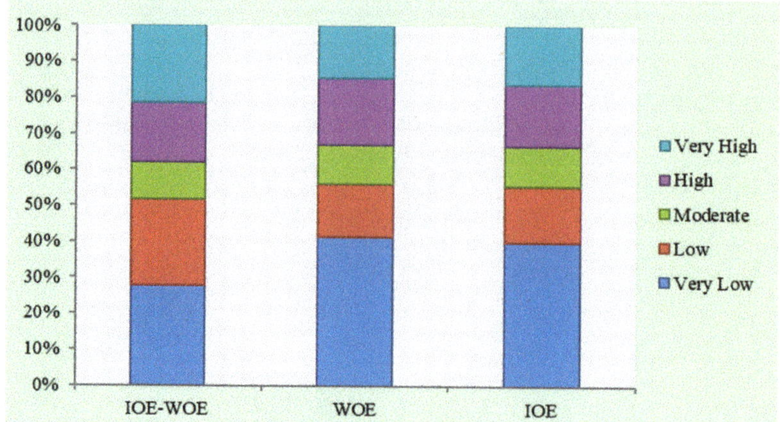

Figure 8. Percentage of each susceptibility classes in three models of the weight of evidence, index of entropy, and WoE-IoE.

3.5. Validation of Results

The AUROC graphs were created for the training dataset (success-rate curve) and for the validation dataset (prediction-rate curve) (Figure 9a,b). These demonstrate that the WoE model had better performance in terms of the success-rate curve (SRC = 0.899) than did the IoE model (SRC = 0.896). The results also indicate that integration of these models improve their performance: the WoE-IoE integrated model had a better prediction-rate curve (PRC = 0.903) and a better success-rate curve (SRC = 0.918) than either the WoE (PRC = 0.877 and SRC = 0.877) or IoE (PRC = 0.899 and SRC = 0.896).

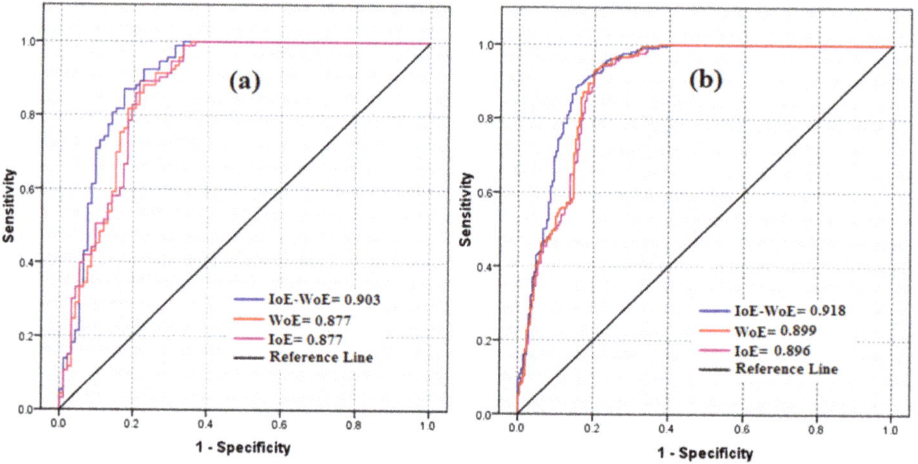

Figure 9. Area under curve values for weight-of-evidence (WoE), index-of-entropy (IoE), and WoE-IoE models: (**a**) validation dataset (prediction rate curve), (**b**) training dataset (success rate curve).

4. Discussion

In terms of the position of the gully in the landscape [71], most of the gullies are on the floor of the valley and in terms of the evolution of the gullies [72], most are continuous. The continuous gully is part of the drainage network, and a discontinuous gully is separated from the drainage network. In terms of gully head-cut plan [73,74], the gullies in the study area are the digitized type, usually found adjacent to rivers and often located at the intersection of their branches. These gullies form in areas of 0% to 5% slope. The most important factors that led to the development of gullies in the study area are: the undercutting of the boundaries of flood flows and gravitational formation of mass fractures; formation of a groove in gullies due to rill erosion and the resulting surface runoff; the influences of tunnel erosion (piping) and underground corridors which have a maximum diameter of 8 meters and a maximum length of 12 m—the expansion of these corridors and the collapse of their roofs cause gully development; and head-cut retrogradation and upward development of gullies.

Determination of the susceptible areas affected by different kinds of soil erosion such as gully erosion is useful for land use planning, mitigating soil erosion, and conservation practices [29]. So far, various methods have been applied for gully erosion susceptibility assessment globally [28–30]. Given the shortcomings and limitations of each of the models, scientists have proposed and developed integrated methods to overcome their disadvantages and increase their efficiency [14]. In this study, two types of statistical methods, namely, IoE and WoE, and their ensemble were applied to produce GESM. The IoE–WoE integrated method eliminates the disadvantages of IoE and WoE individual models and improves their advantages. The main advantages of the WoE method are that it calculates the weighted value of the factors based on a statistical formula and thus avoids the subjective choice of weighting factors. In addition, input maps with missing data (incomplete coverage) can be accommodated in the model and under-sampled data do not significantly impact the results. The main shortcomings of the model are the failure to calculate the relative importance of the parameters, and that the weighted values calculated for different areas are not comparable in terms of the degree of hazard [31]. The main advantage of IoE model is that this model does not require that assumptions be made about the proper distribution of explanatory variables; therefore, several properties can be used and tested. This method also examines the statistical relationships between independent and dependent variables and provides metrics for the significance of variables [30].

Analyses of the spatial relationships between GECFs and gully locations using IoE, and WoE show that areas of low elevation and slope and flat topography, where surface runoff concentrates and where erosion-sensitive evaporation deposits of gypsum or salt have formed, have a high susceptibility to gully erosion. This has been demonstrated in other local studies [69,75]. Areas nearer to streams and roads, with sparse vegetation and higher drainage density than other areas, have more potential for gully occurrence. These findings are in line with References [22,76]. Limestone, sandstone, marl, shale, and red conglomerate geological units have a high susceptibility to gully erosion. These geomorphological features are generally known to promote gully erosion; this is confirmed within the study area and has been reported in other research [77,78].

The IoE model has shown that drainage density, slope degree, and rainfall are key conditional factors for gully erosion in the study area. This is in line with References [16,63,79]. Yesilnacar et al. [79] proved that gently sloping areas are susceptible to surface flow accumulation and gully erosion. Pourghasemi et al. [63] assessed the capability of WoE and FR models for spatial prediction of gully erosion susceptibility in the Chavar region of Ilam Province, Iran. Sixty-three gullies and ten GECFs were used in that study. Their results indicate that the distance to roads, drainage density, and LU/LC are key conditions affecting gully occurrence. Arabameri et al. [16] compared three data-driven models and an AHP knowledge-based technique for gully-erosion susceptibility mapping in the Toroud watershed in Semnan Province, Iran using 80 gully locations and 13 GECFs (including elevation, slope degree, slope aspect, plan curvature, distance from river, drainage density, distance from road, lithology, LU/LC, TWI, SPI, NDVI, and LS). Their results show that lithology, slope, and NDVI are the primary determinants of gully occurrence in the Toroud watershed.

Validation results show that the IoE model performed better than the WoE model. This is consistent with the work of others [16,80,81]. Arabameri et al. [16] States that IoE with an SRC = 0.939 and a PRC = 0.925 better predicts areas prone to gully erosion than does WoE with an SRC = 0.926 and a PRC = 0.921. Wang et al. [81] used IoE and FR models for groundwater qanat potential mapping in the Moghan watershed, Iran and their results depict the excellence of IoE model in qanat occurrence potential estimation. The integration of WoE and IoE has improved upon their separate performances, which is also consistent with References [14,39,82–88]. Arabameri et al. [14] used EBF, LR, and a new ensemble EBF–LR algorithm to spatially model gully erosion at Semnan Province, Iran. Their results show that their ensemble method performed considerably better (AUC = 0.909) than did the individual LR (0.802) and EBF (0.821) methods. Pourghasemi et al. [39] assessed the individual and ensemble data-mining techniques for gully erosion modeling and stated that the ensemble models had a higher goodness-of-fit and predictive power than individual models. Arabameri et al. [84] compared the performance of individual and ensemble models for assessment of landslide susceptibility in Sangtarashan watershed, Mazandran Province, Iran and state that FR–RF integrated model (AUC = 0.917) achieved higher predictive accuracy than the individual FR (AUC = 0.865) and RF (AUC = 0.840) models. Du et al. [85] integrated IV and LR individual models for landslide susceptibility mapping in the Bailongjiang watershed, Gansu Province, China and state that the proposed integrated method was reliable to produce an accurate landslide susceptibility map. [88] Used ensemble RF and EBF models for landslide susceptibility assessment in Western Mazandaran Province, Iran and state that introduced ensemble model can be a powerful tool for landslide assessment at regional scales.

Our research will contribute to achieve a better knowledge of the landscape and to develop sustainable policies to achieve the Land Degradation Neutrality challenge and the United Nation Goals for Land Degradation [89,90]. This is a relevant issue to achieve sustainable land management where gullies must be restored when developed as a consequence of human mismanagement, and for this is necessary to use nature-based solutions [91]. To achieve success in gully erosion control the strategies must find a way to reduce the connectivity of the flows [92].

5. Conclusions

Gully erosion is a common geomorphological problem in arid and semi-arid regions, therefore, it is essential to develop methods for predicting gullying with highly accurate and effective models. Knowing the locations that are prone to or susceptible to gullying can enable ways to avoid casualties and financial losses caused by gully development, and can even promote sustainable development. In recent years, many quantitative and qualitative methods have been introduced for GESM. No approach is believed to be a best-approach, but the general consensus is that each method has its advantages and disadvantages. In this study, two models (the IoE and the WoE) and their integrated offspring (WoE-IoE) were tested for GESM to assess what their advantages and shortcomings were. The most important conclusions to be drawn from this study are:

i.   Based on extensive field surveys and multicollinearity tests, topographical, hydrological, geological, soil characteristics, and environmental factors are significant factors that influence gullying in the study area.

ii. Spatial comparisons of GECFs and gullies using IoE and WoE models show that areas with low elevations, low slopes, and flat topography concentrate surface runoff, and areas near streams and roads, having sparse vegetation, and higher drainage densities have greater potential for gully occurrence.

iii. By using the IoE model to determine the relative importance of GECFs, we have revealed that drainage density, slope degree, and rainfall are key conditional factors for gully erosion in the study area.

iv. Validation showed that integration of the WoE and IoE models improves the performance of either of them individually, but also decreases the disadvantages inherent in each. The WoE-IoE integrated model had higher prediction accuracy than the WoE and IoE models.

v.   Integration of the WoE and IoE models and use of remote sensing data and GIS technique be a powerful tool for GESM and have excellent accuracy.

vi. The novel method introduced in this research is adaptable and can be used in other areas.

vii. Our approach can be used to control the growth of the gullies when human-induced.

Author Contributions: Conceptualization, A.A Data curation, A.A.; Formal analysis, A.A. Investigation, A.A and A.C; Methodology, A.A., and A.C; Resources, A.A and A.C.; Software, A.A.; Supervision, A.A and A.C; Validation, A.A; Writing–original draft, A.A.; Writing–review and editing, A.A., A.C., and J.P.T.

Funding: This research received no external funding.

Acknowledgments: We are grateful to the Editor, Miley Fan, and two anonymous referees for their constructive comments which were valuable to improve our manuscript.

Conflicts of Interest: The authors declare no conflict of interest.

References

1. Peugeot, S.; Cappelare, B.; Vieux, B.E.; Seguis, L.; Maia, A. Hydrologic process simulation of a semiarid endoreic catchment in Sahelan west, model-aided data analysis and screening. *J. Hydrol.* **2003**, *279*, 224–243. [CrossRef]

2. Boardman, J.; Parsons, A.J.; Holland, R.; Holmes, P.J. *Development of Badlands and Gullies in the Sneeuberg*; Catena: Great Karoo, South Africa, 2003; Volume 50, pp. 165–184.

3. McIntosh, P.; Mike, L. Soil erodibility and erosion hazard: Extending these cornerstone soil conservation oncepts to headwater streams in the forestry estate in Tasmania. *For. Ecol. Manag.* **2005**, *220*, 128–139. [CrossRef]

4. Amsler, L.M.; Ramonell, C.G.; Toniolo, H.A. Morphologic changes in the Parana river channel in the li ght of the climate variability during the 20the century. *Geomorphology* **2005**, *65*, 56–70.

5. Marker, M.; Angeli, L.; Bottai, L.; Costantini, R. Assessment of land degradation susceptibility by scenario analysis. *Geomorphology* **2008**, *93*, 120–129. [CrossRef]

6. Zhou, P. Effect of vegetation cover on soil erosion in a mountainous watershed. *Catena* **2008**, *75*, 319–325. [CrossRef]

7. Battagli, S.A.; Leoni, L.; Sartori, F. Mineralogical and grain size composition of clays developing calanchi and biancane erosional landforms. *Geomorphology* **2002**, *49*, 153–170. [CrossRef]

8. Borrelli, P.; Märker, M.; Panagos, P.; Schütt, B. Modeling soil erosion and river sediment yield for an intermountain drainage basin of the Central Apennines, Italy. *Catena* **2014**, *114*, 45–58.

9. Luffman, I.E.; Nandi, A.; Spiegel, T. Gully morphology, hillslope erosion, and precipitation characteristics in the Appalachian Valley and Ridge province, southeastern USA. *Catena* **2015**, *133*, 221–232. [CrossRef]

10. Rafaello, B.; Reis, E. Controlling factors of the size and location of large gully systems: A regression based exploration using reconstructed pre-erosion topography. *Catena* **2016**, *147*, 621–631.

11. Poesen, J.; Nachtergaele, J.; Verstraeten, G.; Valentin, C. Gully Erosion and Environment Change: Importance and Research Needs. *Catena* **2003**, *50*, 91–133. [CrossRef]

12. Frankl, A.; Deckers, J.; Moulaert, L.; Van Damme, A.; Haile, M.; Poesen, J.; Nyssen, J. Integrated solutions for combating gully erosion in areas prone to soil piping: Innovations from the drylands of Northern Ethiopia. *Land Degrad. Dev.* **2014**, *27*, 1797–1804. [CrossRef]

13. Meliho, M.; Khattabi, A.; Mhammdi, N. A GIS-based approach for gully erosion susceptibility modeling using bivariate statistics methods in the Ourika watershed, Morocco. *Environ. Earth Sci.* **2018**, *77*, 655. [CrossRef]

14. Arabameri, A.; Pradhan, B.; Rezaei, K.; Yamani, M.; Pourghasemi, H.R.; Lombardo, L. Spatial modeling of gully erosion using Evidential Belief Function, Logistic Regression and a new ensemble EBF–LR algorithm. *Land Degrad. Dev.* **2018**, *29*, 4035–4049. [CrossRef]

15. Arabameri, A.; Pradhan, B.; Pourghasemi, H.R.; Rezaei, K.; Kerle, N. Spatial Modeling of Gully Erosion Using GIS and R Programing: A Comparison among Three Data Mining Algorithms. *Appl. Sci.* **2018**, *8*, 1369. [CrossRef]

16. Arabameri, A.; Rezaei, K.; Pourghasemi, H.R.; Lee, S.; Yamani, M. GIS-based gully erosion susceptibility mapping: A comparison among three data-driven models and AHP knowledge-based technique. *Environ. Earth Sci.* **2018**, *77*, 628. [CrossRef]

17. Arabameri, A.; Pourghasemi, H.R. Spatial Modeling of Gully Erosion Using Linear and Quadratic Discriminant Analyses in GIS and R. In *Spatial Modeling in GIS and R for Earth and Environmental Sciences*, 1st ed.; Pourghasemi, H.R., Gokceoglu, C., Eds.; Elsevier Publication: Amsterdam, The Netherlands, 2019; p. 796.

18. Nwankwo, C.; Nwankwoala, H.O. Gully Erosion Susceptibility Mapping in Ikwuano Local Government Area of Abia State, Nigeria Using GIS Techniques. *Earth Sci. Malaysis* **2018**, *2*, 08–15.

19. Azareh, A.; Rahmati, O.; Rafiei-Sardooi, E.; Sankey, J.B.; Lee, S.; Shahabi, H.; BinAhmad, B. Modeling gully-erosion susceptibility in a semi-arid region, Iran: Investigation of applicability of certainty factor and maximum entropy models. *Sci. Total Environ.* **2019**, *655*, 684–696. [CrossRef] [PubMed]

20. Nachtergaele, J.; Poesen, J.; Sidorchuk, A.; Torri, D. Prediction of concentrated flow width in ephemeral gully channels. *Hydrol. Process.* **2002**, *16*, 1935–1953. [CrossRef]

21. Saynor, M.J.; Erskine, W.D.; Evans, K.G.; Eliot, I. Gully ignition and implication for management of scour holes in the vicinity of the jabiluka mine, Australia. *Geogr. Ann.* **2004**, *86*, 19–201. [CrossRef]

22. Campo-Bescós, M.; Flores-Cervantes, J.; Bras, R.; Casalí, J.; Giráldez, J. Evaluation of a gully headcut retreat model using multitemporal aerial photographs and digital elevation models. *J. Geophys. Res. Earth Surf.* **2013**, *118*, 2159–2173. [CrossRef]

23. Conoscenti, C.; Angileri, S.; Cappadonia, C.; Rotigliano, E.; Agnesi, V.; Märker, M. Gully erosion susceptibility assessment by means of GIS-based logistic regression: A case of Sicily (Italy). *Geomorphology* **2014**, *204*, 399–411. [CrossRef]

24. Chaplot, V.; Giborie, G.; Marchand, P.; Valentin, C. Dynamic modeling for linear erosion intiation and development under climate and land-use changes in northen Laos. *Catena* **2005**, *63*, 318–328. [CrossRef]

25. Dotter weich, M. High resolution reconstruction of a 1300 year old gully system in northern Bararian. *Holocene* **2005**, *15*, 997–1005.

26. Rescher, N. The Stochastic revolution and the nature of scientific explanation. *Synthese* **1962**, *14*, 200–215. [CrossRef]

27. Derose, R.C.; Gomez, B.; Marden, M.; Trustrum, N.A. Gully erosion in Mangatu Forest, New Zealand, estimated from digital elevation models. *Earth Surf. Process. Landf.* **1998**, *23*, 1045–1053. [CrossRef]

28. Rahmati, O.; Tahmasebipour, N.; Haghizadeh, A.; Pourghasemi, H.R.; Feizizadeh, B. Evaluating the influence of geo-environmental factors on gully erosion in a semi-arid region of Iran: An integrated framework. *Sci. Total Environ.* **2017**, *579*, 913–927. [CrossRef]

29. Arabameri, A.; Pradhan, B.; Rezaei, K. Spatial prediction of gully erosion using ALOS PALSAR data and ensemble bivariate and data mining models. *Geosci. J.* **2019**, 1–18. [CrossRef]

30. Zabihi, M.; Mirchooli, F.; Motevalli, A.; Darvishan, A.K.; Pourghasemi, H.R.; Zakeri, M.A.; Sadighi, F. Spatial modeling of gully erosion in Mazandaran Province, northern Iran. *Catena* **2018**, *161*, 1–13. [CrossRef]

31. Dube, F.; Nhapi, I.; Murwira, A.; Gumindoga, W.; Goldin, J.; Mashauri, D.A. Potential of weight of evidence modeling for gully erosion hazard assessment in Mbire District—Zimbabwe. *Phys. Chem. Earth* **2014**, *67*, 145–152. [CrossRef]

32. Hosseinalizadeh, M.; Kariminejad, N.; Chen, W.; Pourghasemi, H.R.; Alinejad, M.; Behbahani, A.M.; Tiefenbacher, J.P. Gully headcut susceptibility modeling using functional trees, naïve Bayes tree, and random forest models. *Geoderma* **2019**, *342*, 1–11. [CrossRef]

33. Kornejady, A.; Heidari, K.; Nakhavali, M. Assessment of landslide susceptibility, semi-quantitative risk and management in the Ilam dam basin, Ilam. Iran. *Environ. Resour. Res.* **2015**, *3*, 85–109.

34. Rahmati, O.; Tahmasebipour, N.; Haghizadeh, A.; Pourghasemi, H.R.; Feizizadeh, B. Evaluation of different machine learning models for predicting and mapping the susceptibility of gully erosion. *Geomorphology* **2017**, *298*, 118–137. [CrossRef]

35. Amiri, M.; Pourghasemi, H.R.; Ghanbarian, G.A.; Afzali, S.F. Assessment of the importance of gully erosion effective factors using Boruta algorithm and its spatial modeling and mapping using three machine learning algorithms. *Geoderma* **2019**, *340*, 55–69. [CrossRef]

36. Hosseinalizadeh, M.; Kariminejad, N.; Chen, W.; Pourghasemi, H.R.; Alinejad, M.; Behbahani, A.M.; Tiefenbacher, J.P. Spatial modeling of gully headcuts using UAV data and four best-first decision classifier ensembles (BFTree, Bag-BFTree, RS-BFTree, and RF-BFTree). *Geomorphology* **2019**, *329*, 184–193. [CrossRef]

37. Arabameri, A.; Pradhan, B.; Rezaei, K. Gully erosion zonation mapping using integrated geographically weighted regression with certainty factor and random forest models in GIS. *J. Environ. Manag.* **2019**, *232*, 928–942. [CrossRef]

38. Hosseinalizadeh, M.; Kariminejad, N.; Rahmati, O.; Keesstra, S.; Alinejad, M.; Behbahani, A.M. How can statistical and artificial intelligence approaches predict piping erosion susceptibilllity? *Sci. Total Environ.* **2019**, *646*, 1554–1566. [CrossRef] [PubMed]

39. Pourghasemi, H.R.; Yousefi, S.; Kornejady, A.; Cerdà, A. Performance assessment of individual and ensemble data-mining techniques for gully erosion modeling. *Sci. Total Environ.* **2017**, *609*, 764–775. [CrossRef]

40. Shirani, K.; Pasandi, M.; Arabameri, A. Landslide susceptibility assessment by Dempster–Shafer and Index of Entropy models, Sarkhoun basin, Southwestern Iran. *Nat. Hazards* **2018**, *93*, 1379–1418. [CrossRef]

41. Xie, Z.; Chen, G.; Meng, X.; Zhang, Y.; Qiao, L.; Tan, L. A comparative study of landslide susceptibility mapping using weight of evidence, logistic regression and support vector machine and evaluated by sbas-insar monitoring: Zhouqu to wudu segment in Bailong River Basin, China. *Environ. Earth Sci.* **2017**, *76*, 313. [CrossRef]

42. Tien Bui, D.; Shahabi, H.; Shirzadi, A.; Chapi, K.; Alizadeh, M.; Chen, W.; Mohammadi, A.; Ahmad, B.; Panahi, M.; Hong, H. Landslide detection and susceptibility mapping by airsar data using support vector machine and index of entropy models in Cameron highlands, Malaysia. *Remote Sens.* **2018**, *10*, 1527. [CrossRef]

43. Haghizadeh, A.; Siahkamari, S.; Haghiabi, A.H.; Rahmati, O. Forecasting flood-prone areas using Shannon's entropy model. *J. Earth Syst. Sci.* **2017**, *126*, 39. [CrossRef]

44. Al-Abadi, A.M.; Pourghasemi, H.R.; Shahid, S.; Ghalib, H.B. Spatial Mapping of Groundwater Potential Using Entropy Weighted Linear Aggregate Novel Approach and GIS. *Arabian J. Sci. Eng.* **2017**, *42*, 1185–1199. [CrossRef]

45. Arabameri, A.; Rezaei, K.; Cerda, A.; Lombardo, L.; Rodrigo-Comino, J. GIS-based groundwater potential mapping in Shahroud plain, Iran. A comparison among statistical (bivariate and multivariate), data mining and MCDM approaches. *Sci. Total Environ.* **2019**, *658*, 160–177. [CrossRef]

46. I.R. of Iran Meteorological Organization (IRIMO). 2012. Available online: http://www.mazandaranmet.ir (accessed on 12 August 2018).

47. Geology Survey of Iran (GSI). 1997. Available online: http://www.gsi.ir/Main/Lang_en/index.html (accessed on 12 August 2018).

48. IUSS Working Group WRB. *World Reference Base for Soil Resources 2014*; World Soil Resources Report; FAO: Rome, Italy, 2014.

49. Cama, M.; Conoscenti, C.; Lombardo, L.; Rotigliano, E. Exploring relationships between grid cell size and accuracy for debris-flow susceptibility models: A test in the Giampilieri catchment (Sicily, Italy). *Environ. Earth Sci.* **2016**, *75*, 1–21. [CrossRef]

50. Parras-Alcántara, L.; Lozano-García, B.; Keesstra, S.; Cerdà, A.; Brevik, E.C. Long-term effects of soil management on ecosystem services and soil loss estimation in olive grove top soils. *Sci. Total Environ.* **2016**, *571*, 498–506. [CrossRef]

51. Arabameri, A.; Pradhan, B.; Rezaei, K.; Conoscenti, C. Gully erosion susceptibility mapping using GIS-based multi-criteria decision analysis techniques. *Catena* **2019**, *180*, 282–297. [CrossRef]

52. Boreggio, M.; Bernard, M.; Gregoretti, C. Evaluating the influence of gridding techniques for Digital Elevation Models generation on the debris flow routing modeling: A case study from Rovina di Cancia basin (North-eastern Italian Alps). *Front. Earth Sci.* **2018**, *6*, 89. [CrossRef]

53. Wu, C.Y.; Mossa, J.; Mao, L.; Almulla, M. Comparison of different spatial interpolation methods for historical hydrographic data of the lowermost Mississipi River. *Ann. GIS* **2019**, in press. [CrossRef]

54. Gesch, D.; Oimoen, M.; Zhang, Z.; Meyer, D.; Danielson, J. Validation of the ASTER global digital elevation model version 2 over the conterminous United States. *Int. Arch. Photogramm. Remote Sens. Spat. Inf. Sci.* **2012**, *B4*, 281–286. [CrossRef]

55. Zhou, C.; Ge, L.E.D.; Chang, H.C. A case study of using external DEM in InSAR DEM generation. *Geo-Spat. Inf. Sci.* **2005**, *8*, 14–18.

56. Zhang, W.; Wang, W.; Chen, L. Constructing DEM based on InSAR and the relationship between InSAR DEM's precision and terrain factors. *Energy Procedia* **2012**, *16*, 184–189. [CrossRef]

57. Arabameri, A.; Pourghasemi, H.R.; Yamani, M. Applying different scenarios for landslide spatial modeling using computational intelligence methods. *Environ. Earth Sci.* **2017**, *76*, 832. [CrossRef]

58. Arabameri, A.; Pourghasemi, H.R.; Cerda, A. Erodibility prioritization of sub-watersheds using morphometric parameters analysis and its mapping: A comparison among TOPSIS, VIKOR, SAW, and CF multi-criteria decision making models. *Sci. Total Environ.* **2017**, *613–614*, 1385–1400.

59. Arabameri, A.; Pradhan, B.; Pourghasemi, H.R.; Rezaei, K. Identification of erosion-prone areas using different multi-criteria decision-making techniques and GIS. *Geomat. Nat. Hazards Risk* **2018**, *9*, 1129–1155. [CrossRef]

60. Arabameri, A.; Rezaei, K.; Cerda, A.; Conoscenti, C.; Kalantari, Z. A comparison of statistical methods and multi-criteria decision making to map flood hazard susceptibility in Northern Iran. *Sci. Total Environ.* **2019**, *660*, 443–458. [CrossRef]

61. Arabameri, A.; Pradhan, B.; Rezaei, K.; Sohrabi, M.; Kalantari, Z. GIS-based landslide susceptibility mapping using numerical risk factor bivariate model and its ensemble with linear multivariate regression and boosted regression tree algorithms. *J. Mt. Sci.* **2019**, *16*, 595–618. [CrossRef]

62. Arabameri, A.; Pradhan, B.; Rezaei, K.; Saro, L.; Sohrabi, M. An Ensemble Model for Landslide Susceptibility Mapping in a Forested Area. *Geocarto Int.* **2019**. [CrossRef]

63. Pourghasemi, H.R.; Mohammady, M.; Pradhan, B. Landslide susceptibility mapping using index of entropy and conditional probability models in GIS: Safarood Basin, Iran. *Catena* **2012**, *97*, 71–84. [CrossRef]

64. Rahmati, O.; Haghizadeh, A.; Pourghasemi, H.R.; Noormohamadi, F. Gully erosion susceptibility mapping: The role of GISbased bivariate statistical models and their comparison. *Nat. Hazards* **2016**, *82*, 1231–1258. [CrossRef]

65. Conforti, M.; Aucelli, P.P.; Robustelli, G.; Scarciglia, F. Geomorphology and GIS analysis formapping gully erosion susceptibility in the Turbolo streamcatchment (Northern Calabria, Italy). *Nat. Hazards* **2011**, *56*, 881–898. [CrossRef]

66. Conoscenti, C.; Agnesi, V.; Cama, M.; Alamaru Caraballo-Arias, N.; Rotigliano, E. Assessment of Gully Erosion Susceptibility Using Multivariate Adaptive Regression Splines and Accounting for Terrain Connectivity. *Land Degrad. Dev.* **2018**, *29*, 724–736. [CrossRef]

67. Romer, C.; Ferentinou, M. Shallow landslide susceptibility assessment in a semiarid environment A Quaternary catchment of KwaZulu-Natal, South Africa. *Eng. Geol.* **2016**, *201*, 29–44. [CrossRef]

68. Hong, H.; Tsangaratos, P.; Ilia, I. Application of fuzzy weight of evidence and data mining techniques in construction of flood susceptibility map of Poyang County, China. *Sci. Total Environ.* **2018**, *625*, 575–588. [CrossRef] [PubMed]

69. Yesilnacar, E.K. The Application of Computational Intelligence to Landslide Susceptibility Mapping in Turkey. Ph.D. Thesis, Department of Geomatics the University of Melbourne, Melbourne, Australia, 2005.

70. Regmi, N.R.; Giardino, J.R.; Vitek, J.D. Modeling susceptibility to landslides using the weight of evidence approach: Western Colorado, USA. *Geomorphology* **2010**, *115*, 172–187. [CrossRef]

71. Brice, J.B. Erosion and Deposition in Loess-Mantled Great Plains, Medecine Creek Drainage Basin, Nebraska. *Geol. Surv. Prof. Pap.* **1966**, 235–339.

72. Heed, B.H. Morphology of gullies in the colorado rocky mountains. Bulletin of the International Association of Scientific Hydrology. *Hydrol. Sci. J.* **1970**, *2*, 79–89.

73. Ireland, H.A.; Sharpe, C.F.; Eargle, D.H. *Principles of Gully Erosion in the Piedmont of South Carolina*; Tech. Bull. 633.; USDA: Washington, DC, USA, 1939.

74. Hongchun, Z.H.U.; Guoan, T.; Kejian, Q.; Haiying, L. Extraction and analysis of gully head of loess plateau in china based on digital elevation model. *Chin. Geogr. Sci.* **2014**, *24*, 328–338.

75. Le Roux, J.J.; Sumner, P.D. Factors controlling gully development: Comparing continuous and discontinuous gullies. *Land Degrad. Dev.* **2012**, *23*, 440–449. [CrossRef]

76. Nyssen, J.; Poesen, J.; Moeyersons, J.; Luyten, E.; Veyret-Picot, M.; Deckers, J.; Haile, M.; Govers, G. Impact of road building on gully erosion risk: A case study from the northern Ethiopian highlands. *Earth Surf. Process. Landf.* **2002**, *27*, 1267–1283. [CrossRef]

77. Gessesse, B.; Bewket, W.; Bräuning, A. Model-based characterization and monitoring of runoff and soil erosion in response to land use/land cover changes in the Modjo watershed, Ethiopia. *Land Degrad. Dev.* **2015**, *26*, 711–724. [CrossRef]

78. Gellis, A.C.; Cheama, A.; Laahty, V.; Lalio, S. Assessment of gully control structure in the rio Nutria Watershed, New Mexico. *J. Am. Water Res. Assoc.* **1995**, *31*, 633–646. [CrossRef]

79. Golestani, G.; Issazadeh, L.; Serajamani, R. Lithology effects on gully erosion in Ghoori chay Watershed using RS & GIS. *Int. J. Biosci.* **2014**, *4*, 71–76.

80. Wang, Q.; Li, W.; Chen, W.; Bai, H. GIS-based assessment of landslide susceptibility using certainty factor and index of entropy models for the Qianyang County of Baoji city, China. *J. Earth Syst. Sci.* **2015**, *124*, 1399–1415. [CrossRef]

81. Naghibi, S.A.; Pourghasemi, H.R.; Pourtaghie, Z.S.; Rezaei, A. Groundwater qanat potential mapping using frequency ratio and Shannon's entropy models in the Moghan Watershed, Iran. *Earth Sci. Inf.* **2015**, *8*, 171–186. [CrossRef]

82. Umar, Z.; Pradhan, B.; Ahmad, A.; Neamah Jebur, M.; Shafapour Tehrani, M. Earthquake induced landslide susceptibility mapping using an integrated ensemble frequency ratio and logistic regression models in West Sumatera Province, Indonesia. *Catena* **2014**, *118*, 124–135. [CrossRef]

83. ThaiPham, B.; Prakash, I.; Tien Bui, D. Spatial prediction of landslides using a hybrid machine learning approach based on Random Subspace and Classification and Regression Trees. *Geomorphology* **2018**, *303*, 256–270.

84. Arabameri, A.; Pradhan, B.; Rezaei, K.; Lee, C.-W. Assessment of Landslide Susceptibility Using Statistical-and Artificial Intelligence-Based FR–RF Integrated Model and Multiresolution DEMs. *Remote Sens.* **2019**, *11*, 999. [CrossRef]

85. Du, G.; Zhang, Y.; Iqbal, J. Landslide susceptibility mapping using an integrated model of information value method and logistic regression in the Bailongjiang watershed, Gansu Province, China. *J. Mount. Sci.* **2017**, *14*, 249. [CrossRef]

86. Tien Bui, D.; Tuan, T.A.; Hoang, N.D.; Thanh, N.Q.; Nguyen, D.B.; Van Liem, N.; Pradhan, B. Spatial prediction of rainfall-induced landslides for the Lao Cai area (Vietnam) using a hybrid intelligssent approach of least squares support vector machines inference model and artificial bee colony optimization. *Landslides* **2017**, *14*, 447–458. [CrossRef]

87. Dehnavi, A.; Aghdam, I.N.; Pradhan, B.; Morshed, V.M.H. A new hybrid model using step-wise weight assessment ratio analysis (SWARA) technique and adaptive neuro-fuzzy inference system (ANFIS) for regional landslide hazard assessment in Iran. *Catena* **2015**, *135*, 122–148. [CrossRef]

88. Pourghasemi, H.R.; Kerle, N. Random forests and evidential belief function-based landslide susceptibility assessment in Western Mazandaran Province, Iran. *Environ. Earth Sci.* **2016**, *75*, 1–17. [CrossRef]

89. Keesstra, S.; Mol, G.; de Leeuw, J.; Okx, J.; de Cleen, M.; Visser, S. Soil-related sustainable development goals: Four concepts to make land degradation neutrality and restoration work. *Land* **2018**, *7*, 133. [CrossRef]

90. Keesstra, S.D.; Bouma, J.; Wallinga, J.; Tittonell, P.; Smith, P.; Cerdà, A.; Montanarella, L.; Quinton, J.N.; Pachepsky, Y.; van der Putten, W.H.; et al. The significance of soils and soil science towards realization of the United Nations Sustainable Development Goals. *Soil* **2016**, *2*, 111–128. [CrossRef]

91. Keesstra, S.; Nunes, J.; Novara, A.; Finger, D.; Avelar, D.; Kalantari, Z.; Cerdà, A. The superior effect of nature based solutions in land management for enhancing ecosystem services. *Sci. Total Environ.* **2018**, *610*, 997–1009. [CrossRef] [PubMed]

92. Keesstra, S.; Nunes, J.P.; Saco, P.; Parsons, T.; Poeppl, R.; Masselink, R.; Cerdà, A. The way forward: Can connectivity be useful to design better measuring and modeling schemes for water and sediment dynamics? *Sci. Total Environ.* **2018**, *644*, 1557–1572. [CrossRef]

MDPI
St. Alban-Anlage 66
4052 Basel
Switzerland
Tel. +41 61 683 77 34
Fax +41 61 302 89 18
www.mdpi.com

Water Editorial Office
E-mail: water@mdpi.com
www.mdpi.com/journal/water